TORIES

TORIES

Fighting *for* the King

in

America's First Civil War

Thomas B. Allen

HARPER

An Imprint of HarperCollins*Publishers*

www.harpercollins.com

HarperCollins books may be purchased for educational, business, or sales promotional use. For information, please write: Special Markets Department, Harper-Collins Publishers, 10 East 53rd Street, New York, NY 10022.

The painting reproduced on the title page is courtesy of the Yale Center for British Art, Paul Mellon Collection, USA/The Bridgeman Art Library.

Library of Congress Cataloging-in-Publication Data has been applied for.

ISBN: 978-0-06-124180-2

10 11 12 13 14   OV/RRD   10 9 8 7 6 5 4 3 2

To Rob Cowley,
who gave me the idea for this book

Hugh McDonald of North Carolina, a fifteen-year-old soldier of the Continential Army, was on the road near Richmond, Virginia, in the spring of 1777. One day, he joined other soldiers in an act of casual cruelty upon a stubborn Tory. How many other Tories were taunted, tortured, or lynched will never be known.

❧ ❧

While passing through the town, a shoemaker stood in his door and cried, 'Hurrah for King George,' of which no one took any notice; but after halting in a wood, a little distance beyond . . . the shoemaker came to us and began again to hurrah for King George. When the General [Francis Nash] and his aids mounted and started, he still followed them, hurrahing for King George. Upon which the General ordered him to be taken back to the river and ducked. We brought a long rope, which we tied . . . round his middle and sesawed him backwards and forwards until we had him nearly drowned, but every time he got his head above water he would cry for King George. The General having then ordered him to be tarred and feathered, a feather bed was taken from his own house, where were his wife and four little daughters crying and beseeching their father to hold his tongue, but still he would not. We tore the bed open and knocked the top out of a tar barrel, into which we plunged him headlong. He was then drawn out by the heels and rolled in the feathers until he was a sight but still he would hurrah for King George. The General now ordered him to be drummed out of the West end of town, and told him expressly that if he plagued him any more in that way he would have him shot. So we saw no more of the shoemaker.

—McDonald, who was illiterate, dictated his recollections, which became "Revolutionary Journal of Hugh McDonald" in the *Colonial and State Records of North Carolina*, vol. ii, p. 835.

Contents

Preface

—

"*LITTLE LESS THAN SAVAGE FURY*"

One of my earliest childhood memories takes me to Putnam Park, near Danbury, Connecticut. The park was named after Maj. Gen. Israel Putnam. I still remember the cannons and a cave. My mother told me that soldiers who fought in the Revolutionary War spent a cold, hungry winter there. That was my first lesson about the war.

My mother did not tell me about Gallows Hill. On a February day in 1779, while his Continental Army division was in winter camp, General Putnam, infuriated by the number of spies and army deserters who had been brought before him, decided to execute one of each—"make a double job of it," he said. The spy was Edward Jones of Ridgefield, who, as an American supporter of the British was a Loyalist, or Tory. The deserter was seventeen-year-old John Smith, who was accused of planning to join the British Army as a Tory convert. Smith and Jones, ordinary men of ordinary names.

Smith spent a few minutes with a chaplain. Then, within a hollow square formed by the soldiers he wished to fight, Smith's death warrant was read. He was taken off and killed by a firing squad, a few yards from a gallows that soldiers had built on the highest hill in the encampment. Jones was brought to it, and his death warrant was read. A noose around his neck was attached to the beam of the gallows. He climbed a ladder leaning on the beam, looked around at people he seemed to recognize, and swore to God that he was innocent. When he refused to step off the ladder, as one account puts it, he had to be "hurried into eternity," presumably by a soldier, although one report says young boys pushed the ladder.[1]

As that day on Gallows Hill so lethally demonstrated, some Americans wanted to kill other Americans in the Revolutionary War. What had begun as political conflict between politicians called Whigs and their opponents, called Tories, had evolved into a brutal war. Our histories prefer to call the conflict the Revolutionary War, but many people who lived through it called it civil war. Americans who called themselves Patriots taunted, then tarred and feathered, and, finally, when war came, killed American Tories. Americans who called themselves Tories gave themselves a proud new name: Loyalists, a label that had not been needed when all Americans were subjects of the king.

When Brig. Gen. Nathanael Greene took command of the Continental Army of the South in 1781, he wrote to Col. Alexander Hamilton: "The division among the people is much greater than I imagined and the Whigs and Tories persecute each other, with little less than savage fury. There is nothing but murders and devastation in every quarter."[2]

There was also collaboration. When we remember the heroic suffering of George Washington's army at Valley Forge, we forget that only twenty miles away the British soldiers occupying Philadelphia were well housed and well fed because Tories and Tory sympathizers were sustaining them. "I am amazed," wrote Washington to a staff officer, "at the report you make of the quantity of provisions that goes daily into Philadelphia from the County of Bucks."[3] Washington believed that most people in Pennsylvania did not support the war and "the languor of others, & internal distraction of the whole, have been among the great and insuperable difficulties I have met with."[4]

Like most Americans, as a schoolboy and as an adult I had heard about the Tories, but I had not paid them much attention, believing that, as a small minority, they had not played a major role in the war. As a native of Connecticut, I had always thought of my state as a place where all the people fought the British. But soon after I started working on this book, I came across a reference to a Connecticut man named Stephen Jarvis, who had become a Tory soldier and killed other Americans. He was one of many Connecticut people who chose the king's side, and his story is far from unusual. Such Connecticut towns as Stamford, Norwalk, Fairfield, Stratford, and Newtown had such large Loyalist populations that Patriots called them "Tory Towns."[5]

Stephen Maples Jarvis, born in Danbury in 1756, was working on the family farm in April 1775 when he heard the news that British Redcoats and Rebels had clashed at Lexington and Concord in Massachusetts. "My father was one of those persons called Torries," Stephen later wrote, quickly veering in his journal to his own clash with his father. Stephen, going on nineteen, was courting a young woman, Amelia Glover, who was "disapproved of by my father . . . and I was under the necessity of visiting the Lady only by stealth."

To defy his father—and perhaps to impress his girlfriend—Stephen declared that he would join the Rebels' Connecticut militia. When Stephen told his father this, the elder Jarvis "took me by the arm and thrust me out of the door."[6]

At that moment in those turbulent times, when general discontent over British rule had flared into rebellion, the divided Jarvis family mirrored the splitting of families and friends throughout the colonies. Amelia Glover's sister was married to a Rebel. Royal colonial militias overnight became Rebel militias. The militia that Stephen joined, originally formed to serve the king, was commanded by his mother's brother, a Rebel.

In Stamford, thirty miles southwest of Danbury, Stephen's uncle on his father's side, Samuel Jarvis, was the town clerk. Soon after the Battles of Lexington and Concord, the Rebels' Tory-hunting Committee of Inspection summoned Samuel, interrogated him about his Tory beliefs, and condemned him as "inimical to the Liberty of America." The committee also found Samuel's son Munson guilty of "signing a seditious paper, the import of which was that they would assist the King and his vile minions in their wicked, oppressive schemes to enslave the *American* Colonies; and tending to discourage any military preparations to repel the hostile measures of a corrupt Administration."[7]

Samuel and Munson, suddenly aliens in their hometown, began planning how to get out. By the early fall of 1776, they could stand on the Stamford shore, look across Long Island Sound, and on the gray horizon see the low-lying land where the British flag had flown since the British Army drove the Continental Army out of New York. As Samuel Jarvis told the story, he and his wife and four children escaped by boat to Long Island.[8] According to the Rebel version, a mob broke into the Jarvis home late one night, stripped every Jarvis naked, dragged them all into a boat, sailed it across the Sound, and forced them to wade to the British shore.[9] Loyalists became a major Connecticut export.

When Samuel Jarvis reached Long Island, he recruited his son Munson and other Tories into the Prince of Wales's American Regiment, one of more than two hundred Loyalist military units.[10] Samuel and Munson would be among the thousand or so Connecticut men who served in Loyalist regiments, aboard the ships of the Royal Navy, or as Tory privateers.[11] Rich and prominent landowners or royal officials organized and commanded Tory regiments, but the soldiers were usually farmers, laborers, craftsmen, and shopkeepers. Munson Jarvis, like Paul Revere in Boston, was a silversmith.

When Stephen's militia was temporarily released from active service, he deserted, apparently without telling his Rebel uncle. Stephen promised his father that he was through with the Rebels, which was true, and that he was through with Amelia, which was not. Stephen joined the Tories by following his uncle's example. With other young Connecticut men, Stephen rowed across Long Island Sound, went into New York City, and, after service in another unit, joined the Queen's American Rangers. They wore forest green uniforms to distinguish themselves from their comrades in war, the British Redcoats.[12] The Rangers saw themselves as the elite unit among all the Loyalist forces fighting for the king.

Later in the war the Queen's Rangers joined with British forces in an attack on Stephen's birthplace, Danbury. Tories guided the invaders to secret stores of Rebel arms.[13] After the battle, Rebel troops, out for revenge, swooped down on suspected Tories. One was Stephen Jarvis's father. They beat him and pillaged his farmhouse.[14]

Stephen did not take part in the Danbury raid, but he soon was heading for Pennsylvania to begin a long campaign of fighting and killing other Americans. In one battle, he wrote, a Rebel soldier "fired and missed me and my horse and before he could raise his rifle he was a dead man."[15]

After seven years as a Tory soldier, Stephen returned to Danbury, naively expecting to resume a life merely interrupted by war. He and his beloved Amelia planned to be married in an Episcopal church by a clergyman who was a relative. Stephen did not realize that, because the Episcopal clergy's duties included prayers for the king, the Patriots had silenced most Episcopal clergymen in the colonies and forced the closing of their churches. (One Connecticut cleric who defied the Patriots was shot at as he preached. The bullet lodged in the sounding board of his pulpit. He kept on preaching and was not shot at again. Many of his fellow clerics had already fled to England.)[16]

Stephen had to change his marriage plans. After calming a mob that burst into his father's house, he hastily arranged to marry Amelia there: "A clergyman was sent for, we retired to a room with a select party of our friends, and we were united, after which the mob dispersed and had left us."

The next morning the local sheriff, carrying a warrant for Stephen's arrest, forced his way into the bedroom of the bride and groom. Stephen "met him with such a determined and threatening attitude that in his retreat he tumbled from the head of the staircase to the bottom.

He then selected a posse—and surrounded the house. . . . I made my appearance at the window of my bedchamber, spoke to the persons outside, who seemed to look rather ill-natured. I threw them a dollar, desired they would get something to drink the Bride's health, which they did, and before they had finished the bottle I had won them all to my side."

But sometime later another mob stormed the house, attacking Amelia and her father-in-law. Stephen ran away and hid out. The war had not ended for him, and now it had not ended for Amelia. He began to think about leaving America. By then thousands of Tories were continuing a flight from America that had been going on since the first stirrings of the Revolution.

The first self-exiles had sailed to the motherland. "As the Rebellion is general thro' the provinces," a Boston clergyman wrote the arch-bishop of London in August 1775, "the friends of Governmt have no certain place to fly to for safety but to Eng."[17] Clergymen and royal officials began the exodus, which continued throughout the war. Thousands moved to temporary sanctuary in places where Tories ruled, hoping to return home after British victory. Tories jammed New York City; others chose Canada, or Charleston, South Carolina, a Tory town of the South.

Some moved to East Florida, where Britain had established an outpost to discourage Spanish incursions. But the treaty that ended the war handed East Florida over to Spain. So, while northern Loyalists were fleeing to Canada, southern refugees fled from Florida and Charleston to Bermuda and Jamaica. The exodus reached its climax in New York City on November 25, 1783, when a British fleet began evacuating thousands of Americans to Canada. These did not resemble the colonial officials and wealthy Loyalists who had sailed to England at the beginning of the war. The 1783 evacuees' occupations included baker, house carpenter, miller, scrivener, trader, cooper, vintner, breeches maker, and innkeeper.[18]

Royal officials, needing settlers for the Canadian wilderness, sent the Loyalists to harbors along the rocky Nova Scotia coast or up broad rivers. They landed on virgin shores and were handed army rations, tools, lumber, blankets, and cloth for making clothing. New communities sprang up. New lives began.

One of the new Canadians climbed to the top of a desolate hill to watch the sails of her ship disappear over the horizon. "Such a feeling of loneliness came over me," she later wrote, "that, though I had not shed a tear through all the war, I sat down on the damp moss with my baby on my lap and cried bitterly." Her name was Sarah Frost, originally from Stamford. She was the daughter of Patriots and the wife of a Tory who became a notorious raider in an amphibious war waged between Connecticut Rebels and Long Island Tories.[19]

By some counts, about 80,000 Tories left the colonies—proportionally, six times the number of people who fled France during the French Revolution.[20] A larger estimate came from a Tory historian who was in New York when, he said, "not less than 100,000 souls" left the city in a mass postwar exodus.[21] That estimate does not count Tories who left from other places in other times, including large-scale evacuations from Savannah and Charleston. We will never know the total number, but we do have solid knowledge about the flight of thousands of individuals. Stephen and Amelia Jarvis and their infant daughter, for instance, left Connecticut on May 1, 1785. They began their Canadian lives in a settlement newly named Fredericktown, in honor of Prince Frederick, second son of King George III.

Among the exiles who sailed to Canada were some thirty-five hundred black Tories, ex-slaves given their freedom because they had joined the Loyalist cause. In 1792, nearly two thousand of them, bitter over the way they were treated in Nova Scotia, sailed from there in a fleet of fifteen ships to Africa, where they became the founders of modern Sierra Leone. Thus, in ways no one could have imagined in 1776, the Revolution led to the creation not only of the United States but also of a new Canada and a new nation on another continent.

From the battle at Concord to the battle at Yorktown, Patriot troops fought armed Loyalists as well as British troops. By one tally, Loyalists fought in 576 of the war's 772 battles and skirmishes.[22] Relatively few of these Loyalist-Patriot clashes get much mention in military chronicles, and few had an important effect on the outcome of the Revolution. But they did strengthen the solidarity of the Loyalists: They were not merely opposing the Revolution; they were fighting and dying to end it.

In the earliest days of the war Patriots looked longingly at Canada as a potential participant in rebellion.[23] But the Rebels' liberation invasion did not trigger an uprising against the king. Canadian Loyalists fought the American Rebels. Canada became a place that resisted the Revolution—and thus a place where Tories could find refuge.

No one knows how many Tories there were. The Tories themselves consistently believed that they were in the majority.[24] But there is no reliable head count for determining the actual number of Tories, white or black, at any specific time. A modern estimate of Loyalist strength—colonists who fought on the king's side, worked for the British, or went into exile—allots them 16 percent of the total population or nearly 20 percent of the white population.[25] To turn that estimate into a Loyalist head count, however, you need to know how many Americans there were. Estimates of the total American population—based on tax lists, militia musters, and other available records—are as low as 2,205,000 and as high as 2,780,400.[26] So, using the 20 percent figure, there may have been as few as 441,000 or as many as 556,080 Loyalists.

Down the years many historians have cited John Adams as an eyewitness source for an estimate of one-third Tories, one-third Patriots, and one-third indifferent. That view has prevailed because of a consistent misinterpretation of Adams's words. In a letter written in January 1815 to James Lloyd, a forty-six-year-old Massachusetts politician between terms as a U.S. senator, Adams says: *"The middle third, composed principally of the yeomanry, the soundest part of the nation, and always averse to war, were rather lukewarm* both to England and France; and sometimes stragglers from them, and sometimes the whole body, united with the first or the last third, according to circumstances." (Sometimes the Adams quote is cited only as far as "lukewarm.") But Adams was not writing about American reaction to the Revolutionary War. He was giving his judgment about how Americans thought about England and the *French* Revolution when he was president.[27]

Adams did discuss the Tories in another long letter that same year. From 1765 to 1775, he wrote, the British government "formed and organized and drilled and disciplined a party in favor of Great Britain, and they seduced and deluded nearly one third of the people of the colonies." In that letter, to the Reverend Jedediah Morse, an author of geography textbooks, Adams went on to say that "many men of the first rank, station, property, education, influence, and power, who in 1765 had been real or pretended Americans, converted during the period to real Britons." Among them, Adams continued, were "my cordial, confidential, and bosom friends," drawn away to the ranks of the Tories by offers of power and prestige.[28]

Adams's description of the effort to convert Americans to Britons covers only the decade before the war began. He did not speak to the activities of Tories during the war. Nor did he mention the thousands of Loyalists who joined the regiments that were formed to fight the Continental Army, or the Continental Army soldiers and state militiamen who deserted their regiments not because they no longer wished to be soldiers but because they wanted to fight on the Loyalist side. Neither did Adams take up the numbering of what George Washington called "half tories," who secretly aided the Rebels, usually as spies.[29]

Two distinguished historians, Henry Steele Commager and Richard B. Morris, analyzed Adams's one-third thesis and wrote:

> *If by Patriot we mean only those who were ready to fight for the new nation, then Adams' one third is too high; after all, a free population of only 2,000,000 could not put over 25,000 men in the field at once, and a rich and fertile land allowed its soldiers to freeze and to starve. If by Loyalist we mean only those who were actively loyal, and whose loyalty carried them into exile or into British ranks, then again Adams' estimate is too large. But if the term Loyalist is stretched to cover not only those who were actively loyal but also those who were against independence and war, and tried to hold aloof, then the figure of one third is clearly too small. Two things are apparent: that there was always a substantial portion of the*

American population which had no enthusiasm for either the rebellion
or its suppression, and that the number and zeal of Patriots and Loyalists
alike changed constantly with the varying fortunes of the war.[30]

Adams believed in the importance of finding the records of what he called the "intrigues" perpetrated by the British to "divide the people." He also wondered how many incriminating records still existed from the proceedings of Patriot committees that ferreted out Tories and punished them. Until those records were discovered, he wrote, "the history of the United States never can be written."[31]

Many of those records do exist. They are the Loyalists' legacy, and, like the Loyalists, they are scattered. I have found them in Canada, Britain, Scotland, Northern Ireland, in the Library of Congress, and in the archives of the original colonies. I have also seen diaries, letters, and other documents collected by American families who discovered, sometimes to their surprise, the Tories in their family trees. Those documents give faces to forgotten Americans who fought on the losing side.

The Loyalists add a dimension to the Revolutionary War, transforming it into a conflict between Americans as well as a struggle for independence. Paddy Fitzgerald, a historian of Irish migration, once told me, "Every country has a Grand Story, and there are always stories under the grand story." Loyalists lived and died in the Grand Story's underground, fighting to keep America ruled by the king. But they were nonetheless truly Americans, introducing the nation's first generation of politicians to a truth that would endure: Woven into the tapestry to be known as *We the People*, there would always be strands of a defiant, passionate minority.

A note about words and labels. I retain the often peculiar spellings that appear in documents of the era, but I do introduce modern punctuation for clarity. As for labels, people who lived during the Revolution called each other Tories and Whigs, Patriots and Loyalists, Rebels and Friends of the King. "Tory" and "Whig" were political labels.

"Whig" faded away when political disputes evolved into rebellion. But "Tory" endured and became the Rebels' favorite name for their foes. Some people today, particularly the descendants of Loyalists, find the word "Tory" offensive. They also object to "Patriots" for supporters of the Revolution on the grounds that their ancestors were patriots, too: British patriots. Some people prefer "colonials" as a label for everybody in America in those days. But I think a "colonial" is what someone would have been before Americans began calling themselves Americans, whether or not they supported King George III.

I use "Loyalist," "Tory," "Rebel," and "Patriot" not as labels that disparage or commend but as descriptive terms that fit the events and times described. I don't use "American army" or "American" for one side or the other because the Revolutionary War was a civil war, and when Loyalists or Tories fought Patriots or Rebels, everyone in the fight was an American.

Thomas B. Allen
May 21, 2010
Bethesda, Maryland

TORIES

TWO FLAGS OVER PLYMOUTH

MASSACHUSETTS, 1769–1774

*Q. What was the temper of America toward Great Britain before
the year 1763?*

*A. The best in the world. They submitted willingly to the govern-
ment of the Crown, and paid, in their courts, obedience to acts
of Parliament. Numerous as the people are in the several old
provinces they cost you nothing in forts, citadels, garrisons, or
armies, to keep them in subjection. . . .*

Q. And what is their temper now?

A. Oh, very much altered.

—*Colonial agent Benjamin Franklin, before Parliament,
February 1766*

The roar of a cannon resounded through the little Massachusetts
town of Plymouth on the morning of December 22, 1769. And
up a flagpole went a silk flag bearing the inscription *Old Colony 1620*.
The cannon and flag grandly marked a celebration created by the Old
Colony Club.

At lunchtime the members gathered at an inn not far from the rock
where the Pilgrims were said to have landed.[2] Their meal included
whortleberry pudding, succotash, venison, clams, oysters, codfish,
eels, seafowl, apple pie, cranberry tarts, and cheese, all "dressed in the

plainest manner . . . in imitation of our ancestors." The club president sat in a chair that had belonged to William Bradford, who had become governor of the Plymouth Colony in 1621.

The members raised a toast to Bradford and their ancestors in what they hoped to be an annual celebration of Forefathers' Day, commemorating the landing of the shallop that had carried the passengers of the *Mayflower* to shore.[3] As the clock struck eleven that evening, the cannon was fired again; the members gave three lusty cheers and went home.

The Old Colony Club had been founded eleven months before by seven Plymouth men who wished to avoid "the many disadvantages and inconveniences that arise from intermixing with the company at the taverns in this town of Plymouth." They also wished to increase their "pleasure and happiness" along with their "edification and instruction." Five more members, including the owner of the inn, were admitted shortly later.

The club, modeled on gentlemen's clubs in London, became a place where the members, most of them *Mayflower* descendants and many of them Harvard graduates, argued the policies of Tories and Whigs. Tories supported the Crown, the role of the king as head of the church, and the traditional structure of a parliamentary monarchy; Whigs, while certainly not Rebels, sought limited political and social reform. They mischievously noted that "Tory" sounded like the Irish word for "outlaw." The Whigs' name could probably be traced to "whiggamore," the label for seventeenth-century Scottish rebels. Both sides could sometimes agree on such matters as property rights and excessive taxes.

Colonists followed in the steps of the motherland's Whigs, who believed that the Crown, the House of Lords, and the House of Commons should share power. In the colonies a governor was likened to the king, a council to the House of Lords, and a local assembly to the House of Commons. (In Massachusetts the legislature was formally known as the Great and General Court.) But by 1769, in the club and throughout the colonies, the debate was moving toward a sharp division between the Tory champions of the king and the Whig champions of what was politely called "opposition to ministerial measures," a phrase that placed the blame for perceived ill rule on the king's ministers, not on King George III.

"Loyalist" was emerging as the word for an opponent of a "Patriot." There would have been no need to be labeled loyal to the king if the Rebels had not dared to challenge royal authority. And, as violent rebellion neared, Plymouth men and women who called themselves Loyalists saw themselves as the real Americans, the people who descended from the original Americans—the *Mayflower*'s passengers. Here in Plymouth the Loyalists began the tradition that, as descendants of the *Mayflower* voyagers, they had been called by fate, or more likely God, to preserve what the *Mayflower* immigrants had begun. Americans of future centuries would continue the idea that being a *Mayflower* descendant was the ultimate American pedigree. Yet Plymouth's *Mayflower* descendants were British subjects who believed that the future of America lay in royal rule rather than in rebellion.[4]

One club member, Edward Winslow, was the great-grandson of Edward Winslow, who had arrived on the *Mayflower* and later served as a governor of the Plymouth Colony. The Edward Winslow of 1769 had inherited the belief that the rebellion, brewing mostly in Boston, was rooted in the trial and beheading of King Charles I in 1649. After the restoration of the monarchy with the coronation of Charles II in 1660, two of the judges who condemned his father—regicides, as they were known—had fled to America. Puritans in Connecticut and Massachusetts had hidden the judges, thwarting royal pursuers. "The seeds of rebellion were thus sown," wrote a Loyalist historian. ". . . . The Pilgrim fathers of Plymouth were as a rule tolerant, non-persecuting and loyal to the king; but the Puritans . . . were intolerant of all religionists who did not conform to their mode of worship."[5] Religion remained an issue as colonists took sides in the 1770s, when virtually every Anglican clergyman in America became a Loyalist, and Presbyterians were labeled Rebels.

Winslow's leadership, like that of his great-grandfather, would focus on Plymouth. But eventually he would become an important leader of Loyalists beyond his native town. In 1769 he was four years past his playboy days at Harvard and was destined to inherit the posts that had been held by his father: registrar of wills, clerk of the Court of General Sessions, and naval officer of the port (a civil, not a military, post). As a friend of Chief Justice Peter Oliver and Governor Thomas Hutchinson, Winslow would join the Loyalist inner circle in Boston.[6]

Like so many colonists, the Old Colony Club members were changing from Britons who happened to live overseas to Americans who were choosing sides or wondering whether sides really had to be chosen. Tories hailed Britain's imperial power while Whigs argued against what they saw as the excesses of British power: the royal proclamation forbidding settlements west of the Appalachians; increased duties on sugar, textiles, and coffee; the outlawing of colonial currency. As criticism of the Crown and Parliament kindled suspicions of disloyalty, many Tories declared themselves to be Loyalists. Some radical Whigs began calling themselves Patriots.

Among the dozen toasts made at the club on that first Forefathers' Day, the looming crisis was only mildly acknowledged. One wished for a "speedy and lasting union between Great Britain and her colonies." Records of the meal and the toasts survive, but there is no mention of what the twelve members and their guests had to say about the troubles that were clouding their little world of Plymouth, about thirty-five miles from the tumult in Boston.[7] Other records—military muster rolls of Tories and Patriots, Tory petitions to the Crown, proclamations of Tory banishment, land records for exiles in Nova Scotia, pension appeals from Continental Army veterans—show that the futures of these men and tens of thousands of others were caught up in a revolution that was also a civil war.

Winslow and the other Tories in the club aspired to take advantage of their birth and station by gaining posts in the royal colonial government or benefiting from its largesse. This was the core of Tory power—the governors, the judges, the customs officials, and the bureaucrats who served the Crown. Radiating out from that core were Anglican clergymen and their leading parishioners—merchants, shipowners, landed gentry—who supported the idea of a British Empire that drew its supremacy from the Crown and dispensed its benefits upon the chosen in the colonies. They believed most of all in a well-ordered society; they abhorred and, in the 1760s, were beginning to fear a challenging class: the radical Whigs, or the Patriots, as they became known, who envisioned a new kind of society, rooted in America and only loosely tied to Britain.

The Old Colony Club was founded at a crossroads in a revolutionary time. Four years before had come the Stamp Act, so called because colonists had to pay for stamps when buying a newspaper, calendar, marriage license, deck of playing cards, or pair of dice. (Such stamps were in use in Britain; some still are.) Parliament had justified this new tax as a way to finance the maintenance of soldiers sent to the colonies to defend their frontiers against hostile Indians— and to defend British interests in North America. The French and Indian War had ended in 1763 with victory for Britain and the addition of French Canada to British colonial territory. But the war had been costly and worldwide, ranging across the globe from Europe and North America to India. The expanded British Empire needed to pay for its upkeep, and the money would come from taxes paid by colonists.

Since 1675 the colonies had been ultimately governed by a standing committee of the King's Privy Council—the Lords of the Committee of Trade and Plantations, familiarly known as the Lords of Trade. Royal governors reported to the Lords of Trade, and ever since the Stamp Act crisis, accounts of unrest filled the reports. The governors were expected to rule their colonies with the aid of their legislatures. If the legislatures began to veer away from the policies that originated in Britain, governors could dissolve them and assume dictatorial power.

Demands for repeal of the Stamp Act swept through the colonies. Officials were hanged in effigy in British colonies from Nova Scotia to the West Indies. In the Virginia House of Burgesses, twenty-nine-year-old Patrick Henry made his "if this be treason" speech, bringing to life a Patriot doctrine: Only colonial legislatures should have the right to levy taxes on their citizens.[8] A Stamp Act Congress met in New York City, producing a united front, not only to protest the stamps and boycott British imports but also to send a reminder to the king and Parliament in the form of a Declaration of Rights, which declared: "It is inseparably essential to the freedom of a people, and the undoubted right of Englishmen, that no taxes be imposed on them, but with their own consent, given personally, or by their representatives."[9] In Boston and New York City, a secret organization called the Sons of Liberty emerged to fight the Stamp Act through a boycott of British imports.

Parliament repealed the Stamp Act in March 1766. Merchants began selling British goods again, and American tempers cooled. But in 1767 Parliament struck again, this time passing the Townshend Acts, named after Charles Townshend, the Chancellor of the Exchequer. The new laws tightened the Crown's grip on the colonies by setting up a board of customs commissioners in Boston and admiralty courts in Boston, Philadelphia, and Charleston, South Carolina. The courts began to crack down on shipowners and importers who had evaded taxes by smuggling goods into secret harbors or ports manned by corrupt officials.

Honest customs officials searching for contraband were given the specific right to wield writs of assistance, powerful search warrants used in contraband searches even of private homes. The new writs aggravated Americans. Added to long-standing taxes on such imports as wine and clothing were new taxes on imported paint, paper, glass, lead, and tea. The new revenues would be used to pay the salaries of royal colonial officials, taking that power of the purse away from the colonies. Parliament also suspended the New York assembly in punishment for that colony's objection to feeding and housing British soldiers.

In June 1768 enforcement of the Townshend Acts led to the seizure of John Hancock's sloop *Liberty*, which carried a smuggled cargo of Madeira wine into Boston Harbor. Hancock, reputedly the wealthiest man in New England, was a Boston selectman and a leading Patriot with solid connections to the Sons of Liberty. Boston's wharves became a stage for the Sons to tread. They stirred up a small-scale riot, bullied the customs men, and celebrated when the charges against Hancock were dropped.[10] Colonists, reprising their moves against the Stamp Act, again started boycotting British imports. Gangs threatened Tory merchants who defied the ban. Rumors spread that Royal Governor Francis Bernard would be assassinated.[11]

People showed their support for the *Liberty* by singing the "Liberty Song" (to the rollicking tune of "Hearts of Oak," a well-known Royal Navy air):

> *Come join hand in hand, brave Americans all,*
> *And rouse your bold hearts at fair Liberty's call;*
> *No tyrannous acts shall suppress your just claim,*
> *Or stain with dishonor America's name.*

The last verse told anyone who wondered that "Liberty's call" certainly did not mean independence from Britain or disloyalty toward King George III:

> *This bumper* I crown for our sovereign's health,*
> *And this for Britannia's glory and wealth;*
> *That wealth, and that glory immortal may be,*
> *If she is but just, and we are but free.*[12]

The song was written by John Dickinson, a onetime conservative Pennsylvanian who had attended the Stamp Act Congress and become a Patriot. He sent the song to his friend James Otis, Jr., in Boston, who saw that it was printed in the *Boston Gazette*; newspapers throughout the colonies republished it.

Otis, a Tory in a long family line of influential Tories, married the Tory daughter of a Boston merchant. His sister Mercy Otis was married to James Warren, a member of the Old Colony Club. Warren, previously a royal sheriff, became a Patriot and was destined to be a

---

* A cup or glass filled to the brim, especially one used for making a toast.

leader in the Revolution. Mercy Otis Warren would become a play-wright whose works skewered royal officials, especially future gover-nor Thomas Hutchinson.

James Otis set up a legal practice in Boston and became what Patri-ots called a "placeman," a royal appointee given his job as a political reward. Otis had been advocate general of the vice admiralty court when his conversion to Patriot began. Believing that writs of assis-tance violated basic British rights, he quit his royal post to represent merchants complaining about the injustice of the writs. Otis's elo-quent but unsuccessful plea—"A man's house is his castle" was his phrase—impressed young John Adams, who was in the courtroom. Otis's conversion haunted his marriage. His wife remained a Tory. One daughter married a British officer; the other married the son of a Continental Army general.[13]

John Adams's cousin, Samuel Adams, was a radical leader in the Massachusetts legislature. He and Otis composed a circular letter pro-testing the Townshend Acts, sent from the Massachusetts legislature to other colonies. In response the British government ordered the leg-islature to rescind the letter and told Governor Bernard to dismiss the legislature if its members refused. Hancock called a protest meet-ing with a proclamation that lamented "this dark and difficult Sea-son" and asserted the right of "American Subjects" to petition "their gracious Sovereign."[14] Representatives from ninety-six Massachusetts towns attended the meeting and urged the legislators to uphold the defiant act. They did, by a vote of 92 to 17.

Some Sons of Liberty commissioned their fellow Son, Paul Revere, to fashion a silver punch bowl—dubbed the "Liberty Bowl," which honored "the glorious NINETY-TWO . . . who, undaunted by the insolent Menaces of Villains in Power, from a strict Regard to Conscience, and the LIBERTIES of their Constituents . . . Voted NOT TO RESCIND."[15]

Governor Bernard realized he could not rely on militiamen or hardly anyone else in Massachusetts to help him enforce the law. Both sides had lawbreaking leaders. Hancock, who made much of his for-tune as a second-generation smuggler, was nevertheless captain of the Independent Company of Cadets, also known as the "Governor's

Own."[16] (Though the governor admitted a fondness for tax-free Madeira, he had the good sense not to get it from Hancock; Bernard's wine came from a smuggler in Cape Cod.)[17]

In July Bernard sought the aid of Lt. Gen. Thomas Gage, commander in chief of British forces in North America. Bernard sent off a courier with a letter to Gage's headquarters in New York City. "All real Power is in the hands of the lowest Class," Bernard wrote.[18] Gage sent four thousand troops to Boston—a ratio of one Redcoat to every four citizens.[19]

Loyalists welcomed the Redcoats as protectors; Patriots and their supporters in the streets saw the soldiers as an occupation force, sent by Britain to tame or even punish dissent. The first wave of troops and their cannons disembarked in October on Boston's Long Wharf, and, in the words of Paul Revere, they "Formed and Marched with insolent Parade . . . , Drums beating, Fifes playing and Colours flying."[20]

Most troops, unable to find quarters, encamped on the Common.[21] John Hancock, looking down from his mansion on Beacon Hill, could see them going through their drills. More troops arrived in November. The city council barred soldiers from invoking the Quartering Act because there were sufficient barracks in Castle William, a harbor fortress. Rather than post the troops that far from anticipated trouble, officers rented buildings in town as barracks.[22] Many officers were welcomed as long-term guests in Loyalist homes, forming bonds that would have profound effects on the future lives of Loyalist families.

Governor Bernard became the target of seething hatred. Sam Adams denounced him as "a Scourge to this Province, a curse to North America, and a Plague on the whole Empire."[23] What might once have been a political dispute among politicians had festered into street battles and the taunting of soldiers with shouts of "Bloodback!" and "Lobsterback!" as Redcoat patrols marched about the city.

In May 1769, during an anti-Bernard riot in Cambridge, a mob swirled around the Harvard campus and stormed Harvard Hall. The rioters spied a portrait of Governor Bernard hanging in the dining room. Someone whipped out a knife, stabbed the chest of the painted figure, cut out a piece of the canvas, and held it up, screaming that he had removed Bernard's heart.

John Singleton Copley, who had painted the portrait, restored Bernard's heart, to the displeasure of many Patriots, who saw him as a budding Tory.[24] Like many colonists, however, Copley was not taking sides: He painted about as many portraits of Tories as of Patriots. Among Copley's subjects were Tory merchants, along with John Hancock and Sam Adams. One day he painted Paul Revere, and another day he painted General Gage or Gage's beautiful American-born wife.[25]

The Sons of Liberty had once been a secret organization. Now as many as 350 Sons could picnic under sailcloth awnings on an August day near a friendly tavern and sing Dickinson's "Liberty Song." John Adams was there, noting in his diary that James Otis and Sam Adams were "promoting these Festivals, for they tinge the Minds of the People, they impregnate them with the sentiments of Liberty."[26] A month later, Otis was beaten up in a fist-and-cane coffeehouse brawl with a customs commissioner.[27]

The Rebel-controlled legislature charged Bernard with conspiracy to "overthrow the present constitution of government in this colony" and unanimously voted to send King George a petition asking him to dismiss Bernard. The governor, who had often said that he longed for a visit to England, sailed in August. Sounds of citizens' celebration were carried to his ears on the fair wind that sped his ship away from Boston.[28]

A hated royal governor. Sons of Liberty. Street mobs. Redcoats—this was political life in Massachusetts at the end of 1769 as the Old Colony Club celebrated the first Forefathers' Day. Words were hardening, and men were moving toward war. Loyalists were worried about their personal safety. Patriots wanted more power to the people, and there was fear in the air. But so far the idea of independence had not surfaced.

The political options of the club members in 1769 were evolving into dangerous and courageous choices that would determine where and how they would live the rest of their lives. And, in the raging years ahead, similar choices would be made by colonists in every layer of society.

The president of the club and one of its founders, Isaac Lothrop, was a Patriot, as was his brother Thomas, although they were the sons of a royally appointed judge and steadfast Loyalist. Isaac joined Plymouth's Committee of Correspondence, one of numerous such groups that the Sons of Liberty fostered throughout the colonies. By local tradition the idea of such committees had come from Plymouth's James Warren, a member of the club.[29] The duties of Committees of Correspondence ranged from keeping the colonies in touch with one another to exposing secret Loyalists and spies. Eventually some committees demanded that people suspected of Loyalist sympathies swear oaths of allegiance to the Patriot cause.

Another founder, John Watson, although known to be a Loyalist at heart, paid a price to remain on good terms with Patriots in Plymouth: He was one of the thousands of Americans who took a pro-Patriot oath while harboring doubts or secret opposition. Member Oakes Angier, after wavering, became a Patriot. Founder Elkanah Watson, a stauncher Patriot, saw his young son and namesake become a courier for Gen. George Washington.

Thomas Mayhew, Jr., became a lieutenant in the militia and marched off to Boston to serve in the Continental Army under Washington. Alexander Scammell and Peleg Wadsworth were both Harvard graduates and teachers in Plymouth. Their pro-Patriot feelings did not cost them their jobs because the Sons of Liberty were gaining power in Plymouth. Scammell and Wadsworth both joined the Continental Army and rose to the rank of general; Scammell would be killed in the last days of the war.

Elkanah Cushman, like many Loyalists who lived near Boston, sought sanctuary in the city, where Redcoats offered protection from Patriot mobs. Cornelius White joined the British and was lost at sea in 1779 while ferrying supplies from Halifax, Nova Scotia, to British-held New York. John Thomas fled to British Nova Scotia around 1780. Gideon White, Jr., whose great-grandfather, Peregrine White, had been born aboard the *Mayflower*, also chose the Tory side. At least three other Loyalist members would take up arms against fellow Americans.[30]

By the time the Old Colony Club got ready for its second Forefathers' Day celebration in 1770, a new revolutionary wave had swept over the colonies. On January 17, in New York City, British soldiers cut down a Liberty pole erected and cherished by the Sons of Liberty. Bayonet-wielding Redcoats fought Sons and their supporters armed with cutlasses and clubs. Several soldiers and rioters were wounded, but no one died.[31]

In Boston the Sons were protesting import taxes by urging merchants to refuse to deal in British goods. Appeals for support went out to many places, including Plymouth. The Sons published the names of Loyalist merchants who refused to support the Patriots' nonimportation campaign. One, Theophilus Lillie, a dry-goods dealer, responded in the pro-Loyalist *Boston Chronicle*. Lillie used a fundamental Loyalist argument—better to be ruled by a king than by a mob: "It always seemed strange to me that people who contend so much for civil and religious liberty should be so ready to deprive others of their natural liberty. . . . If one set of private subjects may at any time take upon themselves to punish another set of private subjects just when they please, it's such a sort of government as I never heard of before. . . . I had rather be a slave under one master (for if I know who he is I may perhaps be able to please him) than a slave to a hundred or more whom I don't know where to find, nor what they will expect of me."[32]

A gang of boys put up an effigy of Lillie outside his shop and noisily picketed it to drive off customers. On Thursday, February 22, Ebenezer Richardson, a fifty-two-year-old Loyalist, went to Lillie's shop and, by trying to destroy the effigy, drew a raucous crowd. He was well known to Patriots, and his work as a secret informer to royal officials had won him a customs post. Thursday was market day and a half day for schoolchildren, which made it a great day for crowd gathering. Richardson fled to his home, got his musket, and from a second-story window fired at the crowd. Christopher Seider, the ten-year-old son of German immigrants, fell, mortally shot in the head and chest. Another young boy, shot in the hand and legs, survived.[33]

The mob burst into Richardson's house, grabbed him and another man, and probably would have hanged him had not a Patriot leader steered the mob toward a justice of the peace. Richardson was jailed and later tried before Thomas Hutchinson, the royal lieutenant governor and chief justice. Hutchinson himself had seen the wrath of a mob one night in August 1765 when Stamp Act protesters broke into his mansion, nearly demolished it, and "scattered or destroyed all the manuscripts and other papers I had been collecting for 30 years."[34] Hutchinson, who would soon become royal governor, put Richardson on trial. He was convicted of murder, but Hutchinson did not sentence him to execution. (Some time later Richardson received a royal pardon and slipped out of Boston to a new customs post in Philadelphia.)[35]

The Sons of Liberty staged a martyr's funeral for the boy. About two thousand people marched behind his coffin, and former slave Phillis Wheatley, already famed as a black poet, wrote a memorial poem, claiming that "The Tory chiefs" made the boy "Ripe for destruction."[36] The killing and the funeral fired up smoldering resentment against Loyalists and the Redcoats. A week after the funeral, an encounter between a lone British sentry and a rock-throwing mob brought other soldiers and Capt. Thomas Preston to the scene. For tense moments the crowd taunted the Redcoats, whose muskets were loaded and aimed. Without an order from Preston, soldiers fired, killing three men and fatally wounding two others.

Paul Revere quickly produced color prints of "The Bloody Massacre perpetrated in King Street," giving a sensational title to a propaganda image that would further the Patriot cause.[37] In the murder trial of Preston and his soldiers, John Adams successfully defended

Preston and six Redcoats, producing testimony that contradicted Revere's image. Adams also got two other soldiers' murder charge reduced to manslaughter: Each had an M-for-murder branded on his right thumb. Adams won few friends among ardent Patriots by describing the mob as "a motley rabble of saucy boys, negroes and molattoes, Irish teagues and outlandish jack tarrs."[38]

The Boston Massacre, as it would be known in Patriot writings and oratory, produced a quick response in Plymouth, where selectmen unanimously endorsed a report drawn up by a committee that included at least two Rebel members of the Old Colony Club. The report, formally answering the Boston Patriots' request for support, said: "Every man not destitute of the principle of freedom and independence, and that has sensibility enough to feel the least glow of patriotism, must at this time be strongly impressed with a sense of the misfortunes of their country in general and of the town of Boston in particular."[39]

The murder of Christopher Seider—along with the massacre and Plymouth's emotional response—dampened the club's preparation for Forefathers' Day. But in April 1770 came the repeal of all the Townshend taxes except for the one on tea. The repeal temporarily calmed the restive colonies and the Patriots of Plymouth. The second Forefathers' Day went smoothly. Edward Winslow spoke, paying homage to his ancestors but making no direct reference to what was going on around him.

By 1772 no one could ignore the rising rebellion. In June, Rhode Island Patriots seized the Royal Navy's eight-gun schooner *Gaspée*, which had run aground while in pursuit of a suspected smuggler. One of the boarders shot the *Gaspée*'s captain, ordered the crew to abandon the ship, and then set it afire. British officials offered a reward for information about the raiders, but no one came forward. The attack provided Gage with another example of the growing audacity of the Rebels.

Plymouth's revolutionary playwright Mercy Otis Warren in 1772 anonymously published *The Adulateur*, a satire that cast Hutchinson as "Rapatio," a villain intent on raping a colony called Upper Servia. One of Rapatio's henchmen tells of a plan to subdue the citizens:

"... cramp their trade till pale-eyed Poverty
Haunts all their streets and frowns destruction on,
While many a poor man, leaning on his staff,
Beholds a numerous famished offspring round him,
Who weep for bread"[40]

Mercy's anti-Hutchinson theme reflected the mood in Plymouth, where Patriots were in the majority. In Boston events were moving faster than Hutchinson could handle. "I am in a helpless state," he said in a letter to Governor William Tryon of New York.[41] Edward Winslow realized that he could remain aloof no longer. He had to try to keep Plymouth from plunging into the Boston maelstrom.

Parliament's decision to retain the tax on tea had produced a highly effective boycott, pushing the principal source of Indian tea, the powerful East India Company, toward bankruptcy. To aid the well-connected company, Parliament in May 1773 passed the Tea Act, which gave the company a monopoly and allowed a bypass of a previous tax on tea that entered Britain from India. The act set a tea tax so low that the total price was less than a colonist would have to pay a smuggler. Once the tea reached America, it was to be to be entrusted to, and marketed by, special consignees—a new plum for loyal subjects.[42] Parliament and the Crown believed that the Tea Act would end the costly tea boycott. "The ministry believe," wrote Ben Franklin from London, "that threepence on a pound of tea, of which one does not perhaps drink ten pounds a year, is sufficient to overcome all the patriotism of an American."[43]

In August the names of the consignees, ostensibly selected by the East India Company, were revealed. The Sons of Liberty could not have asked for a better example of how the British government dispensed power and riches to a class whose members called themselves Friends of the King. The consignees included Thomas and Elisha Hutchinson, sons of the governor; wealthy merchant Richard Clarke, the father-in-law of both Thomas Hutchinson, Jr., and John Singleton Copley; and Joshua Winslow, one of the many privileged Winslows—and one of Copley's many Loyalist clients.[44]

Copley had been rising in the Boston social world since his marriage to Clarke's daughter, Susannah Farnum Clarke. On her mother's side, Susannah was the descendant of a *Mayflower* passenger who, by family tradition, was so eager to reach shore that she jumped from the shallop and waded through the surf, becoming the first woman of the *Mayflower* to set foot on American soil.[45] Copley had a fine home on Beacon Hill, and he dressed in the fashion befitting a man of the upper class. Now he was beginning to wonder whether he could live—and paint—in Boston without taking sides.[46]

The tea crisis built through November. Patriot leaders, in a move reminiscent of Stamp Act strategy, demanded that the consignees resign, a prudent move taken by consignees in Philadelphia and New York but not in Boston. A mob attacked Clarke's home, yelling threats and breaking windows. All the Boston consignees appealed to Hutchinson for protection. The *Dartmouth*, first of the tea ships, arrived in Boston Harbor. A meeting called the "Body of the People" demanded that the *Dartmouth* leave. Hutchinson suspected that Sam Adams was planning to somehow destroy the tea.[47] He was right.

When a Plymouth town meeting endorsed the happenings in Boston, Edward Winslow reacted by declaring that it was "an affront to the common sense of mankind."[48] Club member James Warren, who attended the town meeting, archly noted: "Little Ned Winslow (one of my Cousins) with a few other Insignificant Tories appeared at the meeting and played their Game by holding up the Terrors of the Governor's Proclamation which rather served us than themselves. From these Gentry in this Town we have little to fear."[49]

The crisis was escalating. "The flame is kindled and like lightening it catches from soul to soul," John Adams's wife, Abigail, wrote to her friend Mercy Otis Warren.[50] Copley tried in vain to negotiate an end to the impasse between the Sons of Liberty and Hutchinson. Two more tea ships, the *Eleanor* and the *Beaver*, arrived in Boston Harbor. No merchant would take the chance of challenging the Rebels by trying to unload the tea. Hutchinson refused to allow the ships to leave until the tea tax was paid.

Then, on the rainy morning of December 16, more than five thousand people gathered at the Old South Meeting House. Many of them waited through the day for word that Hutchinson had changed his mind. Finally, late in the afternoon, when a few candles were feebly lighting the assembly, Sam Adams rose to end the meeting with the words, "This meeting can do nothing more to save the country!" Those words have been handed down as the signal for launching a well-planned attack on the tea. Someone shouted, "Boston Harbor a teapot tonight!"[51]

Scores of men, their number unknown and their identities secret, headed for Griffin's Wharf. Many smeared dust or soot on their faces and wielded hatchets, thinly disguising themselves as Indians. A number of them, according to a Boston blacksmith, were young men like him: "apprentices and journeymen . . . living with tory masters." They boarded the ships, hacked open more than three hundred crates, and dumped their contents into the harbor.[52] "This Destruction of the Tea," John Adams wrote in his diary, "is so bold, so daring, so firm, intrepid & inflexible . . . that I cannot but consider it as an Epocha in History."[53] New York City Patriots also dumped tea into their harbor; tea ships were set afire off Annapolis; in Philadelphia, "the Committee of Tarring and Feathering" warned a tea ship captain to turn back, which he sensibly did;[54] tea shipped to Charleston was unloaded, but Patriots locked it up and prevented it from being sold.[55] All those cities transgressed—but only Boston would be punished.

News and official reports traveled slowly. So it would not be until the spring of 1774 that the punishment would come in the form of what became known as the Intolerable Acts: Parliament closed the port of Boston until the dumped tea was paid for. Parliament gave all the royal governors new powers to ban town meetings and appoint justices and other law officers. Other acts of Parliament moved the capital of Massachusetts to Salem, barred trials in Massachusetts of British soldiers for murder or other capital offenses, required citizens to house and feed soldiers on demand and provide them with carriages "with able men to drive the same." Another act expanded the territory of Quebec, enraging American colonists, who were prevented from settling on land they thought was rightly theirs rather than French-speaking Catholics'.

The Old Colony Club had managed to celebrate Forefathers' Day in 1772. But by 1773 all but Loyalists had quit. The Patriots' Plymouth Committee of Correspondence—some of whose members were also members of the club—invited the club to join with the committee in the celebration. The club, in words probably composed by Winslow, replied that the invitation was "so great an invasion of the liberties and privileges of the gentlemen of the town of Plymouth and the Old Colony Club that we cannot approve or comply with the same."[56] With those words the club was effectively abolished.

At low tide on Forefathers' Day in 1774, a band of Liberty Boys, as some of the Sons of Liberty were also called, escorted a wagon pulled by a long train of yoked oxen to a large rock on the Plymouth shore. The men, under the command of a future Continental Army colonel, dug around the embedded rock, dislodging it with manpower and jacks. As they raised the rock high enough to place it on the wagon, Plymouth Rock split in two. They let the bottom half drop back into its ocean bed and loaded the other half onto the wagon.

The oxen pulled the wagon to the town square. There, half of Plymouth Rock was placed near a large elm and a newly erected Liberty pole from which flew a flag. On this Forefathers' Day the words on the flag were "Liberty or Death."[57]

2

_____

ARMING THE TORIES

MASSACHUSETTS, 1774–1775

But, Oh! God bless our honest King
The Lords and Commons, true.
And if, next, Congress be the thing,
Oh, bless that Congress, too!
　　　—"A Poor Man's Advice to His Poor Neighbors," a ballad[1]

By 1774 Edward Winslow could not see any chance for compromise. To him the Sons of Liberty were enemies of his class—"Sons of Licentiousness," a "sett of cursed, venal, worthless Raskalls."[2] His words earned him a severe rebuke from a radical new local authority, the Plymouth County Convention, which charged that he had "betrayed the trust reposed in him" because of his openly Loyalist sympathies.[3] Winslow, fearing the end of the Plymouth that his dynasty had created and preserved, showed more than mere sympathy to Tory views. He believed that a civil war was breaking out and that Loyalists had to fight the Rebels (a label that the Sons of Liberty despised) to save the king's colonies. Winslow met secretly with Governor Hutchinson, who authorized him to raise and maintain a "Tory Volunteer Company."[4]

This was not yet a Loyalist call to arms. Winslow wanted a security force that could protect citizens from roaming mobs that Loyalists called Rebels even though they often were made up of hoodlums with no greater cause than mischief or anger directed at the upper class by the lower.

The Tory Volunteer Company kept Plymouth "in quiet long after all the towns in the neighborhood were in extreme confusion," Winslow wrote.[5] He credited Hutchinson with approving the idea of mobilizing Loyalists. But it was Hutchinson's successor, Lt. Gen. Thomas Gage, who carried out the plan.

The arming of Loyalists was a reaction to a new wave of protests that had begun to roar across the colony on May 10, 1774, when Bostonians learned that Parliament had passed an act closing the Port of Boston.[6] Three days later, Gage, who had been on leave in England during the Boston Tea Party, arrived in Boston. The city, at least on the surface, and at least on that day, appeared peaceable. He was escorted by the elegantly uniformed Boston Cadets, whose commander was Col. John Hancock, and he was given a noisy welcome that included three volleys of musketry and three rousing cheers.[7]

Gage brought with him new powers bestowed by the king and the Earl of Dartmouth, principal secretary of state for North America. Gage's traditional titles—vice admiral, captain general, and governor in chief—now in reality added up to the implicit title of military governor, for he was simultaneously royal governor and commander of British forces in North America. With the aid of the four regiments of British Army Regulars and a fleet of Royal Navy warships, he would rule over a military occupation of the city.

Although Gage's martial rule encompassed the entire colony, he focused his power on rebellious Boston while trying to keep close watch on the towns around the city. For this, he developed allies among leading Loyalists like Winslow. In early June, Gage moved the customs commissioners and their records from Boston to the relative safety of Plymouth, where Winslow was given a stipend for providing "an Office, fuel, and candles."[8] Winslow's work for the customs bureaucracy earned him additional umbrage from the Patriots.

Gage knew that royal authority was rapidly waning, but he refrained from imposing harsh martial law. He saw no use in trying to silence the Rebel newspapers. Freedom of the press had many defenders in Parliament, and nothing was to be gained by further inflaming the Rebels or giving Parliament another issue to debate. To prevent incidents he had ordered his officers not to wear side-arms. They thought that he was kowtowing to the Rebels. An officer, using the soldiers' nickname for Gage, grumbled in his diary: "If a soldier errs in the least, who is more ready to complain than Tommy?"[9]

Loyalists waited expectantly for him to crack down on the Rebels. Peter Oliver, a wealthy landowner and chief justice of the colony, hoped Gage would swiftly end the incipient rebellion by arresting Rebel leaders on charges of high treason, a hanging offense. But, Oliver added, "unhappily for the Publick, the People were disappointed & the Traitors felt theirselves out of Danger." The reason, Oliver said, was "Timidity, in Suppression of Rebellion."[10] Gage, a longtime soldier trying to be a governor, would try to govern by sheathing his sword and picking up a pen: There would be no immediate arrests, no curfew, no raids on Rebel meeting places.

Gage prorogued—discontinued without dissolving—the Great and General Court of Massachusetts and moved it from turbulent Boston to quieter Salem. Voters boldly responded by electing a provincial congress to replace the General Court. Gage canceled his call for a new legislative session. But the representatives kept meeting as a body. It was recognized by Patriots but not by Loyalists, although men of both persuasions were among the members.

Sam Adams, looking beyond protests and mob action, wanted the legislature to vote to send delegates to a congress* that would take up the Patriot cause. He knew that Gage would dissolve the legislature if his Loyalist informers heard of Adams's plan. On June 17, after secretly revealing his intentions to a few Patriot legislators, Adams moved that the legislature appoint a five-man delegation to a Continental Congress to meet in Philadelphia beginning on September 1.

---

* A convening of delegates who, while lacking legal powers, discussed and voted on resolutions describing their views.

Before any Loyalists could dash for the door, Adams locked it and pocketed the key, only relenting when a Tory member claimed to be ill and needed to leave.

The suddenly cured delegate ran to find Gage, who immediately wrote an order dissolving the legislature and dispatched his secretary to the legislative chamber. When no one would admit him, the secretary read the order to some people standing in the outside hall. Behind the locked door the legislators voted 117 to 12 to send Sam Adams and John Adams to Philadelphia, along with Robert Treat Paine, a Patriot leader in Taunton, a Tory stronghold about thirty-five miles south of Boston; Thomas Cushing, a prominent Boston lawyer; and James Bowdoin, a wealthy Harvard graduate who shared with Ben Franklin an interest in electricity and the phosphorescence of the sea.[11]

The Virginia House of Burgesses in Williamsburg had also been officially shut down. But defiant members, including George Washington, Thomas Jefferson, and Patrick Henry, met at the nearby Raleigh Tavern and declared support for Massachusetts by agreeing not to import British goods. The declaration defied traditional British policy, which frowned on collaboration among colonies. In New York a committee consisting mostly of merchants also responded to Boston's appeal. Committees of Correspondence kept the idea moving through the colonies. The result would be the First Continental Congress, attended by delegates from all the colonies except Georgia, the newest and least populous.[12]

Members of the Massachusetts General Court's council, which served as a kind of upper house, had been elected in the past. Acting under the new law, Gage appointed members instead. The appointees were known as "mandamus councillors," a reference to a royal writ of mandamus, from the Latin for "we command." Gage nominated thirty-six councillors. Twelve immediately declined, and most of those who remained swiftly resigned, fearing retribution from the Sons of Liberty.[13]

Similarly, when Gage appointed royal judges, many declined to serve. And those who did agree to take the bench discovered that they could not empanel juries because men who were summoned, including Paul Revere, would not even enter the courtroom. In some places people blockaded the courthouses.[14] In Barnstable, on Cape Cod, for example, a crowd of about twelve hundred men gathered in front of the Court of Common Pleas and refused to allow the chief justice to enter after he demanded passage in the king's name. The confrontation ended peacefully.[15] But royal governance was disappearing from courthouses, just as it had from the legislature. All "civil Government, both Form & Substance," had ended, Peter Oliver lamented. "The People now went upon modeling a new Form of Government, by Committees and Associations. . . . The wild Fire ran through all the Colonies."[16]

The minutes of town meetings in Worcester, forty miles west of Boston, show how conflicts over the Intolerable Acts were turning neighbor against neighbor during that restive summer of 1774. Back in March, when Worcester held its annual meeting, townspeople voted to boycott tea until the tea tax was repealed. Twenty-six "Royalists" voted against the boycott. Shortly afterward forty-three people, all labeled Royalists, signed a petition to reconsider the vote. After a long and bitter debate in June, the petition was defeated.

Now, with Gage in the governor's chair, the Worcester Royalists were emboldened. Determined to air their beliefs, they filed with the town clerk a long statement that began by lamenting how "sober, peaceable men" in their town had been "deceived, deluded and led astray by the artful, crafty and insidious practices of some evil-minded and ill-disposed persons." The Royalists' statement denounced the tea dumpers in Boston and attacked the Committees of Correspondence, whose "dark and pernicious" actions were leading people toward "sedition, civil war, and Rebellion."

The town clerk, to the astonishment of the Worcester Patriots, allowed the statement to be treated like any other public document. The *Boston Gazette* learned of the statement and published it on July 4. The Worcester Patriots, enraged, plotted retaliation.[17]

On August 22 a large force of men from Worcester and area towns, unarmed but marching "in military order," assembled on the town common. A committee called on Timothy Paine, a mandamus councillor and father of a notorious Tory.[18] Paine, fearing violence, agreed to resign. The marchers—a history of the town puts their number at three thousand—then headed for Main Street and formed a gauntlet that extended from the courthouse to the meetinghouse. A crowd gathered, wondering what was going on. Patriots pulled Royalists out of the crowd and pushed them into the gauntlet with Paine, forcing them all to stop frequently to read aloud their "acknowledgment of error and repentance."[19]

Next, a smaller force—about five hundred men, according to the town history—headed for nearby Rutland and demanded the resignation of another new councillor, James Putnam, a fifty-year-old fifth-generation American. A renowned lawyer, he had served as a major in the French and Indian War. John Adams had boarded in Putnam's home and studied law under him for two years while teaching school in Worcester.[20] Putnam had been the leader of the Worcester group that had signed the anti-Patriot statement. He was not home when the protesters arrived, so one of them handed a member of his family a letter ordering Putnam to publish his resignation in Boston newspapers.[21]

Another mob, armed with clubs and muskets and numbering some fifteen hundred, confronted Daniel Murray, a designated councillor from Rutland, and demanded his promise to resign. They menaced him at first. But he was surprised to see them disperse "without doing the least damage to any part of the estate."[22] Still, the roving mobs frightened area Loyalists, several of whom collected firearms, ammunition, and food and gathered at Stone House Hill in Worcester, setting up defenses and transforming the house into what became known as the Tory Fort. They stayed there for about three weeks, awaiting a Patriot attack that never came.[23]

Putnam and scores of other Loyalists were also reviled as "Addressers" because they had signed printed copies of flowery statements, called "addresses," lauding Gage after his arrival on May 13, 1774, and Hutchinson, prior to his departure on June 1. One address to Hutchinson expressed "the entire satisfaction we feel at your wise, zealous, and faithful administration."[24] An equally flattering tribute was presented to Gage, hailing him for his "experience, wisdom and moderation, in these troublesome and difficult times."[25]

A broadside published by the Patriots identified the 123 Addressers of Hutchinson, including relatives of placemen and artisans, such as jewelers and makers of chaises (open carriages, often with collapsible hoods), whose leading customers were Loyalists. Of the Addressers, 14 were officers of the Crown, and 63 were merchants.

Many Patriot tradesmen refused to serve Loyalists. Forty-three blacksmiths in Worcester County, for example, proclaimed that they would not do any work "for any person or persons commonly known by the name of tories." The blacksmiths urged other Patriots "to shun their vineyards" and "withhold their commerce and supplies."[26]

Some of the Addressers attempted to ward off trouble by apologizing for "that unguarded action of ours," hoping that their acts of contrition would restore them as "Partakers of that inestimable blessing, the good Will of our Neighbours, and the whole Community."[27] Repentance did little good. They were all marked men; even those who rejected Gage's offer were denounced or threatened.

The Patriots declared that no one should deal with mandamus councillors in any way. Jessie Dunbar, of Bridgewater, twenty miles west of Plymouth, defied the edict by buying livestock from Nathaniel Ray Thomas, a councillor and a leading Tory in Marshfield, thirty miles south of Boston. The original Edward Winslow had founded Marshfield in 1640. A road built then, connecting Plymouth with Marshfield, was probably America's first; stretches of it still exist as the Pilgrim Trail.[28] The original Winslow sold the original Thomas a large swath of Marshfield land, and the two families, generation after generation, remained connected there, by road and by heritage.

Dunbar drove the animals he had bought from Thomas to Plymouth, where he skinned an ox and hoisted the carcass up on a rack for sale. Men who identified themselves as Sons of Liberty soon appeared, cut down the carcass, put it in a cart, and forced Dunbar into the ox's sliced-open belly. They "carted" him, as they called the ordeal, for about four miles, charged him a dollar for the ride, and then handed him over to a new mob whose members carted him for several more miles, extracted another dollar from him, and passed him on to a third mob in Duxbury. There his tormentors yanked him from the ox's belly, then reached in to pluck out the beast's entrails, which they used to whip him about his body and face.[29]

Duxbury bordered on Plymouth and had been settled in 1632 by *Mayflower* passengers, including storied Miles Standish. Loyalists in Duxbury reported that the local militia had spawned a number of "Minute Men," volunteers who vowed to be ready to go into battle at a minute's warning. The First Massachusetts Provincial Congress had directed that one-third of each militia consist of men and officers ready and equipped to respond swiftly to an alarm. Other colonies copied the Massachusetts system in various ways.[30]

Until the Revolution militias were royal military units, usually commanded by wealthy and influential officers commissioned by the Crown. As Patriots began rising to power, they formed provisional governments whose revolutionary acts included the abolition of royal militias. Many militiamen became self-appointed enforcers, harassing Tories and policing boycotts. The First Continental Congress urged Patriot leaders to control local militias. The second went further, advocating the election of militia officers and suggesting that militia companies form themselves into regiments controlled by elected provincial governors and legislatures. Loyalists eventually reacted by forming their own military units.

Gage undoubtedly knew about the spreading minuteman movement through his intelligence network of Tory spies and informers. But, without disclosing his own knowledge, he politely replied to a letter of concern from Duxbury justices, promising that he would "take every step" in his power to "secure to them the peceable enjoyment of all their constitutional privileges."[31]

* * *

One of the mandamus councillors who did not resign was Daniel Leonard, a thirty-four-year-old lawyer from Taunton. While representing Taunton in the legislature in the earlier 1770s, he had been an eloquent supporter of the Patriots, even though he was the son of a Loyalist judge. But Leonard the Patriot married an heiress and, as John Adams described Leonard's metamorphosis, "He wore a broad gold lace round the rim of his hat, he had made his cloak glitter with laces still broader, he had set up his chariot and pair. . . . Not another lawyer in the province . . . presumed to ride in a coach or chariot."[32] By 1774 Leonard was a committed Loyalist.

"When I became satisfied that many innocent, unsuspecting persons were in danger of being seduced to their utter ruin, and the province of Massachusetts Bay in danger of being drenched with blood and carnage," Leonard wrote to Gage, "I could restrain my emotions no longer."[33]

Patriots in Taunton had tolerated Leonard until they learned that he had accepted Gage's appointment as a councillor.[34] A huge mob gathered near Leonard's home and began shouting threats. Leonard's father stepped outside the mansion. Saying his son was in Boston, he promised to try to talk Daniel into resigning. The mob dispersed. The next night a smaller group appeared and, seeing a light in a bedroom, believed that Daniel was home. Someone fired a gun. The shot shattered the window of the bedroom window, narrowly missing Daniel's pregnant wife in her bed. When the Leonard baby was born mentally disabled, the family blamed the terrifying shot in the night.[35]

In western Massachusetts, Gage had selected Israel Williams, a longtime legislator, to be a mandamus councillor. He, too, had declined the appointment. Nevertheless, one night a mob kidnapped the sixty-five-year-old. They brought him to a house several miles away and confined him to a room with a fireplace, locked the doors, built a fire, and then blocked the chimney. His captors kept Williams gasping for air until morning, when he stumbled out of the smoke-filled room and signed a paper that pledged his opposition to British authority.[36]

Patriot mobs' "smoking" treatment was so commonplace that John

Trumbull used the term in references to Murray and Williams in a poem. Trumbull, a lawyer who had practiced with John Adams, was a Patriot poet best known for *M'Fingal*, a long poem that satirized Loyalists. It included these lines:

> *Have you made Murray look less big,*
> *Or smoked old Williams to a Whig?*[37]

Smoking and intimidation had mostly replaced tar and feathers, which had been the punishment of choice during Stamp Act days. The practice, which British torturers had been performing since the Crusades, was hurting the Patriot cause.[38] "Americans were a strange sett of people," a member of Parliament remarked in 1774, ". . . instead of making their claim by argument, they always chose to decide the matter by tarring and feathering."[39] After the Stamp Act protests died down, Patriot leaders tried to curb the mobs of Massachusetts; tarring and feathering dropped off, although one incident was well publicized.

John Malcolm, a notorious customs informer, had beaten a Boston man with a cane. On January 25, 1774, avengers put Malcolm "into a Cart, Tarr'd & feathered him—carrying thro' the principal Streets of this Town with a halter about him, from thence to the Gallows & Returned thro' the Main Street making Great Noise & Huzzaing."[40]

Redcoats in Boston chose to use the punishment on a civilian caught in a sting operation when he accepted a soldier's offer to sell his musket. The man was stripped naked, covered with tar and feathers, placed in a cart, and escorted by fifes, drums, and about thirty grenadiers to the Liberty Tree, a venerable elm around which Patriots had been rallying since the Stamp Act protests. A large crowd of scowling citizens gathered and rescued the man as the parade about-faced and rapidly headed back to the British barracks. A protest to Gage went unanswered.[41]

By the summer of 1774, tension between Bostonians and Redcoats was growing and violence was spreading. In July, Jonathan Sewall, the Massachusetts attorney general, talked about the crisis with his

longtime friend John Adams. Their legal duties happened to put
them at the same time in the northern part of the colony that was
called Maine. They climbed a hill in Falmouth (now Portland) and
were buffeted by the breezes of Casco Bay as they spoke.

As one of the Massachusetts delegates to the Continental Congress,
Adams would soon be off to Philadelphia. Sewall urged his friend not
to go and to abandon the Patriot cause. "I answered," Adams wrote in
his diary, "that the die was now cast; I had passed the Rubicon. Swim
or sink, live or die, survive or perish with my country was my unalter-
able determination."[42]

Through the summer Rebel militiamen from several towns withdrew gunpowder from the Massachusetts Provincial Powder House
about six miles northwest of Boston in what is now Somerville. The
towerlike stone structure was topped by one of Ben Franklin's lightning rods.

In some way—probably through a Tory informer—word of the
gunpowder removal reached Maj. Gen. William Brattle, the highest-
ranking officer of the colony's royal militia.

Brattle, a remarkable man of many achievements, had not yet personally acted on the crisis. In one of John Singleton Copley's earliest
portraits, the large, imposing Brattle is presented in a dazzling gold-
trimmed uniform. He lived a comfortable life in one of the mansions
on a quiet street that curved along the Charles River in Cambridge, a
street that became known as Tory Row.

In Cambridge the Tories were a people apart. About 90 percent
of the fifteen hundred residents were descendants of Puritans. They
worshipped in Congregational churches and clung to their opposition
to the Anglican Church. The Tories of Cambridge, many of whom
owned sugar plantations in Jamaica and Antigua, lived on large adjacent estates along Tory Row and entertained one another with lav-
ish dinners. They were all Anglicans who had built their own Christ
Church down the street, near Harvard College. Each Tory family
purchased a pew, allotting space in the rear for slaves and servants.[43]

The church and Tory Row formed a little social world untouched
by Boston. "Never had I chanced upon such an agreeable situation,"

wrote the wife of a Hessian general who had lived nearby for a time. "Seven families, who were connected with each other, partly by the ties of relationship and partly by affection, had here farms, gardens, and magnificent houses, and not far off plantations of fruit. The owners of these were in the habit of daily meeting each other in the afternoons, now at the house of one, and now at another, and making themselves merry with music."[44]

Brattle, who graduated from Harvard in 1722 at the age of seventeen, had had three successful careers—clergyman, physician, lawyer—in his public life. But when rebellion began to brew, he became best known as a military officer. Outwardly he tried to maintain neutrality between Loyalists and Patriots. But he was secretly informing General Gage about Rebel activity in Cambridge. After learning about the steady removal of gunpowder, he passed the information to Gage.

Gage moved quickly. In a smooth, predawn military operation, 260 Redcoats slipped out of Boston Harbor in longboats, were rowed to a landing near the powder house, got the key from a royal sheriff, and carried off 250 half barrels of gunpowder, which they transported through Cambridge, Roxbury, and Dorchester to military headquarters at Castle William on Castle Island in the harbor.[45]

On September 1, after learning what had happened, Patriots set off what became known as the Powder Alarm. Church bells tolled. Drums banged. Thousands of Rebel militiamen mobilized. Express riders carried wild rumors—warships are shelling Boston, Redcoats are on the march—as far as Connecticut, where more musket-carrying men assembled. A mob of four thousand men swarmed into Cambridge and marched to Tory Row.

Somehow Brattle's letter to Gage had become public. In the words of John Rowe, a prominent Boston merchant, the letter "exasperated the country people against Brattle, so that he now takes refuge in Boston."[46] Brattle's sudden desertion from Tory Row began an exodus of Loyalists to the protection of the Redcoats of Boston, establishing the city as a sancuary for Tories.

The mob swarmed down Tory Row, shattering the windows of

Jonathan Sewall's mansion. His fast-thinking wife, Esther, talked
the mob into leaving the house intact in exchange for the contents of
Sewall's well-stocked wine cellar.[47] The next day mobs marched on
the Cambridge courthouse in Harvard Square and successfully ex-
acted resignations from two Tory judges serving as mandamus councillors. Later, Patriots stormed the mansion of Brattle's good friend,
Lt. Gov. Thomas Oliver, demanding his resignation. Oliver picked up
a quill pen and wrote, "My house at Cambridge being surrounded by
about four thousand people, in compliance with their demand I sign
my name."[48] Patriot mobs also hunted down Col. David Phips, the
sheriff who had given up the powder house key, and forced him to
promise he would not enforce the new anti-Patriot laws.

Patriots—"hoodlums," the Loyalists called them—were openly asserting their power. Rowe's diary tracks the swiftly growing crisis:

> Sept. 2—. *A great number of people from the country are collected at*
> *Waltham, Watertown, and Cambridge, occasioned as tis reported from*
> *the behaviour of Genl Brattle. . . . [General Gage] has reinforced the en-*
> *trance at the Neck [the only approach to the city by land]. Commissioner*
> *[of Customs] Hallowell has been insulted in his way through Cambridge;*
> *he fled for shelter to this town [Boston]. Sept 3— [Gage] sent four field*
> *pieces to Boston Neck. . . . Sept. 7—[Gage] has doubled the guards at the*
> *Neck, and I believe designs to fortify it."*[49]

In October a red flag emblazoned with the words "Liberty and
Union" was raised on the town green of Taunton. Enraged Loyalists
reacted by claiming that hoodlums were threatening to take over the
town. Taunton's militia was no longer royal; it was commanded by
Col. David Cobb, a Patriot. His brother-in-law, Robert Treat Paine,
was in Philadelphia, serving in the Continental Congress.

Paine certainly was not a hoodlum. He had been a prosecuting attorney for the Crown in the Boston Massacre trial. Son of a minister
and grandson of a Connecticut royal governor, he had graduated with
honors from Harvard at the age of fourteen. He had served in the
French and Indian War as a Congregationalist chaplain, Though he

possessed many of the attributes of a Loyalist, he had chosen the Patriot path.

To Ned Winslow and many angry and fearful Loyalists like him, that path was leading toward civil war. They believed that a stronger military presence was needed in the towns outside of Boston, and they wrote to Gage asking that a warship be sent to Plymouth Harbor to provide an emergency exit for Loyalists whose escape to Boston would be barred by towns "where inhabitants are notorious factions & malicious."[50]

Gage was obviously aware of Patriot activity in the countryside. Two companies of Redcoats—about one hundred men—patrolled the area around his secondary residence, an estate in Danvers about twenty-five miles north of Boston.[51] But Gage hesitated to commit more of his men to Tory protection duty.

Sometime around the beginning of 1775, two hundred of the "principal inhabitants" of Marshfield, asserting that they had been "insulted and intimidated by the licentious spirit that unhappily has been prevalent among the lower ranks of people in the Massachusetts Government," asked Gage for troops "to assist in preserving the peace, and to check the insupportable insolence of the disaffected and turbulent."[52]

Patriot selectmen from Plymouth and five other towns wrote to Gage protesting proposals to send troops out of Boston to protect Loyalists.[53] Patriots in Plymouth threatened to march on Marshfield and drive the Loyalists from their farms. Marshfield Loyalists reacted by sending a horseman galloping off to Gage with an urgent plea for help. Gage ordered three officers and one hundred men of the Queen's Guards to board two ships that sailed immediately to Marshfield. The ships also carried three hundred "stands of arms," presumably for civilian soldiers. In those days a stand of arms usually consisted of a musket, a bayonet and a cartridge box, and belt.[54]

The Guards and their officers bivouacked on the vast Marshfield estate of Nathaniel Ray Thomas. "The King's troops are very comfortably accommodated, and preserve the most exact discipline," a Loyalist wrote, "and now every faithful subject to his King dare fully utter his thoughts." The commander of the unit, Capt. Nesbit Balfour, and other officers made themselves at home in Thomas's mansion, even having a wine cellar built to Balfour's specifications.[55]

When the Queen's Guards marched into Marshfield, according to the Loyalists' rollicking account of the Patriots' panicky withdrawal, only about a dozen Rebel militiamen appeared—and then quickly disappeared. Gage sent the troops as a show of force against the so-called minutemen (a new, more frightening word than the old and familiar "militiamen") who had begun to roam the country roads beyond the usual reach of his Boston Redcoats. Queen's Guards escorted Loyalists when they set out on those roads. In one clash between the militiamen and a Redcoat patrol, a British sergeant threatened to shoot the militiamen if they failed to give up their arms. They had handed over their weapons.[56]

"Minutemen" became the Loyalists' label for a new brand of Patriots—Rebels carrying muskets. In March 1775 a Loyalist had a harrowing encounter with three minutemen who pounded on his door and announced that they were going to carry him off to jail. When they refused to leave,

> I then took down my gun from where it hung, and . . . all three Sprang upon me, Renched ye Gun out my hand, when my sick wife and all my children stood by, Screeching, Screaming & Crying, my wife begging, as if it was for her Life. I then Ran in to my Shop, and took my Sword . . . and Swore if they did not Leave the House I would Run them through. . . . . I then told my little son to fetch me my horse and put on the saddle, which he did; I then, with my Every Day Clothes, went out of my house and mounted my Horse . . . . I Rode out, fast.[57]

In Marshfield, as in nearby Taunton, the Patriot-Loyalist conflict was tearing the town apart. Some three hundred townspeople, including members of the Winslow family, belonged to the Associated Loyalists of Marshfield. On January 31, 1774, at a town meeting, a majority approved a resolution pledging loyalty to the king. From a town meeting on February 14 came a counterdeclaration pledging that Patriot townspeople would contribute their "last mite for the cause of liberty in the province."[58]

The arrival in Marshfield of Captain Balfour and his Queen's Guards temporarily quieted the town. Balfour, making his acquaintance of prominent Loyalists in the neighborhood, became a frequent guest at Winslow's home. During one visit he offered to station some of his Guards in Plymouth. Winslow declined the suggestion because, he said, local Rebels would be inflamed by the sight of Redcoats on the streets of his town.

When one of Balfour's subordinates did stroll down a Plymouth street, a crowd surrounded him. He darted into an apothecary shop. Some people followed him in and demanded his sword. When he refused, they snatched it from him, broke it into pieces, and departed, satisfied with merely humiliating the officer.[59]

During a dinner at Winslow's home, Winslow laid out to Balfour a plan to march some of the Guards from Marshfield to Plymouth to capture the Rebels who essentially ran free there. Balfour turned to one of the guests, John Watson, a member of the now-defunct Old Colony Club, and asked if the Rebels would fight back. "Yes, like devils," Watson replied. Balfour politely turned down Winslow's plan.[60]

Late one night in August 1774, a mob attacked Col. Thomas Gilbert of Freetown, about thirty-five miles south of Boston. Gilbert, a tough old soldier, fought off his assailants. Then the same mob fell upon Brig. Gen. Timothy Ruggles, who, "by his firm Resolution," also emerged from a scuffle unharmed. But, said Tory chronicler Peter Oliver, "in Revenge" men in the mob cut off the tail and mane of one of Ruggles's finest horses and painted the animal.[61] Another account says that his cattle were maimed and poisoned.[62] Gilbert himself later wrote that several attempts were made on his life.[63]

Gilbert and Ruggles separately began planning the creation of Loyalist military forces that would fight the Patriots and their minutemen without dependence on Gage and his Redcoats. As proud

comrades in battle during the French and Indian War, they preferred
to be addressed by their wartime ranks. Gilbert, who lived in the
Tory-dominated Assonet section of Freetown, in effect took command of armed Loyalists in southeastern Massachusetts. He seems
to have become the quartermaster general of the stands of arms that
had been shipped with the Queen's Guards sent to Marshfield. With
Gage's approval, Gilbert stored muskets, powder, and bullets in his
home and, in his words, "collected and armed about 300 Loyalists,
trained and Exercised them."[64] This would be the first Loyalist military corps raised in the colonies.[65]

A fifth-generation American, Ruggles could claim, through his
wife, a link going back to the beginning of the Plymouth Colony.
President of the Stamp Act Congress of 1765, he had launched his career as a staunch Loyalist. But he had refused to sign the Declaration
of Rights.[66] He became a wealthy man, the owner or five farms. He
kept thirty prized horses and maintained a deer park to entertain his
hunting guests.[67]

Ruggles was distantly related to John Adams,[68] who once wrote
that Ruggles's "grandeur consists in the quickness of his apprehension,
steadiness of his attention, the boldness and strength of his thoughts
and expressions, his strict honor, conscious superiority, contempt of
meanness, &c." After Ruggles became a militant Loyalist leader,
Adams changed his assessment, condemning Ruggles's behavior, accusing him of being governed by "pretended scruples and timidities,"
and claiming that he was "held in utter contempt and derision by the
whole continent."[69]

All around him Ruggles saw that the Patriots—"a banditti, whose
cruelties surpass those of savages"[70]—were organizing and uniting for
war. Ruggles believed Loyalists should quickly do the same, forming a counterforce that he called the Loyal American Association. Its
members would pledge, "with our lives and fortunes," to "stand by
and assist each other in the defence of life, liberty and property, when-
ever the same shall be attacked or endangered by any bodies of men,
riotously assembled upon any pretence, or under any authority not
warranted by the laws of the land."[71] (The First Continental Congress

used similar language, resolving that Americans "are entitled to life, liberty & property.")

Brigadier Ruggles was a stubborn, committed man. When he had been appointed a mandamus councillor and set off to Boston to be sworn in by Gates, he had to cross a bridge on the road out of Hardwick. At the head of a menacing crowd stood his Patriot brother Benjamin, the captain of a royal militia that was rapidly turning rebellious. Town tradition records the scene:

Benjamin quietly faced his older brother and said that if he crossed the bridge he would never be allowed to return.

"Brother Benjamin," Timothy replied, "I shall come back—at the head of five hundred soldiers, if necessary."

"Brother Timothy," Benjamin replied, "if you cross that bridge, this morning, you will certainly never cross it again—alive."

Timothy Ruggles turned, waved his hand to the crowd, and at a brisk military pace crossed the bridge.[72]

Ruggles's proposed Loyal American Association would protect "the good people of the province." Members of the association were to pledge that they would not submit to any rebellious assembly and would instead "enforce obedience" to the king and his laws; they would defend each other if imperiled by "any Committees, mobs, or unlawful Assemblies," and, "if need be, will oppose and repel force with force." Finally, if the person or property of any member of the association was injured and "full reparation" was refused, "we will have recourse to the natural law of Retaliation."[73] Within three weeks, according to an enthusiastic Loyalist, 150 men of Marshfield signed up for the association.[74]

Not long after Ruggles's plan became public, a band of Loyalists in Portsmouth, New Hampshire, signed an agreement that they would "defend and Protect Each other from Mob Riots or any unlawful attacks . . . to the utmost Extremity." Those Loyalists were reacting to the storming of William and Mary Castle, a crumbling fort about three miles offshore, by Portsmouth Patriots. After easily overpowering the British guards, the raiders carried off cannons, muskets, and one hundred barrels of gunpowder. John Wentworth, the New

Hampshire royal governor, appealed to Gage for help. Two Royal Navy warships soon appeared off Portsmouth, and from one of them the governor borrowed twenty stands of arms for his newly formed Loyalist Association.[75] Wentworth boarded one of the warships and, with his wife and infant son, sailed away to Boston hoping in vain that he could persuade Gage to save New Hampshire from the Rebels.[76]

On village greens from New England to South Carolina, Rebel militias were drilling. Patriot couriers were carrying messages from town to town and colony to colony. Georgia, which managed to send only one delegate to the Second Continental Congress, finally showed its solidarity in May 1775, when Sons of Liberty broke into a royal magazine in Savannah, stole gunpowder, and then shared it with South Carolina Rebels.[77] The authority of the Continental Congress was spreading and showing itself, through the creation of provincial congresses and through "resolves" that ranged from military matters to social behavior. In Wilmington, North Carolina, for example, the Committee of Safety decreed that "Balls and Dancing at Public Houses are contrary to the Resolves of the General Congress" and warned a known Tory to withdraw her plans for a ball at her house. She stubbornly went on with the ball, apparently after some negotiating. But the committee issued a general warning against future dancing.[78]

In Virginia delegates to the Virginia Convention met to approve acts of the Congress. To evade the royal governor—John Murray, the Earl of Dunmore—the delegates met in a church in Richmond instead of the royal capital of Williamsburg. On March 23 Patrick Henry proposed raising a Patriot militia and, to critics seeking conciliation, he declared:

> *Gentlemen may cry, Peace, Peace—but there is no peace. The war is actually begun! The next gale that sweeps from the north will bring to our ears the clash of resounding arms! Our brethren are already in the field! Why stand we here idle? What is it that gentlemen wish? What would*

they have? Is life so dear, or peace so sweet, as to be purchased at the price of chains and slavery? Forbid it, Almighty God! I know not what course others may take; but as for me, give me liberty or give me death![79]

In London, Parliament debated conciliation and then voted against the idea, 270 to 78.[80] War now seemed inevitable; indeed, as Patrick Henry said, it had actually begun. Ruggles's Loyal Americans was only a proposal. And Gilbert's three hundred armed Loyalists were not members of a trained armed force. But Ruggles's and Gilbert's plans posed a strategic question for Gage: Would the Loyalists fight against their countrymen?

3

FLEE OR FIGHT

MASSACHUSETTS, JANUARY–APRIL 1775

The flame of Civil War is now broke out in America, and . . .
it will rage with a Violence equil to what it has ever done in
any other Country. . . . I think that the people . . . will adopt the
proverb, which says, when the Sword of Rebellion is Drawn, the
Sheath should be thrown away; Ocians of blood will be shed.
                                        —*John Singleton Copley*[1]

Residents of Cambridge's Tory Row who were harassed during the September Powder Alarm—William Brattle, Col. David Phips, Lt. Gov. Thomas Oliver, and others—fled to Boston, Gage's garrisoned city. The exodus from towns near Boston had begun around the time that Governor Hutchinson sailed away and General Gage arrived. Those events had produced the "Addresser" phenomenon, which identified leading Loyalists, made them certified objects of Patriot wrath, and inspired John Trumbull to give them notoriety in *M'Fingal*—

> *And look our list of placemen all over;*
> *Did heaven appoint our chief Judge Oliver,*
> *Fill that high bench with ignoramus,*
> *Or has it councils by mandamus?*

Who made that wit of water-gruel
A judge of admiralty, Sewall?
And were they not mere earthly struggles,
That raised up Murray, say, and Ruggles?
. . . Or by election pick out from us
That Marshfield blunderer, Nat. Ray Thomas . . . ?[2]

Some fled from real or imagined mobs, taking their families with them, and some fled alone, all hoping that the rising storm soon would pass, that the Redcoats would prevail, that they could return to their homes and reclaim their lives. Israel Williams—the Tory who had been smoked—choose to sail to England, as did Jonathan Sewall, the former Massachusetts attorney general, and, after a sojourn in Philadelphia, former judge Samuel Curwen.

Mrs. Curwen, a descendant of the Winslows of Plymouth, chose to stay in America, because, Curwen archly observed, "her apprehensions of danger from an incensed soldiery, a people licentious and enthusiastically mad and broken loose from all the restraints of law or religion being less terrible to her than a short passage on the ocean."[3] In London, Curwen found "an army of New Englanders." But, he added, ". . . My distress and anxiety for my friends and countrymen embitter every hour."[4]

John Singleton Copley, painter of Patriots and Loyalists, also sailed away, first to Italy for a tour through churches and galleries of great art, and then to London, where he became a member of the newly formed Loyalist Club. His wife sailed later. Copley, an Addresser, followed the family of his wealthy tea merchant father-in-law, who had been driven out of Boston by a mob. His half-brother, Henry Pelham, a committed Loyalist, remained awhile in Boston and started spying for Gage. "My hand trembles while I inform you that [the] Sword of Civil War is now unsheathd," Pelham wrote in a letter to Copley. Soon after, Pelham joined Copley in London. Like many who chose refuge in England, Copley and Pelham never returned to America.[5] Eventually more than seven thousand Loyalists moved to England.[6]

By the early spring of 1775, hundreds of Massachusetts refugees, along with some from New Hampshire,[7] were living in Boston. Most of them thought that Gates and his Redcoats would quickly put down any armed rebellion that might arise. Others, like the recruits for Ruggles's Loyal American Association, believed that Gage would need the aid of armed volunteers. And some spewed invective. One refugee described the Patriots as "more savage and cruel than heathens or any other creatures and it is generally thought than devils." Another "wanted to see the blood streaming from the hearts of the leaders."[8]

Most of the refugees moved their families into the homes of Boston friends. Others rented houses that had been vacated by jittery Bostonians—Loyalists, Patriots, and the undeclared—who had headed elsewhere, hoping that the move was temporary. Some men, such as former Old Colony Club member Gideon White, Jr., struck up social acquaintances with British officers and eventually joined the British Army.

Loyalists who fraternized with British officers put themselves at risk. One was Thomas Amory. Although an Addresser of Gage, Amory was well liked and kept his Loyalist sympathies to himself. For a time he was not harassed. A merchant like his namesake father and grandfather, he lived in Milton, bordering on Boston, in the imposing house of a former royal governor. One night word reached the Patriots that Amory was entertaining some British officers. A brick-throwing mob attacked the house. One of the bricks smashed a windowpane in his young daughter's room and landed on her bed. Amory stepped out to his porch and tried to calm the mob while the officers slipped out the back door and in the darkness made their way down a garden path that ended at the shore of the Neponset River estuary. They boarded a boat tied up at Amory's pier, rowed to Boston Harbor, and went to their lodgings. Amory later moved to Watertown, about eight miles west of Boston.[9]

The well-documented flight of the placemen, the Addressers, and other eminent Loyalists produced for posterity a distorted image of the refugees. Many ordinary families also sought the protection of

General Gage's Redcoats and marines. But documentation is scant about these lesser Loyalists.

John Nutting of Cambridge is an exception. Born in 1739 to a mother who was soon to become a widow, he was apprenticed to a "housewright" in Reading, Massachusetts. After serving in a royal militia during the French and Indian War, he married his master's daughter, built a house—"two story high, three rooms on a floor"—in Cambridge, and started a family. He prospered as a builder and an entrepreneur, importing to Boston timber from land he had bought in the Maine wilderness.

Nutting had become a Tory, and when General Gage decided to empty the Massachusetts Provincial Powder House in September 1774, Nutting somehow helped, perhaps as an informer and guide. This, in his words, "raised the resentment of the populace against him to such a Degree as obliged him to quit his House & Family, & take refuge in Boston."[10] He moved, followed shortly by his wife and six children, at a propitious time, for the chill of autumn was sweeping through the tented bivouacs of the British forces. Bostonians were not about to billet the shivering soldiers. So barracks had to be built, and few carpenters would sign on for the work.

Nutting volunteered to serve as General Gage's master carpenter, becoming, as he later wrote, "the first person of an American that entered into the King's service when the troubles began." Nutting recruited forty or so carpenters from other places, inspiring a Patriot warning that such work would "deem them as enemies to the rights and liberties of America." Taking the warning seriously, the builders walked off the job.[11]

Carpenters imported from Canada and New Hampshire worked on the most crucially needed barracks, which housed the troops assigned to fortifications built on Boston Neck. Boston was then a near-island, attached to the mainland by a slender peninsula. Boston Harbor lay to the east and south, the Charles River on the north and west. Across the river rose the heights of Charlestown, and to the southeast loomed Dorchester Heights.

By November the Boston Neck guardrooms "were not half finished, having neither fire places or Stoves fixed," Lt. John Barker wrote in his diary. But after their stove was repaired, "we were pretty comfortable."[12] The post at the Neck, called "the Lines," was a choke-point for people entering and leaving Boston. Gates and his troops could control the city. But Gates realized that if the Patriots started a war and managed to muster a large-enough army, they could take the high ground of Charlestown and Dorchester and lay an effective siege.

Gage's basic strategy was simple: Deprive the Patriots of an army by depriving them of gunpowder, muskets, and cannons. In October the king had ordered that no gunpowder or arms be exported to America. Patriots learned of this in December and stepped up their efforts to arm their militias.[13] Responding quickly, Patriots in Portsmouth, New Hampshire, attacked Fort William and Mary and made off with its store of gunpowder. Patriots in Rhode Island snatched forty-four guns from Fort George at Newport and openly began supplying arms to militiamen.[14]

Gage reacted slowly. Following up on his September seizure that touched off the Powder Alarm, he ordered a strike against the Rebels of Salem, about twenty-five miles north of Boston. Gage was well informed about Rebel activities by Loyalist spies controlled by his principal intelligence officer, his brother-in-law Maj. Stephen Kemble. Like Mrs. Gage, Kemble was born in New Jersey and was more knowledgeable about Americans than typical British-born officers. Gage's intelligence network served him well. Many of his agents were never identified.[15]

Dr. Benjamin Church, a Boston physician and member of the Patriots' Committee of Correspondence, frequently wrote Patriot prose and poetry. A native of Newport, Rhode Island, he graduated from Harvard in 1754. He later studied medicine at the London Medical College and returned to America with an English wife. He was a member of the Provincial Congress of Massachusetts and the Sons of Liberty. He was also a secret Tory and Gage's best-placed spy, an operative that a modern spymaster would call a mole or penetration agent.[16]

From Church or one of his other agents, Gage learned about the Rebel arsenal in Salem, where, by February 1775, work was ending on a bold project to arm Patriots with cannons. Col. David Mason, as engineer for the Massachusetts Committee of Public Safety, had been a captain of British artillery in the French and Indian War. Somehow he had acquired about a dozen twelve-pounders, which had originally been emplaced at a French fortress in Canada. Now he was supervising their conversion to field cannons. Mason was the Renaissance man of Salem. He gilded and varnished carriages, painted portraits, tinkered with machinery, and gave paid lectures on electricity—a phenomenon he had discussed with his old friend Benjamin Franklin. Mason's interests included the science of explosives.[17]

Mason had turned the fortress cannons and some others over to Robert Foster, a militia captain and a blacksmith. In his Salem workshop, Foster added carriages and modified the cannons with ironwork to Mason's specifications. Mason's wife and daughters were meanwhile sewing five thousand flannel powder bags for the cannons.[18]

Work on the cannons was almost completed when General Gage ordered a raid to seize the arms. He gave the mission to the 64th Infantry Regiment, which was stationed at Castle William. Members of that regiment could avoid being seen on the streets of Boston because they could directly be boarded onto a transport off Castle Island. About 250 men of the 64th sailed off on the night of Saturday, February 26, under the command of Lt. Col. Alexander Leslie.[19]

The troops were to sail to Marblehead, land there, march to nearby Salem, and confiscate the cannons. Paul Revere's counterspies heard rumors of an expedition and sent three men to the waters near Castle Island to investigate. They were spotted, seized, and confined until Monday, so that, a Patriot report said, no warning could be sent "to our brethren at Marblehead and Salem."[20]

The secrecy held until Leslie and his men began their march to Salem on Sunday, the fifes and drums playing "Yankee Doodle," a tune that the British often played to mock the Rebels."[21] John Pederick, a Marblehead man, came upon Leslie on the road. They saluted each other and Leslie courteously parted the marching files to let Pederick through.

Leslie had known Pederick as a royal militia officer who had been cashiered because of his Tory views when Patriots took over the militia.

Pederick had found redemption in the Patriot cause and, when he saw the troops arriving in Marblehead, he knew he had to get past Leslie and warn Salem. After their friendly meeting, Pederick rode slowly past the troops. Then, out of sight, he galloped down the road to Salem. The first man he met was Mason, the cannons' owner. Many people of Salem were at afternoon worship in a meetinghouse when Mason burst through the door and ran down the aisle shouting that the Regulars were coming.[22]

As church bells rang and militia drums thundered a call to arms, Mason rode off to supervise the hiding of the cannons. Men hauled some of them into a dense wood and buried them under a deep carpet of leaves. The bulky carriages were taken elsewhere. Patriots loaded other cannons onto horse-drawn wagons that headed down the road to Danvers, where they were buried in a gravel pit.[23]

By the time the men of Leslie's advance guard reached a bridge between Marblehead and the southern end of Salem, they found the planks ripped up. Soldiers quickly repaired the bridge, but the men hiding the cannons had won a few minutes. The Redcoats marched on, colors flying and drums beating, until they reached the courthouse in the town square. Leslie was soon in muted conversation with three local Loyalists. One was an Addresser. Another was the half-brother of Col. William Browne, a mandamus councillor and one of the notorious "Seventeen Rescinders" of 1768.[24]

Now informed where the cannons were most likely hidden, Leslie led his troops toward an arm of the sea called the North River. Arching over the ship channel was a drawbridge. The troops—and a large, unruly escort of shouting, taunting Salem men, women, and children—were nearing the bridge when the northern leaf of the draw was raised. Everyone stopped at the southern end of the bridge.

By then a growing band of Rebel militiamen was assembling near Leslie's force. On the other side of the water stood armed men. Some climbed up on the leaf of the bridge and, dangling their feet, jeered at the soldiers. Among the men on the bridge was Capt. Richard Derby.

As a privateer during the French and Indian War, he had taken from a French ship many of the cannons now in Salem. "Find the cannon if you can," Derby shouted. "Take them if you can. They will never be surrendered."[25]

Militiamen were stationing themselves along the road to Marblehead, Leslie's potential line of retreat. Militiamen from Marblehead, Danvers, and other nearby towns began heading for Salem. Many of these hard-eyed men were veterans of the French and Indian War and cod fishermen who had earned their strength and tenacity by defying the North Atlantic. Redcoats did not faze them.

Leslie, his sword drawn, threatened to clear the road with a volley of musketry. "You had better not fire," a militia captain told him, "for there is a multitude, every man of whom is ready to die in this strife." (Another version has the captain say, "Fire and be damned! You've no right to fire without further orders. If you fire, you'll all be dead men!")

Near the bridge Leslie noticed three flat-bottomed scows called gungalows, grounded by low tide. He ordered some soldiers to commandeer them for passage. Leslie's men headed toward the gungalows, bayonet-tipped muskets at the ready. Several Salem men got there first, jumped into the scows, and smashed their bottoms with axes. A man bared his chest to the winter chill and defied the Redcoats to stab him. One did, pricking the Salem man's chest. "He was very proud of this wound in after life and was fond of exhibiting it," says an account, which also gives Salem the claim that this was "the first blood of the American Revolution."

The Reverend Thomas Barnard, a former Loyalist and an Addresser who had later publicly begged forgiveness, stepped forward as a mediator. Facing Leslie, Barnard, in his deep, resonant voice, pleaded restraint. Accounts vary, but eventually the consecrated version was this: "You cannot commit this violation against innocent men, here, on this holy day without sinning against God and humanity! The blood of every murdered man will cry from the ground for vengeance upon yourself, and the nation which you represent! Let me entreat you to return!"[26]

Barnard probably came up with the idea that defused the face-off: The bridge leaf would be lowered so that Leslie and his men could cross the bridge in a pantomime of searching for the cannons. He would then march his men to a point about thirty rods (495 feet) beyond the bridge, turn, and march them back to their ship in Marblehead.

Leslie agreed. The bridge leaf was lowered. Leslie marched his men across the bridge and about five hundred feet beyond, where Robert Foster's blacksmith shop—and a line of silent militiamen—stood. Leslie ordered his men to halt, face about, and march back across the bridge. There were shouts and an occasional catcall from opened windows as the Redcoats departed from Salem. They marched on to Marblehead, boarded their transport, and returned to Boston.

The Salem episode, which gleeful Patriots quickly dubbed "Leslie's Retreat," emboldened the Rebel cause and stunned the Loyalists, especially those who were seeking refuge in Boston. Gage, undaunted, continued his disarm-the-Rebels strategy. On February 22 he had ordered two officers, Capt. John Brown and Ens. Henry De Berniere, to disguise themselves as civilians and travel from Boston to Worcester, which, like Salem, had become a Patriot hive.

Gage seemed to be contemplating an attack on Worcester. Writing in August 1774 to Lord Dartmouth, Gage said the Rebels in Worcester "openly threaten resistance by arms, have been purchasing arms . . . and threaten to attack any troops who dare to oppose them. . . . I apprehend I shall soon be obliged to march a body of troops into that township."[27]

Gage ordered Brown and De Berniere, an accomplished artist, to produce "a sketch of the country as you pass." He wanted to know "distances from town to town" and entrances in and out of them. "The rivers also to be sketched out, remarking their breadth and depth and the nature of their banks on both sides, the fords, if any, and the nature of their bottoms." Also, "You will remark the heights you meet with, whether the ascents are difficult or easy; as also the woods and mountains, with the height and nature of the latter, whether to be got round or easily past over. . . . whether the country admits of making roads for troops on the right or left of the main road, or on the sides," and how much food and forage were available.[28]

Brown and De Berniere, donning plain brown suits and tying "reddish handkerchiefs round our necks," thought they could pass as "country men" on their spy mission. But British officers were served by military valets called batmen, and the would-be spies brought one along. When they stopped for a meal at a tavern, they sent their batman to the kitchen to dine with the tavern servants. The officers' dinner "was brought in by a black woman, at first she was very civil, but afterwards began to eye us very attentively; she then went out and a little after returned, when we observed to her that it was a very fine country, upon which she answered so it is, and we have got brave fellows to defend it."[29]

The batman rushed out to the dining room, and suggested that they pay their bill and leave right away. The waitress, he said, had recognized Brown as a British officer from Boston. "This disconcerted us a good deal," De Berniere later wrote, "and . . . we resolved not to sleep there that night."

The two spies did better at their next stop, the Golden Ball Tavern on the Old Post Road in Weston, sixteen miles west of Boston. The owner, Isaac Jones, was such a stalwart Loyalist that he had named a son William Pitt Jones, after the British statesman who was a friend of the colonies but not of armed conflict. Not quite a year before, after accusing Jones of being a Tory, Patriots had raided the tavern and stolen liquor, along with two expensive delicacies, raisins and lemons. Only a month before, the Patriots' anger had cooled and Jones was allowed to keep his tavern open. Jones would later switch allegiance still again and work hauling supplies for the Continental Army.[30]

Jones recommended a Tory inn in Worcester. "The landlord was very attentive to us," De Berniere wrote, "and on our asking what he could give us for breakfast, he told us Tea or any thing else we chose—that was an open confession what he was." The spies knew they had been watched by Patriot counterspies through most of their journey. In case they were stopped and searched, they sent their bat-man on ahead to Boston with their maps and sketches. But, with the aid of the Tory underground—and a snowstorm that covered their tracks—they got back to Boston without incident.[31]

The help that the officers received from Tories suggested that a British military force dispatched to Worcester might get partisan aid. But the detailed report also showed rivers, marshes, and potential am-bush sites along the spies' trek. This convinced Gage that a forty-four-mile march to Worcester could become a disaster.

Gage had an alternative. His leading spy, Dr. Church, had provided extensive intelligence about a Rebel arsenal much closer to Boston: in the town of Concord.[32] On March 20 Gage sent Brown and De Berniere there, not only to study the terrain and roads but also to con-firm the existence of arms caches that Church had reported. He most likely did not know about munitions stored in Congregational meet-inghouses, among the most secret of Patriot hoards. Loyalists claimed that there were so many Congregational clergymen in Patriot ranks that they formed a "black regiment."[33]

On this second mission the two spies were well supplied with knowledge about Rebel arms—and the names of Tory partisans. Of Concord's 250 voters in 1774, town fathers estimated that 52 were To-ries. When Patriots confronted them and demanded that they recant and apologize, the number dwindled to 44. The following year thir-teen persons from the town's most prominent families were added to the list of Tories.[34]

Untrained in espionage, Brown and De Berniere exposed one of their contacts when, within hearing of Concord Patriots, they asked for directions to the home of Daniel Bliss, a well-known Royalist. Rebels tracked down the woman who directed the spies and threat-ened her with tarring and feathering.[35]

Bliss's family, like so many others, was split. His father-in-law had been a British Army officer, as had a brother-in-law. Bliss's two broth-ers were Patriots, and his sister was married to the Reverend William Emerson, pastor of the First Parish in Concord, who preached the Patriot cause and was chaplain to the Rebel-controlled militia.

The previous December, Bliss had spoken about the Boston Port Bill at a meeting in the Concord meetinghouse, criticizing the Patriots for "flouting your King." The colonies, he said, "are England's dependent children" and he warned that "England is a mighty nation" and "Rebellion will lead inevitably to crushing defeat." Bliss, a Harvard graduate and a lawyer who had given counsel to Royal Governor Hutchinson, was not accustomed to having his opinions questioned. But one man at the meeting did speak up: thirty-nine-year-old Joseph Hosmer, an avowed Patriot and a lieutenant in the militia. Hosmer repudiated Bliss's every word with an eloquence that astonished and annoyed Bliss. When another Loyalist asked Bliss who this upstart was, Bliss is said to have labeled Hosmer "the most dangerous man in Concord."[36]

Brown and De Berniere dined with Bliss at his home, which happened to be near a storehouse containing arms. Believing they were under surveillance, the officers slipped out of Concord late at night and, taking a roundabout route, returned to Boston, escorted by Bliss. He was never again seen in Concord,[37] and his family quietly moved to join him in Boston. Later, they went to Canada, where Bliss was commissioned as a British Army colonel and made assistant commissary general.[38]

The spies' visit to Concord convinced Patriots that their town would be the target of a Gage raid. As for Gage, his options were quickly narrowing. For some time his immediate superior, Lord Dartmouth, had been urging action against the Rebels. Gage had responded with dire predictions that if Britain chose armed force the result would be the "horrors of civil war."

But in March Gage wrote a "private letter" to Viscount Barrington, the secretary of war, saying, "It's beyond my capacity to judge what ought to be done, but it appears to me that you are now making your final efforts respecting America; if you yield, I concede that you have not a spark of authority remaining over this country." However, Gage went on, if Britain took the path to war, "it should be done with as little delay as possible, and as powerfully as you are able, for it's easier to crush evils in their infancy than when grown to maturity."[39]

What amounted to an order to go to war came from Dartmouth in the form of a letter he wrote on January 27. He allowed the letter to linger in London, while word of it circulated in Parliament and among ministerial officials and advisers. It finally left England on February 22, on board HMS *Falcon*. Two days later, a duplicate copy was given to Capt. Oliver De Lancey of the 17th Light Dragoons, who was sailing on HSM *Nautilus*. Local storms delayed the sailing of both ships. The *Falcon* did not arrive in Boston until April 12; the *Nautilus*, two days later.[40]

In the letter marked "Secret," Dartmouth said, "Your Dispatches . . . shew a Determination in the People to commit themselves at all Events in open Rebellion. The King's Dignity, & the Honor and Safety of the Empire, require, that, in such a Situation, Force should be repelled by Force." A royal government must replace the illegal Continental Congress, Dartmouth said, and "the first & essential step to be taken towards re-establishing Government, would be to arrest and imprison the principal actors & abettors in the Provincial Congress (whose proceedings appear in every light to be acts of treason & rebellion)."[41]

Dartmouth wanted Gage to launch an offensive immediately rather than wait for promised reinforcements because "the People, unprepared to encounter with a regular force, cannot be very formidable; and though such a proceeding should be, according to your own idea of it, a Signal for Hostilities; yet, for the reasons I have already given, it will surely be better that the Conflict should be brought on, upon such ground, than in a riper state of Rebellion."

"I understand," Dartmouth added, "a Proposal has been made by Mr Ruggles for raising a Corps of Infantry from among the friends of Government in New England. Such a Proposal certainly ought to be encouraged, and it is the King's Pleasure that you should carry it into effect upon such Plan as you shall judge most expedient."[42]

Dartmouth apparently heard about the Ruggles plan from Israel Mauduit, a British political writer and brother of the Massachusetts agent, or lobbyist, in London. In October 1774 Ruggles had written Mauduit about mobilizing "friends of government." Dartmouth's papers show that in February 1775, after writing the Gage letter, he was mindful that Ruggles believed he could raise a Loyalist force from "the majority in New Hampshire, New York, New Jersey, and the Southern Colonies." This is one of the earliest references to the belief, held by many Loyalists, that they were strong beyond the rebellious epicenter of Massachusetts, Rhode Island, and Connecticut.[43]

Capt. Oliver De Lancey was an incarnation of that belief. The fact that Dartmouth entrusted such an important letter to a twenty-one-year-old cavalry officer showed the depth of the connection between royal officialdom and the powerful De Lancey family, the staunchest of New York's Loyalists. Oliver was an American, but he had a partially British biography. Born in New York City in 1752, he was educated in England, where he was commissioned an officer in the British Army.

Oliver was a nephew of General Gage's wife, the former Margaret Kemble, from East Brunswick, New Jersey, who adapted to British ways while clinging to her American identity. Gage's official family included not only Stephen Kemble, Gage's aide-de-camp for intelligence, but also Samuel Kemble, Gage's confidential secretary. After his arrival on the *Nautilus*, Oliver De Lancey became another aide-de-camp to Gage.[44]

The day after getting Dartmouth's letter, Gage received a report from Dr. Church on arms hidden in Concord.[45] Gage moved swiftly—and secretly. He knew that the Patriots in Boston had his troops under close observation. On the afternoon of April 18, Paul Revere and other Patriot leaders began getting reports from watchful Bostonians about signs of a major British operation. Revere had heard such rumors before, but this time he believed they were true. So did cautious Dr. Joseph Warren, a physician who was a member of the Committee of Correspondence and a confidant of Sam Adams. Warren decided to seek confirmation from his most trusted and confidential source, whose name he never revealed.

From that source, another Patriot later reported, Warren "got intelligence of their whole design; which was to seize Sam Adams and Hancock, who were at Lexington, and burn the stores at Concord. Two expresses were immediately despatched thither, who passed by the guards on the Neck just before a sergeant arrived with orders to stop passengers."[46] One of the express riders was William Dawes, a Boston tanner; his business often took him through the Boston Neck checkpoint, where he knew the guards. The other rider was Paul Revere.[47]

We will never know for sure who that source was. But speculation has persisted that Warren's spy was Margaret Kemble Gage. Only a circumstantial case exists. She is said to have told a friend that she felt herself as profoundly torn as Lady Blanche in Shakespeare's *King John*. Blanche, niece of King John of England and daughter of the King of Castile, is caught in a web of power. She laments her divided loyalty, a lamentation that resounded in many a Loyalist heart after the deadly day of April 19:

> *The Sun's o'ercast with blood: fair day, adieu!*
> *Which is the side that I must go withal?*
> *I am with both: each army hath a hand;*
> *And in their rage, I having hold of both,*
> *They swirl asunder and dismember me. . . .*
> *Whoever wins, on that side shall I lose. . . .*[48]

4

"TO SUBDUE THE BAD"

FROM CONCORD TO BUNKER HILL, APRIL–JUNE 1775

The ... sanguinary measures adopted by the British Ministry ...
threatening to involve this Continent in all the horrors of a civil
War, obliges us to call for the united aid and council of the Colony,
at this dangerous crisis.

—New York Provincial Congress[1]

*L*ate on the night of April 18, 1775, a force of about nine hundred
Redcoats and marines[2] boarded Royal Navy longboats at a deserted Boston beach. They were rowed across Back Bay to an isolated
point of Cambridge land, owned by a Tory, where a road led to the
Rebel arms cache at Concord. Long before the next dawn the troops
were on the march, heading for what would be the start of the Revolutionary War.[3] Alongside some of the British forces during that long
and bloody day were a number of Loyalists, anonymous comrades of
the British soldiers.

Some Loyalists joined with the Regulars, straggling along, carrying
muskets, wearing civilian clothes, and drawing taunts from people
who would remember those Tory faces. Other Loyalists, referred to
in accounts simply as "Tory guides," served unseen and remain unknown.[4] Those who are identified include Edward Winslow, well re-

membered in Plymouth as a founder of the Old Colony Club, and his childhood friend George Leonard.[5]

Three others known to have served the king that day were Thomas K. Beaman,[6] Daniel Murray, and his brother Samuel. Beaman, who had been a captain of militia in the French and Indian War, lived in Petersham, about thirty miles north of Worcester. He was reported to be one of Gage's intelligence agents; that would explain his presence as a "guide," often used as a polite term for spy.[7] Family tradition has Thomas Beaman's only brother, Joseph, also heading for battle that day—as a minuteman.[8] John Murray, a mandamus councillor whose home had been attacked in 1774, was the father of Daniel and Samuel. (Murray's daughter was the wife of Daniel Bliss of Concord, who had aided Gage's roaming officer-spies.)

For all five known Tory allies, their work that day began their military careers as Loyalists. Beaman would serve the British Army as a provincial officer handling supply-wagon trains. Daniel Murray would become the captain of a Loyalist volunteer company whose members included his two brothers; later he became a major in the King's American Dragoons.[9] Leonard took command of a small fleet of ships that raided New England seacoast towns,[10] and Winslow would be commissioned as muster master general, responsible for keeping track of all the Loyalist armed forces in North America.[11]

Daniel Murray's war began when he was accompanying Royal Marine lieutenant Jesse Adair at the head of the advance British force on the road to Concord. Around four a.m., near Menotomy (now Arlington), they heard hoofbeats on the road ahead. Assuming the horsemen were express riders spreading reports of the march, Murray and Adair waited until the riders neared, then dashed from the side of the road to seize the horses' bridles. After appropriating the horses, Murray and Adair forced the Patriots, Asahel Porter and Josiah Richardson, to march along with the troops.[12] This was the first action of the day.

General Gage possessed enough intelligence to provide his Concord strike force with a map showing exactly where militiamen had hidden "Artillery, Ammunition, Provisions, Tents, Small Arms and . . . Military Stores." He gave detailed instructions on how to disable discovered cannons and what to do with other stocks: "The Powder and flower [flour] must be shook out of the Barrels into the River, the Tents burnt, Pork or Beef destroyed in the best way you can devise. And the Men may put Balls of lead in their pockets, throwing them by degrees into Ponds, Ditches &c., but no Quantity together so that they may be recovered afterwards."[13]

As the British marched off toward Lexington, the alarm had already been spreading along that same Cambridge road.

The British had learned that Sam Adams and Hancock—two of the "principal actors & abettors" Lord Dartmouth wanted arrested—were somewhere in Lexington. The two Rebel leaders, along with Hancock's fiancée, Dorothy Quincy, and his aunt Lydia, had been staying at the parsonage of the Reverend Jonas Clarke in Lexington. Dr. Joseph Warren, watching over Boston in Adams's absence, had sent Paul Revere and William Dawes, by separate routes, to Lexington to warn Adams and Hancock that the Regulars were coming. (Dr. Church learned of Revere's ride from his wife, Rachel, who wrote Paul a letter—"take the best care of yourself"—with £150 enclosed, and entrusted it to Church to deliver. Church passed the letter on to Gage; what happened to the money is not known.)[14]

Revere and Dawes reached Lexington at just about the time the British had begun their march. Arriving separately, the two riders stopped to warn Hancock and Adams, then set off for Concord.[15] On the road they met Samuel Prescott, a young doctor heading home to Concord after an evening spent courting a Lexington woman. A mounted British patrol stopped the riders. Dawes and Prescott managed to gallop away, Dawes back to Lexington, Prescott to Concord. Revere, his horse taken, was held for a while and later released. Other riders, part of a complex militia warning system, sped the Concord Alarm throughout Middlesex County.

About seventy militiamen were mustered on the Lexington green when the leading British troops came up the road.[16] Asahel Porter, who had been captured by Murray and Adair, was near the head of the column. Adair wheeled his forward companies toward the militiamen. "No sooner did they come in sight of our company," Clarke later wrote, "but one of them, supposed to be an officer of rank"— apparently Adair—"was heard to say to the troops, 'Damn them! We will have them!'" Porter tried to run away from his captors, but he had been doomed by the chance encounter with a British officer and his Loyalist ally. Shot as he ran, he was one of eight Patriots killed at Lexington.[17] What happened next to Daniel Murray is not known. The names and exploits of Loyalist volunteers did not usually find their way into official British battle reports.

Capt. John Parker, commander of the Lexington militia, was a battle-hardened veteran of Rogers' Rangers in the French and Indian War. He knew what to do when he saw the size of the British force and realized how outnumbered he was. "I immediately ordered our Militia to disperse and not to fire," he later wrote.[18] Maj. John Pitcairn of the Royal Marines, who led the British advance guard into Lexington, said in his official account that when his men got within one hundred yards of the militiamen "they began to File off towards some stone Walls on our Right Flank. . . . I instantly called to the Soldiers not to Fire," and "some of the Rebels who had jumped over the Wall, Fired Four or Five Shott at the Soldiers."

After the militiamen fired more shots, Pitcairn wrote, "without any order or Regularity, the Light Infantry began a scattered Firing . . . contrary to the repeated orders both of me and the officers that were present."[19] Militiamen's accounts quote Pitcairn as ordering the Rebels to disperse and then shouting: "Damn you! Why don't you lay down your arms?" Another remembered an officer yelling: "Disperse, you damned Rebels! You dogs, run! Rush on, my boys!"[20] The men kept on shooting until officers managed to get them under control.

"We had a man of the 10th Light Infantry wounded, nobody else was hurt," a British officer later wrote. "We then formed on the Common, but with some difficulty, the men were so wild they could hear no orders." He could not tell how many Rebels were killed "because they were behind walls and into the woods. . . . We waited a considerable time there, and at length proceeded our way to Concord."[21]

Neither Revere nor Dawes had reached Concord, but Prescott had, and his news started the Concord Alarm. "This morning between one and two o-clock we were alarmed by the ringing of the bell," wrote the Reverend William Emerson, whose name was on the Militia Alarm List. Local militiamen began assembling at Wright's Tavern while militias from other towns headed for Concord.[22] A militia officer sent a rider galloping off to Lexington to see if the British were there. He arrived back to report the Lexington firefight.

People in Concord hurried about, finding new hiding places for hoards of ammunition, arms, and supplies. About one hundred and fifty eager young minutemen wanted to confront the British as they approached Concord. But militias were democratic, and a majority voted for a strategic withdrawal to a ridge north of town. Others crossed the North Bridge across the Concord River.

When the British troops arrived, they split into three groups: one at the North Bridge, confronting the minutemen; another in town as a search party for Rebel caches; and a third that crossed the North Bridge to the farm of Col. James Barrett, a French and Indian War veteran who was commander of the militia and overseer of the biggest caches of arms and provisions. Barrett's collection included twenty thousand pounds of musket balls and cartridges, fifteen thousand canteens, a great number of tents and tools, seventeen thousand pounds of salt fish, thirty-five thousand pounds of rice, and large quantities of beef and pork.[23] A great deal of the supplies had been rehidden in nearby towns before the British arrival.

As militiamen began streaming into Concord from other towns, the Patriot force beyond the bridge grew to at least five hundred men, strung out along a ridge overlooking the town center, now populated only by women, children, and men too old or infirm for militia service. British soldiers moved among the glowering residents, dug up three cannons, and knocked off their trunnions, the projections forming an axis on which a cannon pivots up or down; this was the fastest way of disabling the weapon. And they discovered only a small portion of the Rebel stores they had come to find.

Disobeying Gage's orders, the weary searchers did not load their pockets with bullets. Instead they dumped a large amount of bullets into a millpond, from which they would be recovered the next day. The soldiers piled up wooden gun carriages and set fire to them. They cut down the flag-flying Liberty pole and burned that, too. A couple of buildings began to burn, apparently because of carelessness. The soldiers, who had been ordered not to destroy property, joined the citizens in dousing the fires.

On the ridge, militiamen saw the smoke curling up. Barrett, suddenly in command of townsmen and strangers alike, had good reason to hold back from battle. He faced a superior force and lacked any authority to fire on the British, who had not fired upon his men. Barrett's adjutant was the minuteman Joseph Hosmer, whom, the Loyalist Daniel Bliss had not long before called Concord's most dangerous man. Now the fiery Patriot pointed to the smoke rising from the center of Concord, turned to his commanding officer, and said, "Will you let them burn the town down?" Other officers spoke up. Capt. Isaac Davis, who had led his minutemen down the road from Acton, said, "I haven't a man who is afraid to go."

Barrett ordered his men to load their muskets. Many of them remembered that he told them to withhold their fire until the British fired. The militiamen advanced from the ridge, downward toward the North Bridge and the Regulars. The British soldiers pulled back from the bridge, ripping up planks. They formed a volley-firing formation on the other side of the bridge and opened fire, killing Captain Davis and a private.

At a range of about fifty yards the militiamen responded with a volley that killed four of the eight British officers at the bridge. Milling about in panic and confusion, the British began a ragged retreat. Their fat, inept commander, Lt. Col. Francis Smith, managed to get control and stop the retreat. A young American—never positively identified—walked up to a badly wounded Redcoat and, in the words of the Reverend Emerson, "not being under the feelings of humanity, very barbarously broke his skull and let his brains out with a small ax," an auxiliary weapon for many militiamen. "The poor object lived an hour or two before he expired."[24]

This first known atrocity of the Revolution would cast a long shadow. Some Patriots would deny it happened, but it triggered savage acts of hate, revenge, and retribution involving British versus Rebels—and Tories versus Rebels.[25] British soldiers, angry in retreat, saw the bloody head of their comrade. An enduring rumor swept through their ranks: The Rebels are scalping their foes. The rumor even found its way into official reports. One said that Rebels had "scalped and cut off the ears of some of the wounded men who fell into their hands."[26]

The British withdrew from Concord, bearing some of their wounded in commandeered horse-drawn open carriages. Smith led them to the road they had taken from Lexington. "When we arrived within a mile of Lexington, our ammunition began to fail, and . . . we began to run rather than retreat in order," wrote Henry De Berniere in his report to Gage. A month before, De Berniere had been a Gage spy in Concord. On his return to Concord as a soldier, De Berniere survived to report 73 men killed, 154 wounded, and 21 missing.* American losses were estimated to be 50 killed and/or dead of wounds, 39 wounded, and 5 missing.[27]

Some thirty-six hundred militiamen were assembling in, or heading for, Lexington.[28] Around noon, as the British reached a bend in the road, militiamen struck, ambushing the staggering columns. The militiamen fired from behind rocks, trees, and stone walls on both sides of the narrow road. Several British, many of them targeted officers, were killed. Sergeants took over, leading their men about five hundred yards farther down the road—and into another ambush. About thirty men fell, dead or wounded. Survivors struggled on toward Lexington. Militiamen on a hill fired down, wounding Colonel Smith and knocking him from his saddle. Redcoats charged the hill, killing at least one militiaman. Pitcairn's horse bolted, throwing him

---

* Later official British casualties were 65 killed, 157 wounded, and 27 missing.

to the ground. Some men stumbled to the side of the road, unable to march any farther. That morning the British had outnumbered the Rebels. In this murderous afternoon, the British were outnumbered, and, it seemed, doomed.

Nearly twelve hours before, Colonel Smith, suddenly aware that the militias were rising, had sent a courier to Gage in Boston asking for reinforcements. Gage had dispatched a brigade under his adjutant, Brig. Gen. Lord Hugh Percy. As the eldest son of the Duke of Northumberland, Percy would inherit the dukedom and an immense fortune, if he did not die on a battlefield. A soldier for half of his thirty-two years, he had arrived in Boston in 1774 as commander of his regiment, the Fifth Foot. He developed a haughty dislike for Bostonians—"a set of sly, artful, hypocritical rascalls, cruel, & cowards. I must own I cannot but despise them compleately," he wrote to a cousin in England soon after his arrival.[29]

Now, riding at the head of about one thousand men and two field guns, with fifes and drums playing "Yankee Doodle," Percy marched with other Americans—Friends of the King, armed Loyalists in civilian clothes, including Winslow, George Leonard, and Samuel Murray. Winslow later reported that his horse had been shot from under him and that Percy had personally cited him for bravery.[30] Murray was captured that day and taken to a jail in Worcester. On June 15 he was released "to his father's homestead in Rutland," about sixty miles west of Boston.[31]

The other Loyalists who accompanied Percy are not known. But curiously, there is a record showing that on April 19 a number of "gentlemen volunteers" joined the Loyal American Association under the command of Brigadier Ruggles.[32] Some of those gentlemen probably marched off with Percy.

A record in Gage's files states Leonard's role: "George Leonard of Boston deposes that he went from Boston on the nineteenth of April with the Brigade commanded by Lord Percy upon their march to Lexington. That being on horseback and having no connexion with the army, he several times went forward of the Brigade." On a sortie as a spy for Percy about a mile south of Lexington, Leonard met a wounded man who "said Some of our pepol fired upon the Regulars; and they fell on us Like Bull Dogs and killed eight & wounded nineteen." Leonard reported that two other men had told him the same. He added a line that would please Gage, who officially emphasized that the Rebels had fired the first shot at Lexington: "All three Blamed the rashness of their own pepol for fireing first."[33]

Instead of marching along the longboat-to-Cambridge route that Smith's force had taken, Percy led his brigade across Boston Neck, through Roxbury and Brookline, to the Great Bridge that crossed the Charles River between Boston and Cambridge. Patriots had removed the bridge planks. Percy, anticipating this, had brought along tools, planks, and carpenters. But the thrifty Patriots had put the removed planks nearby, so all Gage's craftsmen had to do was walk along the risers, retrieve the planks, and replace them.[34]

In Cambridge, a lieutenant noted, "few or no people were to be seen; and the houses were in general shut up."[35] Percy had hoped for some intelligence from friendly Tories. When he asked a citizen for directions to the Lexington road, the man courteously obliged—and was spotted giving aid to a Redcoat. He was so harassed by angry Patriots that he quickly got out of town.[36]

As Percy's men neared Lexington, they heard a rattle of musketry and then saw, under a cloud of dust up the road, the swirling panic of Smith's force. Percy stopped at Lexington, scanned the scene before him, and acted swiftly. He put his field pieces on two hills and fired on the militiamen, who backed off, stunned. Percy had his men form a hollow square that encompassed the road and stretches of land on either side. The square opened to admit Smith's exhausted troops, "their tongues hanging out of their mouths, like those of dogs after a chase."[37] Within the defensive square were houses. Percy, seeing them as snipers' nests, ordered them burned. Flanking troops forced their way into other houses, and, a lieutenant later wrote, "All that were found in the houses were put to death."[38] Another officer noted, however, that "some soldiers who staid too long in the houses were killed in the very act of plundering by those who lay concealed in them."[39]

On the march back to Boston, Percy later wrote, his men were "under incessant fire, which like a moving circle surrounded and followed us wherever we went." As many as five thousand militiamen hounded the British along the eleven-mile gauntlet. The sun had nearly set when the brigade reached Cambridge, where Patriots waited in still another ambush. Redcoats on the flank found them, killing three. Instead of returning on the morning's route, Percy surprised his pursuers by crossing Charlestown Neck, between the Charles and Mystic rivers, to the hills of Charlestown, protected by the guns of HMS *Somerset*, anchored near the Charlestown ferry slip.

Percy met with the selectmen of Charlestown and negotiated an armistice. He promised to restrain his men, and the town promised not to attack the troops as they took *Somerset* longboats and other craft to reach Boston and the refuge of their garrisons.[40] Percy's appraisal of his foe had changed. "Whoever looks upon them as an irregular mob," he wrote the next day in a letter, "will find himself much mistaken; they have men amongst them who know very well what they are about. . . ."[41]

The battle reverberated the next day in Marshfield, where about one hundred soldiers of the Queen's Guards had been quartered on the estate of Nathaniel Ray Thomas. Reacting to the attack on Lexington and Concord, militias near Marshfield began preparing for a massive attack on the Guards. On the road to Marshfield, a militia officer, Maj. Judah Alden, happened to meet a slave named Cato.[*] The slave told Alden that Captain Balfour had sent him out as a scout to watch for Rebels. Alden told Cato to go back to Balfour and tell him that he had seen a large militia force on its way to Marshfield.

The next morning some five hundred militiamen did march on Marshfield. Crews of fishing boats spotted the marchers and came ashore to join them. But, as the Patriots were preparing to advance on the Thomas estate and attack the Guards, two sloops hove into sight. They anchored off Brant Rock, near Plymouth, and sent off boats for the Guards, who successfully escaped to Boston. The Patriots, primed

---

[*] Major Alden was a descendant of John Alden, the hero of Longfellow's *The Courtship of Miles Standish*.

for a fight, had to be satisfied with collecting the equipment and sup-plies left behind in the Guards' hasty withdrawal.[42]

A week before the battles of Concord and Lexington, Rebel militia-men from the Taunton area had struck. A large force marched on the Tory-dominated Assonet section of Freetown, about thirty-five miles south of Boston, where the Loyalist militant Thomas Gilbert, aided by his brother Samuel, had organized a Tory countermilitia armed by Gage and nicknamed "Gilbert's Banditti" by the Patriots. Gilbert claimed to have raised and commanded three hundred Loyalists. In the melee, Samuel Gilbert later said, he lost the sight of his right eye and was rendered "deaf and Stupified by the Blow" he had received to make him sign "what Paper the Rebels desired."[43]

The attackers took the muskets, powder, and bullets stored in Gil-bert's house and seized twenty-nine of Gilbert's men, who were later released. Some fled to Boston, but most of the Loyalists remained in Assonet to await whatever was going to happen next.[44]

The Gilbert raid was inspired by a letter that he had written in March 1775 to Capt. Sir James Wallace, commanding officer of HMS *Rose*, on station off Newport, Rhode Island. In the letter, which was somehow intercepted by the Patriots and made public, Gilbert said he feared an attack by "thousands of the Rebels" and asked Wallace to dispatch some of his small boats, called tenders, up the Taunton River as rescue vessels if the Gilbert men needed to evacuate.

The Massachusetts Provincial Congress responded to the letter by declaring that Gilbert was "an inveterate enemy to his country, to reason, to justice, and the common rights of mankind" and that "whoever had knowingly espoused his cause, or taken up arms for its support, does, in common with himself, deserve to be instantly cut off from the benefit of commerce with, or countenance of, any friend of virtue, America, or the human race." This withering declaration fore-shadowed the campaign to drive the Loyalists from, if not the human race, at least Massachusetts.

After the declaration was published, Gilbert disappeared and made his way to sanctuary aboard the *Rose*. He had not been at home dur-ing the raid, which focused on his son-in-law, who had hidden in a kitchen oven. He was hauled out, placed backwards on his horse, and led off to the Taunton jail. Before the procession reached the jail, Thomas Gilbert and his slave Pompey appeared. Gilbert, who had hurried back to Assonet when he got word of a looming attack, faced down the crowd and demanded the release of his son-in-law. The crowd, cowed by the old soldier, let his son-in-law go.[45] Gilbert and his family soon joined the growing Tory community in Boston.

The militiamen in Marshfield and Taunton, like those who had responded to the Concord Alarm, acted on the orders—often preceded by a democratic vote—of local commanders. Now that a war had begun, there was no overall strategy, no plans for continuing to resist British troops, no policy for dealing with Loyalists, who overnight had become belligerents in a war. The Committee of Safety, the de facto executive branch of the Massachusetts provincial congress, immediately decided that the militias should be drawn together into an army.

The committee also chartered the *Quero*, a swift American schooner, to race across the Atlantic and secretly deliver the Patriots' version of events to the British public before Gage's official report reached the king's ministers. Fearing that Tory spies would learn of the *Quero*'s mission, Dr. James Warren, chairman of the Committee of Safety, told the ship's captain to sail evasively to "escape all enemies that may be in the chops of the channel," adding: "You are to keep this order a profound secret from every person on earth."[46] The captain, Richard Derby, Jr., was the son of the Salem mariner who had refused to hand over his cannons to British troops in February 1775.[47]

The *Quero*, sailing without cargo, crossed speedily. It boldly passed the Royal Navy base at Portsmouth and slipped into an Isle of Wight harbor. Derby secretly made his way to London on May 28. A letter accompanying Warren's report was addressed to Ben Franklin, who for many years had been acting as an agent for Massachusetts and other colonies. The letter asked Franklin to spread the news. But Franklin had left London and sailed for Philadelphia on March 25; he would arrive in Philadelphia on May 5.[48] Nevertheless the news did spread. English newspapers published the report; the *London Evening Post* even reprinted the *Gazette*'s account. Gage's terse report, dated April 22 and borne by a heavily laden merchant ship, arrived in London two weeks later.[49] Many British readers of Gage's official report gave more credence to the Patriot version, which made Britain the instigator of the attacks.

Back in Massachusetts, the militiamen assembling in Cambridge, Roxbury, and Charlestown got an overall commander selected by the Committee of Safety: Maj. Gen. Artemas Ward, a big rugged man who owned a prosperous farm in Shrewsbury. He was a classic citizen soldier. After graduating from Harvard in 1748, he entered politics, serving in the General Court and for a time appearing to be on the Tory path. During the French and Indian War, he became a lieutenant colonel in a regiment that took part in a campaign to capture Fort Ticonderoga. After the war he was appointed colonel of his royal militia regiment and a judge in the Court of Common Pleas.

As soon as Ward joined the opposition to the Stamp Act, Massachusetts royal governor Francis Bernard revoked his militia commission. Ward later said that he told Bernard, "I consider myself *twice* honored, but more in being superseded, than in having been commissioned . . . since the motive that dictated it is evidence that *I am*, what *he is not, a friend to my country.*"[50] Bernard's successor, Governor Thomas Hutchinson, called Ward "a very sulky fellow, who I thought I could bring over by giving him a commission in the provincial forces . . . but I was mistaken."[51]

When an express rider arrived with the news that the British were on their way to Concord, Ward was in bed, suffering from bladder stones. At daybreak he painfully eased himself out of bed and rode thirty-five miles to Cambridge.[52] Within days he was in command of an army of about twenty thousand armed men from all over New England.[53]

There was no way to know how long any of the sudden soldiers would stay. Although they were not able to launch an attack on the British, geography made them into besiegers of a city that was a near-island, linked to the rest of Massachusetts by Boston Neck. Militiamen in Cambridge confronted Gage's troops across the Neck; those in Charlestown faced the British across a narrow stretch of water. Rebel troops formed a semicircle around the city.

On Gage's fortified and garrisoned Boston Neck, Patriots hastily raised their own fortifications. The Neck became the border between the king's friends and the king's foes. Tories fled into Boston and Patriots poured out, driven not so much by fear of neighbors as by fear of war or by a desire to join other Patriots.

Notorious Tories who had tarried too long sped toward Gage's protection. On the day of the attacks on Lexington and Concord, Dr. Josiah Sturtevant of Halifax galloped the six miles to Boston so fast that it was said he lost his saddlebags. He was commissioned a British Army captain and put in charge of a military hospital.[54] Abijah Willard, a mandamus councillor who had been attacked and imprisoned, entered Boston after the Battle of Lexington, offered his services to Gage, and was commissioned a captain in the first company of Brigadier Ruggles's paramilitary Loyal American Association.

Cambridge became the nerve center of the Patriots. Huge vacated homes along Tory Row filled with Patriot officers and their men. The colonnaded mansion of John Vassall, who had already lost his hay and stables to the Patriots, became the headquarters of General Ward. Here was Tory Row at its grandest: The vestibule's massive mahogany doors, "studded with silver," opened into a wide hall, "where tessellated floors sparkled under the light of a lofty dome of richly painted glass," a peacetime visitor wrote. "Underneath the dome two cherubs carved in wood extended their wings, and so formed the center, from which an immense chandelier of cut glass depended. Upon the floor beneath the dome there stood a marble column, and around it ran a divan formed of cushions covered with satin of Damascus of gorgeous coloring. . . . All the paneling and woodwork consisted of elaborate carving done abroad."[55]

Troops set up tents and shelters on the greensward of Cambridge Common. They occupied the brick buildings of Harvard, whose classes would soon be suspended to make room for militiamen, by order of the Committee of Safety. Militiamen also made a barracks of Christ Church, abandoned by its fleeing Loyalist congregation.

Ward's makeshift army had little artillery and consisted of men who had come to fight a battle, not a war. Many of his soldiers had been summoned by a piece of paper handed out by twenty-three-year-old Israel Bissell of East Windsor, Connecticut, who rode off from Watertown, near Boston. Tradition has him shouting "To arms, to arms, the war has begun!" on his four-day journey. He carried a concise report written on the morning of the battles by a member of the Committee of Safety; the report politely requested that the news be spread and that people supply Bissell with fresh horses.

After alerting General Ward in Shrewsbury, Bissell kept on riding down the Boston Post Road. His first horse died in Worcester. He got another horse and galloped on. At each stop he handed the report to a Patriot leader, who quickly hand copied it and distributed more copies in his area. The ritual continued in town after town until Bissell got to Philadelphia, where a bell, soon to be called the Liberty Bell, pealed, drawing thousands to assemble and learn the news.[56]

When Bissell had reached New Haven, Connecticut, early on the afternoon of April 21, a town meeting was hastily called. The majority, which included a number of suspected Loyalists, voted against sending armed aid to the Massachusetts Committee of Safety. This did not go down well with Benedict Arnold, the thirty-four-year-old captain of an elite Connecticut militia, the Governor's Second Company of Foot. He immediately sent runners off to round up members and volunteers for a march to Cambridge. On the morning of Saturday, April 22, before a large crowd, Arnold paraded his men, resplendent in uniforms of scarlet coats with buff facings, white breeches and stockings, and black half leggings.

Arnold sent an aide to the town selectmen, who were meeting in Hunt's Tavern. In Arnold's name the aide asked for gunpowder and bullets from the town magazine. When the selectmen refused, Arnold marched his company to the tavern and sent in a messenger for the keys. Rebuffed again, he said that if he did not get the keys he would order his men to break down the door to the magazine. Arnold got the keys, supplied his men with ammunition, and marched them north to Cambridge.[57]

In 1758 Arnold had been a militiaman in the futile British-American attack on Fort Ticonderoga, which commanded the strip of land between Lake Champlain and Lake George on the route to Canada. Artemas Ward had also fought at Ticonderoga, and when Arnold reached Cambridge, he suggested an attack on Ticonderoga to seize its cannons. Ward approved the plan. The Massachusetts provincial congress commissioned Arnold as a colonel for "a secret service," provided him a small amount of money and supplies, and on May 3 he was off. Seven days later, after an odyssey that brought Ethan Allen and his Green Mountain Boys into the war, Ticonderoga fell without a shot, as did Crown Point, a smaller fort a few miles north.

Arnold's bold move extended the war into New York and its disputed northern frontier—without any authority beyond the Massachusetts Provincial Congress and the Committee of Safety. Coincidentally, Ticonderoga fell on the same day that the Second Continental Congress assembled in the State House in Philadelphia with delegations from every colony but Georgia. The delegates did not learn about Ticonderoga until May 18, and the reaction was more embarrassment than joy. The New York and Connecticut delegations denied responsibility. Congress, uneasy about the sudden warfare, ordered Arnold and Allen to keep the king's cannons and other property safe until that time when "the former harmony with Great Britain and these colonies so ardently wished for" is restored.[58]

In Boston there was no hint of harmony. In early May, British soldiers, beginning to feel the paranoia of a siege, heard rumors of a Rebel plot to slip assassins into the city. According to the rumor, the Rebels would stage a feint that would sound an alarm, officers would rush from their lodgings—and be shot. Gage may have believed the rumor because he ordered officers to move into barracks, an order that violated a long tradition of separate quartering for officers and enlisted men. Civilians were also growing fearful. "Numbers of People are quitting the Town every day with their families and Effects," a

British lieutenant wrote; "it's a distressing thing to see them, for half of 'em don't know where to go to, and in all probability must starve."[59]

On May 25 HMS *Cererbus* arrived in Boston, delivering three major generals: William Howe, Henry Clinton, and John Burgoyne. Gage was still officially the field commander, but he now had the three generals looking over his shoulder and urging action. Burgoyne, a flamboyant part-time playwright whose nickname was "Gentleman Johnny," made a comment that was widely circulated: "What! Ten thousand peasants keep five thousand king's troops shut up? Well, let *us* in, and we'll soon find elbow room!" The remark became so well known that Burgoyne got the additional nickname "Elbow Room."[60]

The siege was well into the month of June when Gage, in his role as royal governor, issued an excessively wordy proclamation, written by Burgoyne, declaring martial law and offering pardons to "those in arms and their abettors" who should lay down their weapons. The only Rebels ineligible for a pardon were Sam Adams and John Hancock, "whose offences are of too flagitious a nature to admit of any other consideration than that of condign punishment." Left unsaid was the fact that the fitting punishment for treason was execution.[61]

The Patriots' intelligence network had yet to discover that Dr. Benjamin Church was spying for the British. On May 12, as chairman of a subcommittee of the Committee of Safety, he signed a report recommending that a defensive system be established at several places on Charlestown Neck, including Bunker Hill.[62] Oddly, Church seems not to have reported this to Gage because the British apparently had no knowledge of the Patriots' imminent plans. Patriot agents had learned that Gates, goaded by Howe, was about to break out a force that would occupy the heights on the mile-long Charlestown Peninsula, separated from Boston by a narrow stretch of water.[63] In a preemptive move the Committee of Safety authorized the taking of Bunker Hill, the peninsula's 110-foot high point.

Walking alone on the moonlit night of June 16, Major General Clinton was reconnoitering, looking across the dark water to the peninsula where British troops would soon be landing. He thought he heard some odd noises drifting across the water. He came upon a British sentry. Yes, the sentry said, he had also heard what Clinton himself thought he had heard: unusual muffled sounds on those hills that loomed over Boston. From those heights, both general and sentry knew, the Rebels could fire down on British ships and British troops. Clinton hurried back to headquarters and told Howe and Gage that he was sure the Rebels were moving onto the high ground. He urged Gage to land a force next morning. But Gage wanted to wait until dawn, when he could see what the Rebels were doing.[64]

The first British eyes to see what the Rebels were doing belonged to a lookout aboard HMS *Lively*, anchored off the southern shore of the peninsula. In the light of dawn he saw a large number of Rebels building a redoubt atop seventy-five-foot Breed's Hill. *Lively* quickly opened fire. Soon, on the orders of Adm. Samuel Graves aboard HMS *Somerset*, all the British ships in the harbor began firing. But the broadsides had little effect on the Americans because the warships had trouble elevating their guns at a high-enough angle to reach the hilltop.[65]

The lookout had spotted a few of the thousand militiamen who had assembled in Cambridge the night before. Their original orders had sent them off, silently and by lantern light, to Bunker Hill under the command of Col. William Prescott. There, throughout the night, they were to dig and build a redoubt. But at the last minute the objective was changed: Breed's Hill, which was more accessible, would be fortified first. (Confusion over the hills' names stems from the fact that both together were often called "Bunker Hill" while the southern slope was locally known as "Breed's Hill.")

Col. Richard Gridley, chief engineer of the Massachusetts forces, objected and was overruled. He then laid out a plan for a redoubt on Breed's Hill; it would be about forty yards square, with a six-foot parapet on which gun platforms would be mounted. Two flanking entrenchments were designed so that attackers would be hit with enfilading fire about twenty yards beyond the face of the redoubt.

The soldiers grabbed spades and began digging into soil still hardened by frost.[66] "Cannon shot are continually rolling and rebounding over the hill," wrote an American officer, "and it is astonishing to observe how little our soldiers are terrified by them."[67]

In Boston after the cannonading, Gage, his aides, and a Loyalist adviser went to Beacon Hill for a better view. Gage picked up his telescope and studied the redoubt that was emerging from Breed's Hill. "Who appears to be in command?" he asked, handing the telescope to Col. Abijah Willard, the wealthiest landowner in Lancaster, forty-seven miles northwest of Boston. Willard had been commissioned a colonel during the French and Indian War, when he led a regiment in three major campaigns. Willard recognized Colonel Prescott, a tall man whose head and shoulders jutted from the rampart. He was a friend, a fellow officer in the previous war, and the husband of Willard's sister Elizabeth.

"Will he fight?" asked Gage.

"I cannot answer for his men," said Willard. "But Prescott will fight you to the gates of hell!"[68]

It was a poignant moment for Willard. He had been made a mandamus councillor, had publicly recanted his appointment, had kept his own beliefs private, and had remained out of the strife. By his wealth and social position he was a Loyalist, just as Prescott could have been. But Prescott had chosen sides. So, finally and obliquely, had Willard. On the morning of April 19, he had filled his saddlebags with seeds for his farm in Beverly, north of Boston, and ridden along a route that would have taken him through Concord. When he came upon long columns of militiamen heading for Concord and Lexington, he realized he could not join them and had to oppose them. He turned toward Boston and sought refuge with Gage.[69]

In the morning's light Gage saw fortifications that had, in a British officer's words, "appeared more like majick than the work of human beings."[70] But still Gage did not believe the motley American army was a match for the two thousand men he would send on a frontal assault up Breed's Hill, with General Howe leading the attack.

Early on the afternoon of June 17, Howe landed on the southern shore of the peninsula and first hit what appeared to be a weak spot along a narrow beach. Americans waited until the British were within fifty yards and, in a burst of fire, riddled the line of Redcoats, killing ninety-six. The rest of the attackers pulled back. Howe called up soldiers and marines held in reserve. A second assault was also turned back. Dueling against snipers in houses in Charlestown, British artillery in Boston fired red-hot cannonballs and balls called "carcasses," which were full of burning pitch. Much of the town went up in flames.

The third surge, led by General Clinton and aimed directly up the hill, reached the redoubt through diminishing fire because the Americans were running out of ammunition. Some were throwing rocks as British survivors stormed the redoubt. Attackers with bayonets and defenders using muskets as clubs fought in the bloody redoubt. Prescott, who battled bayonet-wielding Redcoats with his sword, later showed the bayonet holes in his waistcoat and banyan, a kind of robe he wore over his uniform.[71] Under covering fire from reinforcements, the Americans retreated up Bunker Hill—which would give its name to a battle not fought there—and across Charlestown Neck. The British gave up pursuit.

Of the 1,500 Americans who defended the peninsula, about 450 were killed and wounded. Of the 2,400 British troops who landed on the peninsula, about 1,150 were killed or wounded; the Queen's Guards who had slipped out of Marshfield were all but wiped out. Among the Rebel dead was Dr. Joseph Warren, who, though commissioned a major general, had volunteered to fight as a soldier. British marine major John Pitcairn, the savior at Lexington, was mortally wounded and carried to his deathbed by his wounded son, marine lieutenant William Pitcairn.[72] "A dear bought victory, another such would have ruined us," General Clinton later wrote.[73]

Clinton, Howe, Burgoyne, and Gage now knew the answer to Gage's impertinent question—"Will he fight?" But that question continued to hover over the relationship between British officers and their Loyalist allies. Veterans like Willard and Brigadier Ruggles could advise Gage. Lesser Loyalists could spy for him. Ruggles's Loyal Americans could tear down abandoned wooden buildings and salvage the lumber for firewood.[74] But few Loyalists were given a chance to fight alongside British soldiers.

One of them was Gideon White, Jr., a charter member of the Old

Colony Club.[75] Another was John Coffin, son of the Nathaniel Coffin
who, as the king's cashier of customs at Boston, had one of the most
lucrative royal posts in America. At eighteen John had been chief mate
on a ship that in 1774 had brought reinforcements from England. Coffin met General Howe, who took a liking to the brash young man.
Coffin joined the force that Howe led to the Battle of Bunker Hill.
He survived and later raised the Orange Rangers, a mounted Tory
rifle corps.[76]

After Bunker Hill, Loyalists emerged as an auxiliary occupation force,
helping the British tighten their control of the city. Sometime later a
British general would declare that what started in Concord was a civil
war: "I never had an idea of subduing the Americans. I meant to assist
the good Americans to subdue the bad."[77] The assistance began on the
streets of Boston when General Gage reacted to his losses at Bunker
Hill by starting to purge the city of Patriots.

THE WAR FOR BOSTON

Motives of patriotism and private interest prompt me to hazard my fortune in this noble conflict with my brethren in the provincial army. . . . My friends afford me no encouragement, alleging that, as this is a civil war, if I should fall into the hands of the British, the gallows will be my fate.

—Dr. James Thacher, Continental Army surgeon[1]

" T he day, perhaps, the decisive day, is come," Abigail Adams wrote on April 18 in a letter to her husband, who was in Philadelphia as a delegate to the Continental Congress. Like so many people who lived near Boston, she feared that vengeful British troops were on their way from Bunker Hill, marching to wherever they could find Rebels to slaughter. "It is expected," she wrote, "they will come out over the Neck to-night, and a dreadful battle must ensue. Almighty God, cover the heads of our countrymen, and be a shield to our dear friends!"[2]

The British, tending their dead and wounded and recovering from the shock of Rebel ferocity, had no intention of leaving the relative safety of Boston. The decisive day had another consequence. Before Bunker Hill there was a chance that Concord and Lexington had been encounters like the Boston Massacre—bloody but not acts of war. Bunker Hill, however, was far more than an encounter. A war had begun, and people already saw it as a civil war: There was no room in Gage's Boston for Rebels and there was no room for Tories in the Patriots' Cambridge, or in Roxbury, or in fire-gutted Charlestown.

On April 27 Gage announced that anyone could leave the city, and a new stream of refugees flowed out of Boston.[3] Days later he ordered the printing of permits that allowed people to leave "between sunrise and sunset." Whatever they had with them was examined because, the pass said, "No arms nor ammunition is allowed to pass."[4] Permits were sold illegally, and people paid more and more to buy them. Then, beginning in July, anyone "desirous of leaving the Town of Boston" could do so only after presenting his or her name to James Urquhart, the British officer serving as town major. Within two days more than two thousand Bostonians registered with Urquhart.[5]

Many others hesitated to depart because they knew that the homes they left behind would almost certainly be plundered by British soldiers. Urquhart ruled that each refugee could possess no more than five pounds in cash and that no one could carry away silver-plate valuables.[*] But women sewed treasures into their garments, and families found ways to smuggle their silver spoons, bowls, and teapots in the goods that were piled on their carts.[6] Urquhart issued a special secret pass that allowed Henry Pelham, John Singleton Copley's half-brother, to "take a plan of the towns of Boston & Charlestown and the Rebel works round these places." Pelham was also assigned a sergeant and some soldiers to help him when he was in Boston. When he prowled as a British agent beyond Redcoat control, he acted like an artist making sketches. His spying gave the British detailed cartographic intelligence about Rebel fortifications—and gave posterity a magnificent map of the greater Boston area.[7]

In a letter to Copley in England, Pelham sketched Boston in words. The "Sword of Civil War is now unsheathd," he wrote. "Thousands are reduced to absolute Poverty. . . . Buisness of any kind is entirely Stop'd. . . . We find it disagreeable living entirely upon salt Meat. . . .

---

* This generally meant real silver.

Almost every shop and store is shut." At a dinner in August with General Howe, Pelham had "the only bit of fresh Meat I have tasted for very near four Months past. And then not with a good Conscience, considering the many Persons who in sickness are wanting that and most of the Convenency of Life. The usual pleas now made by those who beg a little Bacon or Saltfish is that its for a sick person."[8]

Loyalists responded to the blockade by promising "to contribute our Aid to the internal Security of the Town." They formed a voluntary association, saying that they would do whatever Gage deemed necessary or by helping "to raise a Sum of Money for promoting this salutary Purpose."[9] Gage, drawing from a pool of about two hundred Loyalist volunteers, assigned forty-nine to patrol the streets each night.[10]

Immediately after Bunker Hill, Gage found a mission for mandamus councillor Abijah Willard. He was put in charge of one hundred Loyalists assigned to find food for Gage's troops, a demanding task in hungry Boston. Details on exactly how Willard's victuallers did it are lacking; one report credits them with delivering one hundred oxen and sheep. James Putnam of Worcester, another councillor who had been targeted by Rebels, took command of a company of Brigadier Ruggles's Loyalist volunteers.[11] Three hundred Loyalists—with more signing up every day—formed a significant portion of the population. As people poured out of Boston, the number of residents dropped to 6,753. Gage had to feed 5,000 British troops and their families: 18,600 wives and children.[12]

At this early point in the blockade Gage did not anticipate firefights inside Boston. Selectmen had assured him that all the residents had been disarmed. But Gage did not trust the selectmen, for good reason. They were all pro-Rebel, and at least one was a Son of Liberty. Presumably Gage's many spies had reported widespread possession of firearms by lingering Boston Rebels. His proclamation ordered all firearms to be delivered immediately to a Loyalist-controlled courthouse. All who held on to firearms "should be deemed enemies to his majesty's government."[13] People turned in 1,778 "firearms" (presumably muskets), 973 bayonets, 634 pistols, and 38 blunderbusses.[14]

In Cambridge, Loyalists and Patriots sparred but did not fight. Friends of departed Loyalists barricaded the entrances to vacated houses and nailed doors closed to seal off rooms piled high with furniture and other bulky valuables, stored for the day of victorious return. Patriots, who outnumbered their opponents, forced the guardians to take down the barricades and then plundered the houses.[15] One of the houses belonged to John Nutting, the busy carpenter who had supervised the building of British barracks in Boston. His house served up a bit of irony: According to a Tory neighbor, it "was made a Barrack for the american Souldiers and much Damaged thereby."[16]

During the lull in Boston and Cambridge, Patriots in Maine—the sparsely populated northernmost part of the Massachusetts Colony— sparked a confrontation with the Royal Navy. In Falmouth, Thomas Coulson, a wealthy Loyalist merchant, was outfitting his ship, the *Minerva*, and had ordered sails and rigging from Britain in defiance of the Patriot embargo on British goods. When the ship carrying Coulson's goods arrived, Patriots ordered it to go to sea without unloading its cargo. Coulson objected, claiming that the ship was unfit for a return voyage and needed repairs. But when the repairs dragged on, Patriots decided that the ship had tarried too long and repeated their demand.

William Tyng, Falmouth's Tory sheriff, had been given a colonel's commission by General Gage. Expecting waterfront trouble, Tyng asked for help. Gage sent HMS *Canceaux*, a merchantman converted into an eight-gun Royal Navy warship, under the command of Lt. Henry Mowatt. The *Canceaux*'s appearance happened to coincide with the arrival of news of Lexington and Concord. Fearing that emboldened Rebels would move against them, Coulson and Tyng sought refuge on the warship, along with the Reverend John Wiswall, a Church of England clergyman, and some of his flock.

As the standoff crisis seemed to ease, sixty militiamen from Brunswick, led by Col. Samuel Thompson, arrived to support the Patriots. Mowatt, Wiswall, and the ship's physician had disembarked and were strolling onshore when, apparently on impulse, some of Thompson's men pounced on them. Both Patriots and Loyalists were shocked. Mowatt's second in command, learning of the kidnapping, threatened to shell the town if Mowatt was not released. He fired a couple of blanks as a warning.

Panic swept through the town. As one witness wrote, "Our women were . . . in tears, or praying, or screaming; precipitately leaving their houses, especially those whose husbands were not at home, and widows, hurrying their goods into countrymen's carts . . . and carrying their children either out of town, or up to the south."[17]

Thompson, pressured from both sides, released his captives "under parole." Wiswall came ashore, was interrogated by Patriots, promised to return to his parsonage—and sneaked back to the *Canceaux*. The crisis worsened as some six hundred Patriot militiamen from nearby towns poured into Falmouth. A few militiamen, rebelling against their own officers, seized them and demanded rum and food as ransom. Other militiamen threatened to attack local Loyalists—or even the *Canceaux*.

The crisis ended when Coulson's *Minerva*, the ship that started it all, used its illicitly imported sails and rigging to head for Boston, carrying off Falmouth's first Loyalist refugees—Tyng, Wiswall, and Coulson. Their families would follow later.[18]

The next confrontation between Maine Rebels and the Royal Navy flared into the first naval battle of the Revolution. Maine provided Boston with much of its lumber. Ichabod Jones, who had built one of the first sawmills in the little town of Machias, had made his fortune by delivering Boston goods to Machias in exchange for lumber. Machias, a river town up Machias Bay, was so far north and so isolated that its people associated themselves more with Nova Scotia than with Boston. The town did not have a militia, but it did have a Liberty pole and a Committee of Correspondence and Safety.[19]

On June 2 Jones sailed up the Machias River aboard a British warship, the armed schooner HMS *Margaretta*, ready for trouble. Accompanying the warship were two coastal trading sloops that were to deliver goods to Machias and carry off lumber. The sloops tied up at the wharf while the *Margaretta* anchored offshore. Jones and Midshipman James Moore, the young captain of the *Margaretta*, went ashore. Jones announced that he was there to deliver supplies to Machias—if the town provided lumber. No lumber, no supplies. Jones did not identify his customer, but people knew it was the British Army, in need of lumber for barracks and floorboards for entrenchments and other earthworks.

Midshipman Moore had something to add: If the town did not take down its Liberty pole, he would bombard Machias.

Dublin-born Jeremiah O'Brien, a Patriot leader, stepped forward to say that a town meeting had voted for the Liberty pole, and only a town meeting could vote to take it down. O'Brien, knowing that town Tories were on Moore's side, was stalling for time to rally Patriot support. He knew that the provincial congress had "strongly recommended" that no one supply the British troops with goods, including, specifically, timber, boards, spars, pickets, and tent poles.[20]

O'Brien also knew that if Machias did not get Jones's goods, the town would not starve, not while there was game in the woods and fish in the sea and in the Machias River. No lumber sale meant only a shortage of rum, snuff, tobacco, and luxuries not many could afford anyway. Besides, anything could be smuggled, for a price, from Nova Scotia.

The town meeting on Saturday, June 10, voted to keep the Liberty pole. Jones reacted by distributing provisions only to those who voted against it.[21] O'Brien persuaded Moore and Jones to wait for a second vote on Monday. On Sunday, Moore and his officers were invited to the meetinghouse, which served as the Machias church. The Reverend James Lyon, a Presbyterian minister, was the pastor. As the chairman of the Patriots' Committee of Safety and Correspondence, Lyon was undoubtedly aware that O'Brien and his Patriots had a reason for wanting their visitors to go to church.

On Sunday, while listening to Lyon's sermon, Moore looked around and noticed that there were few able-bodied men in the congregation. Through an open window he spotted a band of men approaching the church. Signaling his officers to follow, he leaped through the window and led them racing down the road to the wharf. Jones, knowing the area, fled to the woods and hid there for several days, undoubtedly aided by Loyalist sympathizers.

On board the *Margaretta*, an alert petty officer saw his officers being chased by a mob. He ordered a boat launched and fired a swivel gun, a small cannon mounted on a deck railing. The ball missed the pursuers but slowed them down. Moore and his officers made it to the boat, which was rapidly rowed back to the ship.

The next day O'Brien, his five brothers, his black servant, and about twenty other men swarmed aboard the *Unity*, one of the trading ships, while another group of Rebels boarded the second trading ship, the *Polly*. The two crews had among them about twenty muskets, along with an assortment of clubs, pitchforks, and timber axes. One of the men lugged a large-caliber gun aboard the *Unity*.

As the two ships set sail, the *Margaretta* slipped its anchor cable, hoisted sail, and headed down river toward the bay, six miles away. During the chase the *Polly* ran aground. When the *Margaretta* reached the bay, it fired on the fast-closing *Unity*, killing a man who had propped the big gun on a gunwale. Another man stepped forward and fired, killing the *Margaretta* helmsman. Musketry from the *Unity* raked the British warship. The two ships were so close that Moore, from the quarterdeck, could hurl small, powder-filled cannonballs with lighted fuses down on the *Unity*, mortally wounding one man and wounding several others.

O'Brien rammed the *Unity* into the *Margaretta*. His men lashed the ships together and leaped upon the *Margaretta*'s deck. Moore, cutlass drawn, led his crewmen and marines against the boarders. In a fierce fight, Americans swung muskets, clubs, axes, and pitchforks against British bayonets and Moore's cutlass. A lethal musket shot felled Moore, and the crew shortly surrendered. O'Brien leaped to the halyards and hauled down the British ensign. Four Americans were killed and ten wounded; the British suffered ten killed and ten wounded. O'Brien and two of his brothers later became privateers in the war against the British at sea.[22]

As for Ichabod Jones, he was caught in the Machias woods. Condemned as "a known enemy" by the provincial Committee of Safety, he was taken to a Massachusetts town four hundred miles from Machias and released under a large bond. His property was confiscated, a retribution that would befall more and more Tories.[23]

Widespread confiscation of Loyalist property got its start in Cambridge. The Committee of Safety decreed that hay belonging to John Vassall, an Addresser from Cambridge, was to be cut and used by the Patriots. They took over his abandoned Tory Row mansion and his two-hundred-acre estate, even ordering that his stables be reserved for Patriots' horses. Vassall and his family soon went aboard one of the six Loyalist-laden ships that left Boston for England that summer.[24]

Gage stationed about two hundred of his men on Boston Neck, the front line between British and American troops. He also strengthened the "Rebel redoubt" on Breed's Hill, and fortified the Boston-to-Charlestown ferry stops and Moulton's Point, where Howe's troops had landed. Soldiers used wood salvaged from the ashes of Charlestown for fortifications, tent floors, and fuel.[25] General Ward's militiamen, numbering about sixteen thousand, threw up earthworks in a line from Cambridge to the Mystic River and built forts in Roxbury. The strongpoint in the Rebel line was on Prospect Hill in Charlestown (today's Somerville), which loomed over Boston and the British fleet.

Propaganda broadsheets compared life for Rebel sentries on Prospect Hill in Charlestown with life for British sentries on Bunker Hill:[26]

PROSPECT HILL		BUNKER HILL	
I	Seven dollars a month.*	I	Three pence a day
II	Fresh provisions and in plenty.	II	Rotten salt pork.
III	Health.	III	The scurvy
IV	Freedom, ease, affluence, and a good farm.	IV	Slavery, beggary, and want.

---

* Seven dollars had about the same purchasing power as about $200 in 2009.

Management of the Massachusetts militiamen was shifting to Philadelphia, where the Continental Congress voted to create the Continental Army—and, shortly after, adopted the "Olive Branch Petition." The document was so effusive in its homage to the "Most Gracious Sovereign" from "your Majesty's faithful subjects" that it could have been written by a Loyalist.

The delegates said they felt "required by indispensable obligations to Almighty God, to your Majesty, to our fellow subjects, and to ourselves, immediately to use all the means in our power not incompatible with our safety, for stopping the further effusion of blood. . . . We therefore beseech your Majesty, that your royal authority and influence may be graciously interposed to procure us relief from our afflicting fears and jealousies." The king ignored the olive branch, declaring that the colonies were in "an open and avowed rebellion" and that "all our Officers, civil and military, are obliged to exert their utmost endeavours to suppress such rebellion."[27]

While waving the olive branch, however, Congress asked Virginia, Maryland, and Pennsylvania to raise ten companies of riflemen. They would be the first soldiers to be mustered into the new army, and they would be carrying a marksman's weapon. The rifleman's gun, unlike a militiaman's smoothbore musket, was rifled—long spiral grooves inside the barrel engaged the ball, starting it spinning along an axis at right angles to the direction it was shot. Stabilized instead of tumbling, the bullet could fly straighter and farther.

Congress next selected George Washington, a Virginia plantation owner, as commander in chief of the Continental Army. He was still remembered as a hero in the French and Indian War, when he had been a fearless twenty-three-year-old Virginia militia officer—"4 bullets through my Coat, and two Horses shot under me" in one battle.[28] And his appointment gave to the South military power that would offset New England political power.

Traveling northward from his Virginia home, he took the Alexandria ferry across the Potomac River. A ferry going in the opposite direction carried an old friend, the Reverend Jonathan Boucher, an Anglican clergyman. "Some patriots in our boat huzzaed," Boucher later wrote, "and gave three cheers." He confined himself to doffing his hat. Washington asked the ferries to stop, and he spoke for a few minutes to Boucher. Washington knew him not as an avowed Tory but as the learned man who had tutored Washington's stepson and often dined at Mount Vernon. Boucher bluntly told Washington that Americans would soon declare for independence. Washington denied that he was "joining in any such measures," Boucher recalled. Like most of his supporters, the new general believed that the colonies were seeking a redress of grievances, not independence.

Soon Boucher would be committed to his own civil war, preaching with a pair of loaded pistols resting on a cushion in his pulpit. He had "given notice that if any man, or body of men, could possibly be so lost to all sense of decency . . . to drag me out of my own pulpit, I should think myself justified before God and man in repelling violence by violence"—stating what was the moral basis for the Loyalists' decision to arm themselves. Boucher would sail for exile in England nine months later, before the Continental Congress declared the independence that Washington had not yet realized he was actually fighting for.[29]

The first armed Loyalists to patrol the streets of Boston were members of General Ruggles's Loyal American Association. Ruggles's men did not have uniforms; a patroller simply identified himself by tying a white scarf around his left arm. They patrolled "the District about Liberty Tree & the Lanes, Alleys & Wharves adjacent" from sunset to sunrise to "prevent all disorders within the district by either Signals, Fires, Thieves, Robers, house breakers or Rioters."[30] Included among them were Edward Winslow and George Leonard, veterans of the Lexington-Concord expedition.

One patroller, Job Williams, led an attack on the Liberty Tree, a great elm on which Sons of Liberty had been hanging effigies of Tory officials since Stamp Act days. Williams and his cohorts chopped down the tree. By one account one man was killed in a fall while attempting to cut off a limb of the tree. Williams went on to become a captain in a Loyalist regiment.[31]

An early muster roll of the patrollers lists several men who represented the upper ranks of Massachusetts society.[32] Richard Saltonstall, Harvard '51, whose family had lived in Massachusetts for six generations, had been a superior court judge and a colonel in the French and Indian War. After a short term of service in the association, he decided he could not take sides in the war and sailed to England, never to return.[33] Thomas Aston Coffin, Harvard '72, came from a similarly illustrious family whose members included the holder of one of the most lucrative of royal posts: Receiver General and Cashier of His Majesty's Customs.[34] John Lindall Borland, Harvard '72, was born and bred on Tory Row. He went from patrolling Boston to serving as a lieutenant colonel in the British Army.[35] Zebedee Terry later joined a Loyalist regiment,[36] as did Benjamin Pollard.[37]

The chaplain of the Loyal American Association was the Reverend John Troutbeck, who in 1753 had been sent to America as a missionary of the Society for the Propagation of the Gospel in Foreign Parts. He became the assistant rector of King's Chapel, Boston's oldest Anglican church.[38] Harried as an Addresser, he left for England in 1776. He was remembered, in Rebel doggerel, as having been both a clergyman and a distiller:

His Sunday aim is to reclaim
Those that in vice are sunk,
When Monday's come he selleth rum,
And gets them plaguey drunk.[39]

Although Ruggles was one of the most militant Loyalists, Patriot neighbors treated him with restraint. After confiscating the arms and ammunition found in his home, the Patriots confined him to one of his farms, about ninety miles west of Boston, allowing him to leave only on Sundays, "fast days, or some other public days," and charging him for the pay of his guards. After a short time members of the town meeting voted to allow him "to go to Boston, and live there, if he pleases."[40] He did, arriving in time to give Gage advice prior to the battle of Bunker Hill. His association became a model for two other Loyalist groups later formed in Boston: the Royal North British Volunteers Formation, made up of "North British Merchants,"* and the Loyal Irish Volunteers, consisting mostly of Irish merchants.

Royal North British Volunteers, each wearing a blue hat bearing a Saint Andrew's Cross (the blue-and-white emblem of Scotland), were to "take into Custody all Suspicious & Disorderly Persons found in the Streets at improper Hours." Similar duties were assigned to the Loyal Irish, who attached white cockades (rosettes or knots of ribbon) to their hats as badges.[41]

Gage did not bring the volunteers into the regular military establishment because that was not the way the British Army worked at that time. Officers came from the noble ruling class. They purchased and sold their commissions. Americans had become aware of this practice during the French and Indian War, when "provincial" regiments went into battle side by side with British forces but were not integrated, and American officers were not given equal treatment with their British counterparts. One of the provincials who discovered this was Col. George Washington of the Virginia Regiment. After learning that he could not be given a British Army commission higher than captain, he resigned.[42]

When Washington arrived in Cambridge, he first boarded at the home of the president of Harvard, who was a Son of Liberty. Washington then moved, setting up his headquarters in the Tory Row mansion that General Ward had taken over.[43] In Cambridge and Boston, Washington found not warfare but stalemate. Loyalists fled to Boston and Patriots left, all seeking safety. Soldiers fought spasmodi-

---

* After the Acts of Union in 1707, joining the kingdoms of England and Scotland to form the United Kingdom of Great Britain, "North British" was the preferred promonarchy name for Scotland.

cally. The British made thrusts across the Neck or tried to drive the Rebels off Prospect Hill. The Rebels slipped into enemy territory to burn a barn or kill a Redcoat. In the Rebel ranks were snipers and Indians armed with bows and arrows. On one June day, two Indians killed four Regulars.[44]

The Indians, members of the Stockbridge tribe of Massachusetts, joined the Patriots' ranks after a propaganda drive aimed at gaining allies from several eastern tribes.[45] The Massachusetts Provincial Congress sent a letter to Indian leaders, telling them that the British "want to get all our money . . . and prevent us from having guns and powder to use and kill deer, wolves and other game with or to send to you for you to kill your game and to get skins and furs to trade with us for what you want."[46] Gage reacted by telling Lord Dartmouth in a letter that "we need not be tender of calling upon the Savages, as the Rebels have shewn us the Example."[47] Gage's remark foreshadowed the British use of Indians as allies, especially along the New York–Canadian border.

The arrival of riflemen sharpened the close combat along the lines. The British were stunned by the way their sentries were picked off by these backwoods marksmen. A British newspaper, quoting a letter from Norfolk, Virginia, described their uniforms as "something like a shirt, double caped over the shoulder, in imitation of the Indians; and on the breast, in capital letters, is their motto, 'Liberty or Death!'" One captured rifleman was taken to England as a curio, much like the American Indians displayed in previous times.[48]

"Never had the British Army so ungenerous an enemy to oppose," a British officer complained. They "send their riflemen (five or six at a time), who conceal themselves behind trees, &c., till an opportunity presents itself of taking a shot at our advanced sentries; which done, they immediately retreat."[49]

The British Army did not yet see its ungenerous enemy as soldiers worthy of decent treatment as prisoners. Washington, who knew Gage from the French and Indian War, viewed him at first as an honorable opponent playing by the accepted rules of war. But when Washington learned that Americans captured in the Bunker Hill battle were not being treated as prisoners of war, he wrote to Gage saying he was shocked that the men had been "thrown indiscriminately into a common jail appropriated for felons" and that some of them, "languishing with wounds and sickness," had undergone amputations in the jail. Gage responded by saying that he viewed the captives as criminals "whose lives by the laws of the land are destined to the cord"—the noose.[50]

Gage, representing British authority, believed he was dealing with a rebellion and a civil war. He saw local Loyalists as counterrebels who could aid him. Gage had a limited view, for militant Loyalists looked beyond Boston. They saw themselves as having more to contribute to the war than service as a home guard for the British garrison. Tory civilians were beginning to realize they were caught in an unforeseen war. "We are in the strangest state in the world, surrounded on all sides," wrote a Tory woman in Boston. "The whole country is in arms, and intrenched. We are deprived of fresh provisions, subject to continual alarms and cannonadings, the provincials being very audacious, and advancing near to our lines."[51]

Gage called on the Royal Navy to help feed his hungry soldiers and civilian wards. The British Army had grown to about six thousand men. Its supply line, which stretched to England, was menaced by Patriot privateers. Adm. Samuel Graves, commander of the Royal Navy's North American Squadron, judged that the rebellious acts of colonists had "rendered our dependance for Fuel and fresh provisions very precarious."[52]

Capt. Sir James Wallace, commanding officer of the *Rose*, on station off Newport, Rhode Island, gathered provisions by having his sailors and marines go ashore and take what they needed. When farmers moved their livestock inland, Wallace sailed to Bristol, Rhode Island, where he sent an officer ashore in a longboat. When it came alongside the wharf, the officer demanded that a small delegation of town officials accompany him back to the *Rose* to meet with Wallace. A magistrate gruffly told him that if his captain wanted to talk, he should come ashore himself.

The officer returned to the *Rose* and reported to Wallace. Back in

March, Wallace had had experience with cheeky Rebels, who had intercepted a letter sent to him by Col. Thomas Gilbert, leader of the Tory militia in Freetown. Now, infuriated by another act of disrespect, Wallace opened fire on the town.

After a few cannonballs struck, the town leaders agreed to go aboard the *Rose*. Wallace demanded two hundred sheep and thirty head of cattle. The townsmen balked, saying that much livestock was unavailable. Negotiations ended when Wallace made his final offer: "If you will promise to supply me with forty sheep, at or before twelve o'clock, I will assure you that another gun shall not be discharged." The townsmen agreed. *Rose* crewmen loaded the sheep on board, and Wallace sailed off with his fleet to Boston.[53]

A few days later Admiral Graves issued an order to Lt. Henry Mowatt, captain of HMS *Canceaux* and recent victim of abduction by Falmouth Patriots. "My Design," Graves wrote, "is to chastize Marblehead, Salem, Newbury Port, Cape Anne Harbour, Portsmouth [New Hampshire], Ipswich, Falmouth in Casco Bay, and particularly Mechias where the *Margaretta* was taken . . . and where preparations I am informed are now making to invade the Province of Nova Scotia."[54] Graves was right. Machias was the secret headquarters of American Rebels and anti-British Nova Scotians who were planning an invasion. One of the planners was the Reverend James Lyon, the Presbyterian clergyman who had been a confederate of the Machias captors of the *Margaretta*.[55]

The *Canceaux* and three other warships left Boston on October 8, 1775. Supposedly, bad weather forced Mowatt to change course and as a result Falmouth became the first port he would "chastise." But Mowatt obviously had a personal reason for picking Falmouth. On October 17 he sent an officer ashore with a warning that he would bombard the town in two hours. He also invited local Loyalists to come aboard the *Canceaux*. None did, fearing that they would be shot as they rowed to the warship.

A committee that included a leading Loyalist won a delay until the next morning, when Mowatt ordered all five ships to begin a cannonade with shells and the incendiary balls called carcasses. The bombardment, in his words, "continued till six, by that hour the body of the town was in one flame." Then Mowatt sent some men ashore to finish the mission by setting fire "to the vessels, wharfs, storehouses, as well as to many parts of the town that escaped from the shells and carcasses."[56] The bombardment destroyed about three-quarters of the town's buildings, including about 130 dwellings, many of them housing two or three families; a church, the courthouse, the library, and almost every store and warehouse. Sheriff Tyng's mansion was spared, spotlighting his Tory status and guaranteeing his self-exile.[57] Fourteen ships were burned and several others seized.[58]

As autumn settled on Boston, Gage made one of his last decisions in America, for he was scheduled to sail to England on October 10. Knowing that his successor, General Howe, would soon go into winter quarters, Gage wanted his soldiers—many of them living in tents—to have decent shelter by occupying some of Boston's many abandoned houses. To find someone to supervise the emptying of the houses and the storing of their contents, he reached beyond Massachusetts to the Tory stronghold of New York. He chose Crean Brush, a striving, self-promoting Dublin-born Loyalist.

In the words of Gage's commission, Brush would become the official receiver of "large quantities of Goods, Wares, and Merchandize, Chattles and Effects of considerable value left in the Town of Boston by persons who have thought proper to depart therefrom." To give the looting a veneer of legitimacy, Gage ordered townspeople to provide Urquhart, the town major, with their names and addresses so that someday their property could be returned. Then, in a conflicting proclamation, he made a leading Loyalist the sole auctioneer for selling off the goods to the highest bidder.[59]

Brush affected a military-style wardrobe and, claiming to be a former British Army officer, demanded to be addressed as "Colonel." He had arrived in northeastern New York about a dozen years before. As a favored Loyalist in New York, he became wealthy by simultaneously serving in placeman posts, becoming a member of the legislature, and acquiring land.

Most of Brush's holdings were in disputed territory claimed by New York, New Hampshire, and a land speculator named Ethan Allen. Brush solved the dispute by introducing a bill that outlawed Allen and his lieutenants in the Green Mountain Boys, a guerrila force originally formed to fight for disputed land bordering on Vermont. By 1775 he was the owner of twenty thousand acres. But Patriots took over the New York legislature, which wiped out Brush's claims and made him a public enemy.[60] In 1777 most of his New York land would become part of the Republic of Vermont.

In Boston, Brush used his newfound connection with Gage to ask a favor: When the general arrived in London, would he please present a petition to King George III? The petition asked royal approval of Brush's plan to "raise a Body of Volunteers," a three-hundred-man unit modeled on the Royal Fencible Americans, an early Loyalist regiment formed in Nova Scotia.* Brush promised that he and his men would "open a Line of communication" from the Connecticut River westward toward Lake Champlain.

Brush said his "intimate Knowledge of that Frontier" would enable him to track down the "dangerous Gang of Lawless Banditti, who, without the least pretext of Title, have, by Violence, possessed themselves of a large Tract of Interior Territory." That territory, of course, was land that Brush claimed.[61] On the New York frontier, British officials did depend on Loyalist allies but did not need any help from Crean Brush.

Brush's emptying of houses was only part of the British pillaging of Boston. Dragoons turned the Old South Church into a riding school, hacking and carrying away the pulpit and pews and spreading dirt and gravel on the floor. An exquisitely carved pew, with silk seats, was taken off to become a hog-sty. A stove was installed, the

---

* A "fencible soldier" was eligible for home service only. The British interpreted this
  to mean that fencibles could be deployed anywhere in the colonies.

church library's books and manuscripts providing the kindling. Another church became a barracks, and its steeple was dismantled for firewood, as was the entire church. More than one hundred houses were torn down for firewood. Some of the work was done by members of Ruggles's Loyal American Association.[62]

So much looting was going on that General Howe directed his provost marshal "to go his rounds, attended by the executioner, with orders to hang up on the spot the first man he should detect in the fact without waiting for further proof for trial."[63] Because of the food shortage, Howe, loosening the rules, allowed families to leave in increasing numbers. Loyalists appealed to him to stop the exodus because they feared that the Rebels would burn down the town if its only population were Tories and soldiers. Howe agreed, after letting some three hundred people leave on November 25. When they got past the checkpoint, one was dead and two were dying. Washington, fearing a smallpox epidemic, barred the refugees from the American camp.[64]

Soon after taking over from Gage, Howe formed quasi-military "companies," commanded by officers he named, and directed by members of the mandamus council. Volunteers for the companies, he proclaimed, "shall be properly armed, and an allowance of fuel and provisions be made to those requiring the same, equal to what is issued to his Majesty's troops within the garrison."[65]

By December members of another Loyalist home guard unit—three hundred Royal Fencible Americans—were patrolling the streets, on the watch for fires and burglaries while Crean Brush was busily emptying houses.[66] Generals Howe and Washington, meanwhile, were planning their strategies for the months ahead. Howe looked to New York as the next battleground. Washington looked to Canada.

6

INTO THE FOURTEENTH COLONY

*My friends and Fellow Subjects—The unhappy necessity which
subsists of dislodging the Ministerial Troops obliges me to carry on
Hostilities against your city. . . . I find myself reduced to Measures
which may overwhelm you with Distress!*

—Brig. Gen. Richard Montgomery,
letter to the people of Quebec[1]

*W*ashington was on his way to Cambridge, as commander in
chief of the Continental Army, when General Ward and his
council of war decided to invade Canada. First Congress and then
Washington ratified the decision, for there was reason to believe that
Canada, Britain's fourteenth colony, would join the other thirteen in
rebellion. Hope was strengthened by fresh news of revolutionary fervor to the north.

On May 1, 1775, only twelve days after the Battles of Lexington
and Concord, an anonymous rebel in Montreal defaced a bust of
the king and hung on it a mock rosary, with potatoes for beads. Attached was a sign in French that said, "Here's the Canadian Pope
and English Fool."[2] The desecration triggered widespread clashes
between supporters and opponents of British rule. When word of
the incident reached Congress, invasion enthusiasts believed that a

potential Son of Canadian Liberty had spiritually joined the American revolt.

The symbolic attack on the king was inspired by an uproar over the Quebec Act, which went into effect on May 1. American Patriots detested the act because it barred western expansion. Canadians were infuriated by the act's acceptance of Catholics as royal officeholders and its legalization of previously outlawed priests. So there was a tenuous anti-British link between Canadians and Americans.

But congressmen who backed the invasion nevertheless believed that if New England and the other colonies wanted freedom from British oppression, then Canadians would certainly feel the same. Enthusiasts ignored the fact that Canadian Catholics, mainly in Quebec, distrusted Protestant New England. Among *Canadiens* (French-speaking Canadians), the common name for New Englanders in particular and Americans in general was *Bostonais*, a derogatory term which came to mean anything that was terrible and violent.[3]

The congressmen saw a march northward as a friendly act of liberation from British tyranny, not as an unprovoked attack on a neighbor. Washington, sidestepping the ideology, endorsed the congressional invasion proposal on strategic grounds: The invasion would block a possible British thrust into New York from Quebec. That peril was real, for British officials and American Loyalists along the New York frontier were already rousing Indians against the Patriots. The British Department of Indian Affairs—an agency more military than political—had dozens of agents in northern New York recruiting Indian allies.[4] Throughout the war a British-Indian-Loyalist alliance would fight the Patriots along the New York–Canada frontier.

Congress launched the campaign against Canada by instructing Maj. Gen. Philip J. Schuyler to lead a strong force to Ticonderoga, destroy any vessels that the British could use for an invasion of New York, and then launch an invasion of his own. He was ordered to seize St. John's on the Richelieu River north of Lake Champlain, a gateway to Montreal—and then take Montreal itself. Congress added an odd condition that showed American sensibilities toward the neighbor to the north: Invade only if the invasion "will not be disagreeable to the Canadians."[5]

Washington, after approving the Schuyler expedition, ordered Col. Benedict Arnold to lead a second march through the Maine wilderness to Quebec. Washington gave Arnold copies of an appeal that was to be handed out to the "Friends and Brethren" of Canada, asking them to join in the rebellion against British rule. In a bid for Catholic support, Washington wrote, "The cause of America and of liberty is the cause of every virtuous American Citizen Whatever may be his Religion or his descent."[6]

For a long time the focal point in Canada for Patriot activities was Nova Scotia. In 1768, when the Massachusetts legislature sent a circular letter protesting the Townshend Acts to all the colonies, Nova Scotia was included as a sister colony. The letter so distressed Nova Scotia's lieutenant governor that he felt compelled to send a message to London, assuring the minister for colonial affairs that "no temptation, however great, will lead the inhabitants of this colony to show the least inclination to oppose acts of the British parliament."[7]

In October 1774, the Continental Congress invited Quebec to send delegates to Philadelphia for the next session. "We do not ask you, by this address, to commence acts of hostility against the government of our common Sovereign," the invitation said. "We only invite you to consult your own glory and welfare, and complete this highly desirable union."[8] Congress translated the invitation into French and sent two thousand copies to Thomas Walker, a Montreal merchant who was an ex-Bostonian and a friend of the Rebels. He distributed the invitations to French-speaking Canadians.[9] Quebec declined the invitation.

Walker, once a justice of the peace, was a notorious foe of British rules about the billeting of Redcoats. In 1764 he and four other magistrates had imprisoned a British officer in a dispute over his lodgings. Four masked men avenged the officer by breaking into Walker's house, beating him severely, and hacking off one of his ears. Four soldiers were arrested for the assault, put on trial, and acquitted. A subsequent trial of four others ended the same way. The acquittals added to Walker's rage. By 1773 he was the leader of a small group of radical Canadians—and a spy for both Benedict Arnold and Ethan Allen. A year later, he was a full-fledged revolutionary, eager to bring a fourteenth colony into the Revolution.[10]

Colonial officials in Canada, reacting to the whiff of Revolution from across the border, began to fear that opposition to royal rule could be contagious. Chief Justice Jonathan Belcher, Jr., of Nova Scotia founded the Association for Loyal Allegiance, whose members aided authorities who were mobilizing to repel the invaders. Martial law was later decreed, and strangers entering the province had to report to a magistrate or be treated as American spies.[11] American Loyalists were fleeing from northern New York into Canada—and pro-Patriot Canadians were crossing into New York or Maine.[12]

This was the setting when General Schuyler, a wealthy Albany entrepreneur, assumed command of the invasion army. He had fought at Ticonderoga during the French and Indian War. Now he was back at Ticonderoga, the terminus for the long-established waterway path to Canada. Schuyler, ailing and overly cautious, took his time, spending weeks gathering supplies and building boats for the trip northward down Lake Champlain.[13] He was soon eclipsed by his younger and brasher second in command, Brig. Gen. Richard Montgomery, an Irishman turned American Patriot.

Montgomery had sold his captain's commission in the British Army, immigrated to America in 1771, and bought a sixty-seven-acre farm at King's Bridge, about a dozen miles up the Hudson from New York City. While stationed in New York during the French and Indian War, he had met Janet Livingston, daughter of the ruling land baron of the most powerful Patriot family in New York. Later, back in New York as a wealthy, charming immigrant, he courted Janet Livingston. They were married in July 1773 at Clermont, the family's thirteen-thousand-acre Hudson Valley estate.[14] One of his brothers-in-law was Robert R. Livingston, Jr., who would be a member of the "Committee of Five" that drafted the Declaration of Independence.[15]

When Montgomery volunteered to go to war, his wife had a nightmare in which she saw him die of wounds inflicted by his brother. Entwined in that nightmare was her vision of the civil war that had struck even Clermont: Three of Montgomery's officers would carry the Livingston name into Canadian battles. But, while the Livingstons fought for the Patriots, most of their tenants, turning against their landlords, became avowed Tories.[16] The phenomenon, repeated on baronial estates all along the Hudson Valley, was inspired by the tenants' belief that after the war the victorious British would distribute land to Americans who had supported the king.[17]

On September 5 Schuyler issued a proclamation that said "the Grand Congress" had sent the army to "Expell" British troops "which now, acting under the orders of a Despotic Ministry, would wish to Enslave their Countrymen." He also vowed that he would do no harm to Indians, known to be mostly pro-British.[18]

Schuyler, who suffered from gout and rheumatism, stayed behind as Montgomery led about one thousand men in a fleet of boats to Fort St. John's, on the Richelieu River north of Lake Champlain. They landed a mile and a half from the fort and were slogging through a marsh when they were ambushed by Capt. Gilbert Tice of the British Department of Indian Affairs, who commanded a combined force of Loyalists and Indians. In the thirty-minute firefight, Tice lost ten men, Montgomery eight. Tice, an American Tory who had run a tavern in Johnstown, New York, had been recruiting for some time in the border area.[19]

Schuyler, too ill to go on, put Montgomery in command and returned to Ticonderoga. Montgomery, realizing that his force was too green and disorganized for an attack on the fort, began a siege. He also sent small detachments northward to recruit Canadians. Ethan Allen, acting on his own as usual, declared that he would seize Montreal with the aid of Canadians who would rise up and join the American side.[20] Instead, British troops, aided by Canadian Loyalists, captured Allen and forty of his Green Mountain Boys. Allen was shipped to England and would not be exchanged until 1778.[21]

Fort St. John's finally surrendered on November 2, and Montgomery headed for Montreal, which had a small British garrison. Another
fort, north of St. John's, fell quickly. Among the British officers captured was Lt. John André, who was destined to play a much larger
role in the Revolution. The colors of André's regiment were furled
and sent to Congress as a trophy of a victory on the path to Quebec.[22]

Arnold, meanwhile, had started off with 1,050 men on his epic expedition, which began on September 13 with a march from Cambridge
to Newburyport, where the men boarded ships for a three-day voyage
to a port up the Kennebec River. There a fleet of more than two hundred newly built bateaux awaited the troops for the next leg of their
journey through the wilds of Maine. To penetrate these seemingly
trackless forests, Arnold had a copy of a map that showed a paddle-
and-portage route to Quebec. He expected to reach the city in three
weeks.

The map was made during the French and Indian War by John
Montrésor, a brilliant young British Army engineer. How Arnold obtained the map is not known, but he had been to Quebec as a merchant during that war and had known Indians who had made the
journey. He may have met Montrésor, for Arnold also had a copy of
Montrésor's handwritten journal describing his two round-trip journeys from Quebec to the Maine coast.[23]

Arnold plunged into the unknown with faith in the map and his
handpicked men, who included keen-eyed Virginia riflemen famous
for their sniping of British sentinels in Boston. Their commander was
Capt. Daniel Morgan, a future hit-and-run fighter against Redcoats
and Tories in the South. One of Arnold's other officers was nineteen-
year-old Aaron Burr, accompanied part of the way by a beautiful
young Indian woman whom soldiers called Golden Thighs.[24]

Knowing that he needed to update Montrésor's map, Arnold commissioned a Maine surveyor to make copies for his officers and add
information about the depths and speeds of the Kennebec. The surveyor, who had worked for a Loyalist-owned land company and was

almost certainly a Loyalist himself, deliberately introduced errors into the map. It probably had some inaccuracies already because British Army clerks routinely put errors on copies of maps likely to fall into enemy hands.[25]

The doctored map produced lost days, lost miles, and a loss of embittered men who wandered off, hoping to find a way home. The bateaux were difficult to maneuver and, in Arnold's words, "very badly built."[26] Day after day men paddled and portaged, carrying the four-hundred-pound boats on their chafing shoulders, working up hills, slogging through swamps and bogs, soldiering on for nearly six hundred miles. In the fifty-one-day struggle against storm, disease, hunger, and despair, Arnold lost 40 percent of his men to death or desertion.[27]

On November 12 Montgomery reached Montreal, which was defended by only about 150 men. Attempting to evacuate the city, they embarked in a fleet of small boats. Headwinds and cannon fire stopped the exodus. But their commander escaped and set out for Quebec. He was Gen. Guy Carleton, both governor of Quebec and commander of Canadian forces, under General Gage.

As Carleton found himself defending his province against invasion, he drew upon his experience as a courageous officer in the French and Indian War—and his discovery that he possessed political courage. He had been involved in the writing of the Quebec Act, which he viewed as a bulwark against a French insurgency. And when the possibility of an American invasion loomed, he had taken the unpopular step of placing the province under martial law.[28]

Carleton was ready when Arnold and more than six hundred starving and exhausted survivors of his march reached the southern side of the St. Lawrence River, directly across from Quebec. Arnold immediately began to gather boats and canoes for a crossing. And he found some obvious supporters of the invasion: people willing to help his men make scaling ladders for an assault on the city's walls.[29]

Arnold also got the help of Thomas Walker, the Montreal radical. When the invasion began, Carleton had arrested and jailed Walker

for recruiting Canadians to fight against the British and their Loyalists. As Carleton was escaping from Montreal, he put Walker on
a Quebec-bound ship, which Americans captured.[30] As soon as
he was freed, Walker made his way to Arnold.[31] Another ally was
Philadelphia-born David Franks, a civilian volunteer who served as
Arnold's aide-de-camp. A prominent Jew in Catholic Quebec, Franks
had been accused of the potato-rosary desecration. Attacks on him
inspired his decision to aid Arnold.[32]

British Army colonel Allan Maclean arrived in Quebec on November 12 with reinforcements—220 Royal Highland Emigrants.
Maclean and his men, marching toward Montreal, had been summoned to Quebec by Carleton. A few months before, Maclean had
made a secret recruiting tour of Scots-Irish communities in the Carolinas and in the Mohawk Valley, along the New York–Canada border.[33] Most of his recruits were recent Scottish emigrants to America
and Canada. Many of them were veteran Highlanders tested in combat—stalwart men transplanted from the rugged, rock-strewn land of
gorse and heather to the New World, where their skill at arms would
challenge the Continental Army.[34]

Maclean was born a Highlander on the Hebridean island of Mull in
1725. In an act of clan fidelity, he joined the Jacobites* in support of
Charles Edward Stuart—"Bonnie Prince Charlie." Rebellion flared
when Prince Charlie made his futile attempt to regain the throne that
his grandfather, King James II, had been forced to abandon. In 1746
Maclean fought the British in the disastrous defeat at Culloden that
ended the rebellion. He fled to Holland and enlisted in the Scots brigade of the Dutch army. In 1750 he returned from exile after King
George II granted amnesty to Jacobite officers who swore allegiance
to the Crown.

---

* From the Latin *Iacōbus*: James. Jacobites supported a deposed seventeenth-century
  king, Britain's last Catholic monarch—James II of England (who also ruled Scotland as James VII).

Six years later Maclean accepted a commission in the Royal Americans, an oddly formed new regiment that included foreign officers who had deserted from their armies; rejects of Irish regiments; and colonial recruits drawn primarily from Swiss and German emigrants in Pennsylvania.[35] Sent to America, Maclean fought gallantly in the French and Indian War and raised a Canadian regiment called Maclean's Highlanders. When the war ended, many of his Highlanders settled in Canada on land granted by the Crown but the twice-wounded Maclean chose to return to Britain.

As rebellion simmered in America, Maclean suggested to superiors the idea of the Royal Highland Emigrants.[36] On April 3, 1775—two weeks before the Battles of Lexington and Concord—King George III approved the proposal, authorized the regiment's name, and directed General Gage and the royal governors of North America to assist Maclean.[37]

A half century before, King George I had been warned about the Highlanders: "Their Notions of Virtue and Vice are very different from the more civilised part of Mankind. . . ."[38] But the Highlanders of America were now led by chieftains loyal to the king and hopeful of reward when this civil war ended in royal victory. In Scotland the clan system had been breaking down; landowners were continually raising tenants' rents. "It is a grief to our spirits to leave our native land and venture upon such a dangerous voyage," an immigrant wrote; "but there is no help for it. We are not able to stand the high rents and must do something for bread or see our families reduced to beggary."[39]

Maclean sailed to Boston, where Gage made him, as a lieutenant colonel, commander of the regiment, which he divided into two battalions. Recruits for one of them were to come from Nova Scotia and St. John's Island (now Prince Edward Island). The other battalion, which Maclean would directly command, was to enlist Loyalists in Canada and New York—a limitation Maclean ignored so that he could include Highlanders in the Carolinas.[40]

* * *

Maclean's Royal Highland Emigrants regiment would be the first organized Loyalist military unit to fight in a major battle of the Revolution. When Maclean arrived in Quebec, he found, to his surprise and rage, that pro-American *Québécois* were expecting to surrender the city to Benedict Arnold. Carleton arrived next and reacted swiftly, proclaiming that every able-bodied man who refused to bear arms was to leave Quebec within four days. The proclamation drove out many families known to be pro-American. But Carleton knew that others remained, and he still feared that if the invaders managed to get within the walled city, they would attract sympathizers.[41] He tried to stifle rebellion among Catholics by getting the local Catholic hierarchy to decree that anyone helping the invaders would be denied the sacraments.[42]

Carleton assembled an army of assorted defenders. The garrison could muster only about one hundred men because most of Canada's British Regulars had been sent to Boston as reinforcements for Gage. Added to the garrison troops were four hundred sailors and thirty-five marines of two Royal Navy warships in Quebec Harbor, along with fifty masters and mates of merchant ships. There were also nine hundred local militiamen—some British, some *Canadien*—and a number of Catholic seminary students.[43] Carleton handed military command to Maclean, who strengthened defenses in the two-tier city, which consisted of a vulnerable lower town open to the harbor and a walled, fortresslike upper town.

On November 14 Arnold's men paddled and rowed across the river and encamped on the Plains of Abraham. Here had been the climactic battle of the French and Indian War, when Gen. James Wolfe defeated the Marquis de Montcalm. The fall of Quebec meant the conquest of Canada. Arnold, well aware that history could repeat itself here, boldly sent an officer toward the city under a flag of truce with a demand of surrender. Maclean's guns fired on the officer. The next day Arnold tried again and got the same response. Prudently he withdrew about twenty miles from Quebec and awaited Montgomery.[44]

Montgomery appeared with about three hundred men, including pro-American Canadian militiamen he had picked up along the way. He also brought spoils from British river forts that had fallen to him: ammunition, guns, and Redcoats' winter uniforms. Most of Arnold's men were half naked, shivering in the cold shortening days of December. They eagerly donned the uniforms, attaching twigs of hemlock to their hats to advertise their true identity.[45]

Montgomery took command and established a camp close to the city. He immediately sent a letter to Carleton demanding surrender. Carleton refused to accept it. So Montgomery recruited an old woman to deliver a second letter, which said that the Americans came "with the professed Intention of eradicating Tyranny and giving Liberty and Security to this oppressed Province."[46] Carleton burned the note without reading it, had the woman arrested, and after she spent a night in jail, he put her outside the gates.

Montgomery began a siege. But the enlistment of many of his men ended on New Year's Day, and he knew he had no choice but an attack. At 2 a.m. on the last day of 1775, his men tramped through a blizzard to their places in Montgomery's attack plan: two feints, then—on the signal of two rockets flaring in the darkness—two attacks on the lower town. Once it was taken, the combined force would advance on the upper town. All of this was known by the defenders because a deserter had slipped into the city and revealed the plan. So they were ready when they saw the rockets rising in the black sky and the dots of lanterns advancing in the whirling snow.

Arnold, leading six hundred men into the lower town from the north, reached as far as a barricade that erupted with fierce fire. Shot in the left leg, he leaned against a wall and, waving his sword, urged his men on until, his boot filling with blood, he was taken off, still gripping his sword.[47] He turned command over to Daniel Morgan, who fought his way to the rendezvous site but, instead of linking up with Montgomery, was surrounded by two hundred of Maclean's men and was captured with most of his force.[48]

On the south side of town Montgomery, leading his advance guard, charged a fortified house. He was ten yards from the house when a blast of cannon and musket fire killed him, two other officers, and ten nearby men. Others were wounded.[49] Suddenly leaderless, the rest turned and ran. Many of them would be among the four hundred prisoners taken that day.[50] About eighty in the invading force were killed or wounded.[51]

The next day British troops saw a hand sticking out of a drift of bloodied snow. They dug and found the body of Montgomery, his sword at his side. Had he not been killed, Arnold wrote in tribute, "the town would have been ours."[52]

Arnold, propped up in a hospital bed in a Quebec suburb, commanded about six hundred men who had managed to escape and regroup outside the city. They lived from day to day on diminishing rations while men deserted and Arnold sent couriers to Montreal, still in the invaders' hands, carrying urgent requests for troops and supplies. *Canadiens* drifted into Arnold's camp, hoping to arm themselves and somehow fight the British. Carleton did not attack, for he knew that in spring reinforcements would come in Royal Navy warships.

On April 1, 1776, Arnold, promoted to brigadier general, was ordered to Montreal. There, to his surprise, he met Ben Franklin and two other commissioners dispatched by the Continental Congress in quest of the fourteenth colony. The commissioners had been told to find a way "to promote or to form a union between the United Colonies and the people of Canada." The commissioners stayed in what Franklin called "the best built and the best furnished house in the city"—the mansion of Thomas Walker, the one-eared Canadian rebel. The commission gave Arnold permission to supply the struggling invasion force by seizing goods from the warehouses and ships of Montreal Tories. He paid in worthless Continental currency rather than gold. The authorized plundering was the sole accomplishment of the commission, which fled Canada in May.[53]

Several Continental Army regiments made their way to Quebec, adding to the remnants of the invasion force. But there was little chance for a second thrust into Quebec, for in May and June, British reinforcements arrived, and the Americans and their Canadian allies began to withdraw. Carleton pursued, leading a force of eight thousand Regulars and three thousand German mercenaries, collectively called Hessians. The Patriots, burning or seizing anything of value, retreated up the Richelieu River to Lake Champlain, marshaling finally at Crown Point, just above Ticonderoga.[54]

"The junction of the Canadians with the Colonies," Arnold wrote to General Schuyler, ". . . is at an end. Let us quit then and secure our own country before it is too late." Arnold took command of the retreat as Carleton began building warships at the north end of Lake Champlain to speed up his chase. To fight for control of the lake and thwart or delay a British invasion, Arnold rounded up carpenters and shipwrights to build a fleet of smaller ships near Crown Point, at Skenesborough (now Whitehall, New York).

Skenesborough was named after a powerful local Tory, Col. Philip Skene, who was away on business in England. Arnold seized Skene's trading schooner, the *Katherine*, and renamed her *Liberty*. Arnold's men, searching out Tories, captured Skene's twenty-two-year-old son Andrew, other members of the family, a dozen slaves, and fifty tenants. All but Andrew were released. He was jailed as a military prisoner, but he escaped and made his way to Quebec to become part of the growing Loyalist military presence in Canada.

Arnold's men took possession of Skene's mansion and the large stone fort that he had built to guard his fiefdom. In the cellar of his house they found the body of Skene's wife, preserved, it was said, so that he could continue to receive an annuity provided for him while his wife remained above ground. The soldiers buried the body at the base of what is still called Skene's Mountain.[55]

The sixteen vessels of Arnold's little navy consisted mostly of flat-bottomed craft about fifty-three feet long propelled by oars and a fore-and-aft sail. Each carried a few small cannons. The British fleet, which sailed south in October 1776, dwarfed Arnold's. The British could fire one thousand pounds of shot for every six hundred pounds fired by Arnold's boats.[56] In a ferocious hide-and-seek battle around

Valcour Island near the western shore of the lake, Arnold lost eleven of his sixteen boats but thwarted Carleton's hope for a swift drive southward.[57]

Arnold took charge of a rear guard to shield the rest of his retreating force from Carleton's advance guard. He set Skenesborough afire and got his men to the boats that would save them. As the British came into sight, Arnold galloped to the shore of Lake Champlain, shot his horse to deny it to the enemy, took his saddle, and climbed into a boat destined for Crown Point. He had vowed to be the last man out of Canada, and he was.[58]

During his retreat he had stripped the land of food, impeding and slowing down a British advance toward the Hudson River, the waterway to New York City. Carleton, in the chill of a frustrating October, postponed his planned invasion and decided that the campaign season had ended. He retired to winter quarters, giving the Rebels a hard-won chance to regroup.[59]

The failed invasion did not end the American fascination with the fourteenth colony. Even as Arnold was retreating, another invasion was about to begin. The leader was Jonathan Eddy, one of the many Massachusetts colonists who had migrated to Nova Scotia.

In 1755 Eddy had joined a force formed by John Winslow (Edward Winslow's uncle) to seize the French fort Beauséjour at the head of the Bay of Fundy. The taking of the fort began the campaign to rid British Canada of Acadians. The British distrusted the Catholic, French-speaking Acadians, most of them fishermen and farmers who declined to join Britain in its war against France.

Winslow, at the head of a regiment consisting mostly of Massachusetts volunteers like Eddy, was one of the officers who carried out the expulsion of the Acadians, a tragedy immortalized by Henry Wadsworth Longfellow in his poem *Evangeline*. Winslow began by ordering "both old men and young men, as well as all the lads of ten years of age" to assemble in a Catholic church "that we may impart to them what we are ordered to communicate . . . no excuse will be admitted on any pretence whatever, on pain of forfeiting goods and chattels, in default of real estate."[60]

Winslow shut the doors of the church and addressed the 418 men and boys in English, which few Acadians understood. He told them that they were all "the King's prisoners" and that their "lands and tenements, cattle of all kinds and live stock of all sorts, are forfeited to the Crown." Five days later, as their burned-down homes and barns smoldered behind them, more than one thousand men, women, and children were herded onto ships, with little regard for keeping families intact. They were the first of more than ten thousand Acadians who were expelled between 1775 and 1783, mostly to other American colonies.[61]

Eddy served at Beauséjour—renamed Fort Cumberland—from 1759 until 1760, and after his discharge returned home to Massachusetts. Three years later, drawn by the offer of cheap land, he was among the New Englanders who settled on what had been Acadian farmland. As a member of the new landed gentry he was appointed a member of the Nova Scotia legislature.[62]

As reports of rebellion arrived from Massachusetts, Eddy felt a kinship with the Rebels in his native colony. Resurrecting his old wartime title, Captain Eddy sailed to Massachusetts, leaving his family behind. In March 27, 1776, Eddy met with George Washington in Cambridge and urged him to support Patriots in Nova Scotia. Washington gave Eddy his blessing—and money to travel to Philadelphia and present his proposal to Congress.[63]

Congressmen politely told him they could not give him any assistance. The failed invasion had dampened congressional enthusiasm about actions in Canada. A month before, Congress had secretly turned down a plan to put Halifax, Nova Scotia, to the torch. The arson raid had been proposed by Jeremiah O'Brien, hero of the HMS *Margaretta* seizure in Machias, Maine, and another man, presumably a privateer like O'Brien.[64]

Eddy next went before the Massachusetts legislature, which rejected his request for troops but did give him ammunition and some supplies. He chartered a small ship and crew and sailed to Machias, where he found some followers, and then went on to Nova Scotia for more, particularly around Fort Cumberland. His grand plan was to take the fort and march on to Halifax, his force swelling with more and more devotees of the fourteenth colony. When he was ready to attack the fort, he later claimed, he had an army of 180 men, including Indians, Acadian exiles, and Maine Patriots. On November 10 he appeared before Fort Cumberland, and, under a flag of truce, as "Commanding Officer of the United Forces," demanded surrender.[65]

The fort was manned by Loyalists, the Royal Fencible Americans. The fort's commander replied with a letter ordering Eddy to "disarm yourself and party Immediately and Surrender to the King's mercy." After the Loyalist force easily repulsed two feeble attacks, Eddy withdrew to await reinforcements that never came. Later in the month additional defenders arrived at the fort, including more Loyalists—the Canada-based battalion of the Royal Highland Emigrants.[66] After some desultory sparring between defenders and would-be attackers, the commander of the fort ended the invasion by issuing a general amnesty that pardoned nearly all the Canadians, dissolving Eddy's army.[67]

Eddy fled to Machias, which soon was targeted for reprisal. Four British warships sailed up the Machias River to shell the town. Townspeople and Indian allies defended from both shores. To move a cannon to a strategic spot for firing on one of the ships, militiamen staged a funeral procession, with the cannon disguised as a bier draped with a blanket. When the bier began spewing cannonballs, the warship pulled away, beginning the British withdrawal.[68] The raid was a lesson for British strategists, who realized that the real danger from the Rebels was on the sea. The Royal Navy sharpened its vigilance along the Nova Scotia coast, dueling with privateers like Jeremiah O'Brien.

Washington continued to yearn for the fourteenth colony, but never again would he approve an invasion, even though he knew that a Canada in British hands would always be a menace. Ten days after getting from General Schuyler "the melancholy account" of the failed attack on Quebec, Washington wrote to Arnold, saying of Canada, "To whomsoever it belongs, in their favor, probably, will the balance turn. If it is in ours, success I think will most certainly crown our virtuous struggles. If it is in theirs, the contest at best will be doubtful, hazardous, and bloody."[69] By the time he wrote this, in January 1776, he knew that the war had to be fought not in Canada but where he was, in Cambridge, or where the British and their Tories were, in Boston.

THE FAREWELL FLEET

BOSTON, JANUARY 1775–JULY 4, 1776

*This [Rebel] army, of which you will hear so much said, and see
so much written, is truly nothing but a drunken, canting, lying,
praying, hypocritical rabble, without order, subjection, discipline, or cleanliness; and must fall to pieces of itself in the course
of three months.*

—*A Royal Navy surgeon in Boston*[1]

Loyalists, scanning Dorchester Heights for cannon muzzles, feared
that the Patriots would bombard the British troops as they left
the city. Patriots, looking at the gun ports on the Royal Navy warships
strung around the harbor, expected that the British would give the
city a flaming farewell. Boston's selectmen, all of them Patriots who
had remained in the city, decided to make a last-minute effort to save
Boston from either fate. Three Loyalists joined them in a rare show
of unity.

The Loyalists were Thomas and Jonathan Amory and their friend
Peter Johonnot. Months before, Thomas Amory had faced down a
crowd that attacked his home because he was friendly with British
officers. Soon after the attack he planned that his childless brother
Jonathan would sail to England with family members while Thomas
stayed behind. But, when Boston was threatened, Thomas Amory

forgot the past and sought out the Patriot selectmen. The third Loyalist, Johonnot, a wealthy distiller, had also planned to leave. Instead he agreed to work with the Amory brothers.[2]

The save-the-city delegation went first to General Howe, who ordered a general on his staff to respond. Then, on March 8, the Loyalists, under a flag of truce, went to the fortified checkpoint at Boston Neck. Although borne by Loyalists, the petition was bipartisan; it carried the signatures of the selectmen as "a testimony of the truth." A colonel on Washington's staff accepted their petition, which said that Howe had given them, through his general, assurance "that he has no intention of destroying the town, unless the troops under his command are molested during their embarkation, or at their departure, by the armed force without"—meaning the guns of Dorchester.

Now both sides wanted a similar assurance from Washington. If those cannons were fired on Howe's departing troops, the appeal continued, "we have the greatest reason to expect the town will be exposed to entire destruction. Our fears are quieted with regard to General Howe's intentions. We beg we may have some assurance that so dreadful a calamity may not be brought on . . ."[3]

Washington's colonel duly took the petition to Washington's headquarters on Tory Row. The next day he returned to the checkpoint with a response: "Gentlemen,—Agreeably to a promise made to you at the lines yesterday, I waited upon his excellency General Washington, and presented to him the paper handed to me by you, from the selectmen of Boston. The answer I received from him was to this effect: 'That, as it was an unauthenticated paper, without an address, and not obligatory upon General Howe, he would take no notice of it.' I am, with esteem and respect, gentlemen, your most obedient servant."[4]

The curt answer left the Loyalists anxious. "His excellency" General Washington seemed to be ignoring the plight of the city because his rank and command status had not been recognized by Howe. In fact, without setting forth an acknowledgment in words, Howe had agreed not to torch the city, and Washington had agreed not to shell Howe's men and ships.

Still, during the night of March 9, Washington ordered more cannons entrenched low enough on a Dorchester's Nook's Hill to increase the threat to Howe's troops and ships. The men planting the cannons built a warming fire, which British sentinels spotted, touching off a cannonade that killed five Continentals. Washington answered the British cannons with a bombardment of his own. During a long, terrifying night, artillerymen on both sides fired more than eight hundred shots.

The day after the cannon duel, Howe issued a proclamation and had it distributed in handbills throughout the city. It ordered "all good subjects" to turn in all linen and woolen articles because they were "much wanted by the Rebels, and would aid and assist them in their Rebellion." There were no exceptions for clothing or blankets. People unable to turn in their goods at army headquarters were told to deliver them to Crean Brush, in care of the *Minerva*, which was tied up at a wharf. Brush would give receipts for the goods "and will oblige himself to return them to the Owners, all unavoidable Accidents accepted." Anyone who did not turn in the goods, Howe warned, "will be treated as a Favourer of Rebels."[5]

Howe's proclamation considerably extended Brush's earlier commission for emptying Boston houses and other buildings. Patriots later told of seeing Brush, escorted by soldiers, using the proclamation as a warrant for breaking into stores and hauling the goods to the *Minerva*, the well-known ship whose imported sails and fittings had touched off the Falmouth troubles.

Brush, said a witness, was stripping shops "of all their goods, though the owners were in town. There was a licentious plundering of shops, stores and dwelling houses, by soldiers and sailors, carrying destruction wherever they went; and what they could not carry away, they destroyed."[6] He accumulated so much loot that he needed another ship, the brig *Elizabeth*. For seven consecutive days, Brush said afterward, "my own assiduity was so great that I did not in any one night allow myself more than two hours sleep."[7] While approving Brush's pillage, Howe issued an order saying that "the first soldier who is caught plundering will be hanged on the spot."[8]

Washington kept his troops on alert, wary of British deception. On March 10, as dragoons led their horses toward the docks for herding onto transports, there could be little doubt that evacuation had begun. But Washington was still skeptical, and Selectman Timothy Newell was still worried: He noted in his diary on March 12: "The Inhabitants [are] greatly distressed thro' fear the Town would be set on fire by the Soldiers."[9]

The soldiers, however, were focused on leaving, not burning. Work parties loaded ships with the matériel of war, leaving behind at wharves and in warehouses hundreds of pounds of salt, five thousand bushels of wheat, six hundred bushels of corn and oats, one hundred bundles of iron hoops for barrel making, 157 pack saddles, hundreds of blankets, and paraphernalia too heavy to load rapidly: anchors, lumber, nails, four-wheel carriages. The many abandoned cannons were spiked and had their trunnions broken off, but a great deal of ammunition was abandoned, to be retrieved later by the Continental Army. The British would also leave behind a score of ships, including a brig with all its guns and another laden with a full cargo of molasses. Many of the ships were scuttled at the wharf and salvageable, as were their intact or partially destroyed cargoes.[10]

Logistics officers struggled with the complex task of assigning shipboard space to unaccustomed cargo: the goods of civilian passengers. Down the wharves came carts, wagons, and wheelbarrows overflowing with clocks, tables, chairs, chests of drawers, and other possessions.[11] More and more names were added to the list of the Loyalists sailing in the ships of the fleet.

Howe had known for some time that if he had to evacuate his army he would also be obliged to take along all the Loyalists who wanted to leave. "If we are driven to the difficulty of relinquishing Boston," Lord Dartmouth had written him, "care must be taken that the officers and friends of the government be not left exposed to the rage and insult of rebels, who set no bounds to their barbarity." Howe had to accept Dartmouth's order, but he did remind Dartmouth about the problem of simultaneously rescuing soldiers and civilians: "A thousand difficulties arose on account of the disproportion of transports for the conveyance of property, and the quantity of military stores to be carried away."[12]

Boston Loyalists were overwhelmed by the news of Howe's decision to abandon the city. Suddenly they lost their hope for an easy triumph over the Rebels. "Not the last trump[et] could have struck them with greater consternation," Washington noted, using the biblical allusion to the end of the world.[13]

Many Boston Loyalists were well aware that they would become targets of vengeance when their Redcoat protectors left. They knew they had to flee, and they wanted to take their treasures with them. "The people of the town who were friends of the government," a witness later wrote, "took care of nothing but their merchandise, and found means to employ the men belonging to the transports in embarking their goods; by which means several of the vessels were entirely filled with private property, instead of the king's stores." Another Bostonian complained that Loyalists were outbidding each other to purchase the labor of moving men.[14]

For Howe's troops the evacuation was a complex military maneuver. For the Loyalists the exodus from Boston was a panicky flight inspired by fear, despair, and rage. The Loyalists were abandoning home, possessions, and, for many, a cause that was lost. They had expected first Gage, and then Howe, to crush the rebellion. Instead the rebellion had morphed into a civil war. The city that had been the pride of some and the temporary refuge for others had become a place seething with hatred. Neighbors became enemies. Friends turned against friends. As the Loyalists headed for the Halifax-bound ships, they passed silent glares or heard a taunting chant:

The Tories with their brats and wives
Have fled to save their wretched lives.[15]

Many Loyalists, believing that someday they would return, left friends or relatives behind to guard abandoned homes. Others were rich enough to commit to a permanent move; for some Halifax would be a way station en route to England. But in England few would find happy or productive lives. Whatever position they had attained in the colonies, in England they were simply Americans, aliens in their ancestral homeland.

Even clergymen, expecting to be welcomed into the fold of the Church of England, found a chill reception. The Reverend Henry Caner, seventy-six-year-old rector of King's Chapel in Boston, later told of hearing "many compassionable expressions" from the Archbishop of Canterbury and the Bishop of London—but no offer of an appointment to an English parish. He said he was told, "We can't think of your residing here. We want such men as you in America." English clergymen, Caner decided, looked upon their American colleagues "in no better light than as coming to take Bread out of their mouths."[16]

Like most Loyalists, Caner had only short notice when he was told he could embark for Halifax. He led a final service at King's Chapel on Sunday, March 10, and then boarded a ship for Halifax, taking with him the communion service, the church's register of baptisms, marriages, and burials, and a book of early records.[17] About one-third of the King's Chapel congregation would be gone by the end of 1777, for "Anglican" and "Tory" were almost synonymous.

Caner's name appears on a list of departing Loyalists compiled for the evacuation.* The list was published in 1881 in *Proceedings of the Massachusetts Historical Society*. A note says that the names were "Taken from a paper in the handwriting of Walter Barrell," but how Barrell obtained and recorded the names is unknown. He was inspector general in the Boston Customs House, and he would have had the responsibility of monitoring the manifests of the evacuation ships.

The son of a merchant who was already in England, Barrell had been a member of Ruggles's street-patrolling Loyal American Association.[18] Two of his brothers were Patriots. Barrell's own name is on

---

* The entire list can be found at ToriesFightingForTheKing.com.

the list, which shows that he brought five people, apparently family members, with him to Halifax. The Barrell family was one of about seventy that went from Boston to Halifax and then on to join the growing Loyalist population in England.[19]

One section of the list, devoted to thirteen mandamus councillors, showed that they did not sail alone. Richard Lechmere took eleven family members to England. Foster Hutchinson, brother of a former royal governor residing in England, took a dozen relatives with him to Halifax. Customs official Henry Hulton took eleven to Halifax and soon sailed on to England.[20]

All names are followed by a number indicating how many people (unnamed) are leaving with that person. Many names are followed by an occupation, such as "clerk" or "merchant." And sometimes the same surname is listed several times in a row, each with a number following it, as in the enumeration of the renowned Winslow family, whose total strongly suggests that rank had its privilege in the allotment of ship space:

Winslow, Isaac	11
Winslow, Pelham	1
Winslow, John	4
Winslow, Mrs. Hannah	4
Winslow, Edward, Collector, Boston	1

Hannah Loring Winslow (subject of one of John Singleton Copley's portraits) was the thirty-three-year-old widow of Joshua Winslow, a cousin of Copley's wife. Joshua, as one of the consignees who was to resell imported tea, was a small player in the events leading to the Boston Tea Party. He was the father of six children when he died in March 1775. His widow left for Halifax, listing three children. (The other children may have traveled with other relatives.)[21]

Edward Winslow took with him the large, carved, wooden royal coat of arms that had hung on a wall of the council chamber in Boston. As he later wrote, the lion and the unicorn in the arms "are of course Refugees and have a claim for residence" in Nova Scotia.[22]

Other placemen who sailed to Halifax also carried off official documents in the belief that they would bring them back when the unpleasantness was over.

The list accounts for either 926 or 927 passengers, depending upon interpreting what was originally in Barrell's handwriting. (Included on the list are virtually all the hundred-odd Loyalists mentioned thus far in this book.) Besides those expected refugees there were 105 residents who lived in rural areas or towns and who were not placemen; 213 merchants, and 382 farmers, smaller traders, and "mechanics," the overall term for artisans and people who worked with their hands.

The largest group of refugees was made up of men who most feared retribution—those who managed or worked for the customs house. They included "tidesmen," customs officers who boarded ships to get payments of duties, and "tidewaiters," who examined cargoes in search of contraband.[23] Anglican clergymen were members of another class that left virtually as a bloc, headed by Caner.

The grand total of the Halifax-bound "refugees," as they were called, is estimated at about thirteen hundred men, women, and children. But before and after the evacuation others made their way to Halifax, bringing the probable total to about two thousand. Among these unlisted refugees are a few who get named in other ways. Jolley Allen, a Boston merchant, for example, left posterity an account of his attempt to reach Halifax. And, after the war, many previously anonymous refugees told of their voyages to Halifax in petitions seeking land grants or financial aid from the Crown for their loyalty.

Also not named on the list was John Wentworth, the royal governor of New Hampshire. Wentworth, fearing the rise of Rebels, had mustered a force of Loyalists as a personal guard. He then sought refuge, with his wife and their infant son, on board the HMS *Scarborough*, which took them to Boston. At the time of the evacuation he sailed to Halifax in a schooner with a number of New Hampshire friends. His wife and child went on to England. Wentworth joined Howe and sailed with him to New York. He recruited a Loyalist military unit known as Governor Wentworth's Volunteers.[24] They were described as "Persons of Education and reputable Families who have personally suffered variety of Persecutions, and appear now firmly determined to give additional Proofs of their Attachment to the Cause of Government by exerting their utmost Endeavours to suppress the Rebellion in America."[25]

Another independent traveler was Isaac Royall, one of America's wealthiest Loyalists and one of Massachusetts' few slaveholders. A trader in sugar, rum, and slaves, Royall was a mandamus councillor. He lived in a mansion in Medford, a few miles north of Boston. On his six hundred acres were quarters for twenty-seven slaves. He found himself in Boston on the eve of the Lexington and Concord battles, en route to the family plantation in Antigua. After a while he sailed to Halifax, where he awaited other members of the family, including George Erving, who was married to Royall's daughter, and Erving's brother John. The Royalls and the Ervings all went on to England.[26]

Two Marblehead Tories who immigrated to Nova Scotia and later continued on to self-exile in England were Thomas Robie, an Addresser, and his wife. They were given a raucous farewell by Patriots who went to the dock and shouted taunts. Mrs. Robie, seated in the boat that would take her and her husband to the Halifax-bound ship anchored offshore, shouted back what many departing Loyalists felt in their bitter hearts: "I hope that I shall live to return, find this wicked Rebellion crushed, and see the streets of Marblehead run with Rebel blood."[27]

Many Loyalists had left before the evacuation, and many, many more would leave in the years to come. We do not know most of their names. Lorenzo Sabine, a nineteenth-century historian who collected information on thousands of Loyalists, lamented at how little the future was to know about them: "Men who, like the Loyalists, separate themselves from their friends and kindred, who are driven from their homes, who surrender the hopes and expectations of life, and who become outlaws, wanderers, and exiles—such men leave few memorials behind them. Their papers are scattered and lost, and their very names pass from human recollection." Some could not write, and those who did, consumed by their shattered lives, lacked the time— and sometimes even the paper—to record their hopes and fears.[28]

At least three hundred families either went with the British fleet and did not give their names to Barrell or chose to make their own arrangements. Those who did sail independently probably paid passage on ships that for decades had regularly shuttled between New England and Nova Scotia. During the war that connection worked both ways. So many pro-Rebel inhabitants of Nova Scotia wanted to trade illegally with the enemies of the king that the Massachusetts legislature issued passes to them, directing commanders of American warships and privateers to allow safe passage.[29]

In 1776 Washington had seen Tories merely as disillusioned people, calling them "Unhappy wretches! Deluded mortals!"[30] But his attitude toward them gradually hardened. Writing to his brother, John Augustine, Washington was uncharacteristically nasty about the Loyalist departure: "All those who took upon themselves the style and title of government-men in Boston, in short, all those who have acted an unfriendly part in the great contest, have shipped themselves off in the same hurry. . . . One or two have done, what a great number ought to have done long ago, committed suicide."[31]

One of the Loyalists preparing to leave on his own was the merchant Jolley Allen. He had arrived in Boston from England in 1755, set up shop on a fashionable street, and prospered by selling luxury goods. "Just imported from LONDON, by Jolley Allen," said an ad in the *Boston Gazette*. "A very large Assortment of English and India GOODS, fit for all Seasons, too many to be enumerated separately in an Advertisement."[32] He also ran a pawnshop and offered lodgings for people and stables for horses.[33] Allen suffered financially and personally when the Sons of Liberty began to boycott British imports in the campaign against the Stamp Act.

When the campaign focused on an embargo of tea, Allen surreptitiously bought two chests of tea from the sons of Royal Governor

Thomas Hutchinson. An informer told the Sons of Liberty. Some Sons called on Allen and threatened to tar and feather him, he claimed. By his account, he outwitted them, for they were unable to connect him with English tea. His sales, however, "declined from that time" because he was marked as a Tory. His customers dwindled to "friends of Government and the Army," including General Gage and other officers.

Early in March 1776, as the city was plunged into the evacuation panic, Allen believed he had to leave, for he feared that if the "Americans"—his word—found him, "knowing the Disposition of the people so well for above Twenty-two Years I had lived amongst them," they would hang him and everyone in his family. On the waterfront he found Robert Campbell, a captain willing to take aboard his ship, the *Sally*, Allen, his family, his household goods, and the contents of two warehouses of merchandise. Allen had married a young Irishwoman and "had seventeen children by her plus five miscarriages in thirty-seven years of marriage." Only six children were alive in 1776.[34]

Allen planned to follow the ships of the evacuation fleet wherever they sailed. Most of the available ships were under the control of the Royal Navy. But there were others, like the *Sally*, whose captains were ready to sail under the fleet's protection and would evacuate anyone who could pay the cost.

Opportunistic captains like Campbell added another complication to British Army and Royal Navy officers who were dealing with a harbor full of evacuation ships. Officers also had to cope with the plundering and heavy drinking of their men. "It was not like the breaking up of a camp, where every man knows his duty; it was like departing your country . . . ," an officer wrote.[35] Officers who had bought furniture and other items could neither take them along nor sell them. Some, in frustration, smashed their American acquisitions.[36]

Army and navy officers were experienced in transporting soldiers and their supplies. But for this operation there was additional need for space to carry and feed an unknown number of Loyalist families. Their possessions had to be stowed. To make room for the civilians, soldiers had to cut back. Officers ultimately enforced a strict order: "All Household furniture and other useless Luggage to be thrown over Board."[37]

"Nothing can be more diverting than to see the town in its present situation," a Patriot wrote, "all is uproar and confusion; carts, trucks, wheelbarrows, handbarrows, coaches, chaises, are driving as if the very devil was after them."[38] Selectman Newell wrote in his diary for March 13 that Bostonians of all persuasions were "in the utmost distress thro' fear of the Town being destroyed by the Soldiers, a party of New York Carpenters with axes going thro' the town breaking open houses, &c. Soldiers and sailors plundering of houses, shops, warehouses—Sugar and salt &c. thrown into the River, which was greatly covered with hogsheads, barrels of flour, house furniture, carts, trucks &c. &c."[39]

The tempo of looting increased. Soldiers and civilians broke into shops and carried off their plunder. The soldiers packed away the portable loot. The civilians bore theirs to the wharves, where ships awaited. Crean Brush's gang loaded the *Minerva* and *Elizabeth*. One hundred fifty hogsheads of rum (a hogshead had a capacity of about fifty-five gallons) were loaded onto one of the ships under British Army control.[40] The harbor was filled with transports and warships, signal flags whipping up and down the halyards. Scavengers prowled the shore, picking up the flotsam of discarded freight borne on the incoming tide.

Men hired by Jolley Allen shuttled merchandise and household goods from his shop to the wharf where Robert Campbell's ship was tied up. Allen allowed Campbell to take twenty-one more passengers aboard—"to put more Money in his pocket," Jolley wrote in his account, which sputters with indignation.[41]

On Saturday, March 16, Washington sent a small force to Nook's Hill to fortify the guns. During the night, British guns fired at the hill. Washington's men did not return the fire. To Howe the silence was ominous. Either through intelligence reports or intuition, he concluded that Washington's patience was running out. He ordered the evacuation to be concluded the next day.

British Army lieutenant John Barker, in his diary entry for March 17, recorded the soldiers' last day in Boston: "At 4 oclock in the Morn. the Troops got under Arms, at 5 they began to move, and by about 8 or 9 were all embarked, the rear being cover'd by the Gren[adie]rs. and L[igh]t. Inf[antr]y. The Rebels did not think proper to molest us. We quitted Boston with a fair wind. . . ."[42]

Within half an hour Washington sent more than one thousand men into the city—selecting only those who had had smallpox and were thus immune. Washington himself bore the facial scars of a severe case of smallpox, which had struck him at the age of nineteen. He had not only suffered the scourge, but he had also learned about its prevention. He would later insist on inoculations for every soldier in the Continental Army.[43]

Washington forbade any officer or enlisted man to enter the city without permission because, he said, "the enemy with a malicious assiduity, have spread the infection of the smallpox through all parts of the town." A "person just out of Boston" told Washington that "our Enemies . . . had laid several Schemes for communicating the infection of the smallpox, to the Continental Army, when they get into the town."[44] No one yet knew that smallpox had already sailed from Boston, borne by Loyalists and Redcoats. An epidemic had begun in Halifax months before the evacuation.[45]

Some of the first of Washington's men to enter Boston headed for the fortifications at Bunker Hill and saw what appeared to be sentinels. Two men were sent forward to see how many British remained. Laughing, they waved their comrades to advance. The sentinels were dummies. The redoubt was empty, a symbol of the military situation in the American colonies: There were hardly any British soldiers left on American soil.[46]

While liberating the city, Washington kept watch on the fleet lingering off the Nantasket Peninsula, which formed the southeast side of Boston Harbor. "From Penn's Hill," Abigail Adams wrote to John, "we have a view of the largest fleet ever seen in America. You may count upwards of a hundred and seventy sail. They look like a forest."[47] The ships off Nantasket bothered Washington. Even burdened with the civilians, Washington believed, the nine thousand troops might be heading for New York.

There was also a small chance that the evacuation was a feint and Howe might send an invasion force back into the harbor and begin bombarding the city. Washington ordered the city's redoubts manned and strengthened. And, in case Howe did suddenly veer south, Washington ordered five regiments to march to Norwich, Connecticut, and sail from there to New York. On March 27 most of Howe's fleet weighed anchor and sailed off to Halifax.[48] But the British left a few vessels off Nantasket for more than two months, to the annoyance of American officers in Boston.[49]

Washington ordered a swarm of American privateers to follow the British fleet, which consisted of about 120 ships. The commander of "Washington's Fleet," as the privateer ships were dubbed, was John Manley of Massachusetts. His flagship was the *Lee*, a schooner transformed into a warship with the addition of six carriage guns and ten gunwale-mounted swivel guns.[50] Manley spotted, pursued, and captured a major prize: Crean Brush's loot-filled *Elizabeth*, which he took back to Boston. It contained a "great part of the plunder he rob'd the stores of here," wrote a Bostonian. ". . . I immagine she must be the richest vessell in the fleet."[51] Brush and several henchmen found on the *Elizabeth* were jailed in Boston while Patriot officials drew up a case against them.

Jolley Allen aboard the *Sally*, meanwhile, was making little headway toward Halifax. While the *Sally* was still in offshore waters, Captain Campbell twice collided with other ships in the British fleet and then ran aground. When Allen asked the captain whether the tide was coming in or going out, the captain said he did not know because he did not have an almanac aboard. Under way again, Campbell turned the wheel over to a lad who was on his first sea voyage. The captain pointed to a ship ahead and told the young helmsman simply to follow it.

Soon a squall came up, partially tearing the mainsail off the mast. The sail fell over the side, starting to fill with water, sink, and take the ship with it. Trying to recover the sail, Campbell hauled and tied a line, but—"for want of knowing how to Tye a Sailor's knot," according to an exasperated Jolley—the line skittered out of its belay.

Below, the ship started taking on water. Campbell tried one pump, which did not work, and set off, unsuccessfully, to look for the other. Heavy seas carried off a cask containing all the water on board, along with the large container that contained all the food. Allen went below, "took my Wife by the hand as she lay in bed in the Cabin, and "laid myself down by her to think I should die in her Arms along with her."

His grim reverie was interrupted by frantic sounds and a knocking on the cabin door. The additional passengers, who had been consigned to the hold, pleaded—"for God's sake"—to be allowed to enter. A plank, they said, "had given way, and the Sea was pouering in and the Vessel was sinking, and they beged that they might be permitted to stay in the Cabin till we all went to the Bottom together."

Allen, wondering how far the ship was from land, went on deck and asked Captain Campbell their position. He replied, Allen later recorded, that "it was impossible for him to tell, for he had not kept any reckoning, and the reason he gave me for it was that he had forgot to bring pens Ink and paper." When Allen pointed out that he had those items, the captain told him "he had never learnt Navigation, and that he never was on Salt Water before, but he did know to row a Boat in a River."

As the sun rose, "with our Sails and Rigging torn in ten hundred thousand Pieces," the *Sally* was riding a strong shoreward tide and heading for a landfall that Campbell could not identify. Campbell was for heading out to sea. But Allen, taking command of the foundering ship, ordered everyone on deck and insisted that the ship head toward shore. It scraped across several sandbars and, on the late afternoon of March 28, dropped anchor off Provincetown, Cape Cod, tattered sails signaling its distress.

People ashore, fearing that the strange ship might be carrying smallpox victims, organized a rescue by selecting rowers who had survived the disease. When Allen, his family, and the other passengers were all safely ashore, someone asked where they were from. Allen answered, "From Boston with the Fleet"—marking them all as fleeing Tories.

Allen wrote that he and his family were taken to a "Hog Sty," a half-wrecked cottage with broken windows and a leaky roof. He did not report what happened to the others, but presumably they were confined elsewhere while the local Committee of Safety decided what to do.

Townsmen scrambled aboard the *Sally*, running it aground so they could get at the cargo, which consisted almost entirely of Allen's furniture and merchandise. He was told he would be charged a large sum for the unloading and storage of his property. According to his account, he soon learned—probably from a secret Loyalist—that people were fighting over the cargo, stealing some articles and hurrying away with them, burying others in the sand for retrieval later. Someone, he was told, had torn apart his "crimson Silk Damask Bed which cost One hundred fifty pounds Sterling."

He and his family were still being held in April when his wife died at the age of fifty-two. Allen, fifty-eight, attributed her death to the stress of their misadventures. In May the town selectmen gave him a pass to Watertown, where the legislature was sitting. His seventeen-year-old son was allowed to go with him on the 120-mile journey. His other children and what was left of his property were held as a guarantee of his return.

On his way to Watertown he stopped at his home and found it occupied by his barber, who charged the pair eight shillings sterling for one bed for two nights. When Allen and his son reached Watertown, the legislature ordered them held for further investigation. He was told that the lawmakers were considering a plan to separate his children, order them kept fifty miles apart, and apprentice them to strangers to earn their room and board. He also was told that some people were looking for a tree big enough for the hanging of him and his children—with "me upon the highest Branch."

Finally Allen's brother Lewis Allen learned of Jolley's ordeal and petitioned the House and the Council, which functioned as a kind of senate, saying that he would assume care of the children, that his brother would stay in Lewis's home on his farm in Shrewsbury, about thirty-eight miles east of Boston, and that Jolley would not "hold Correspondence with any Person knowing them to be Enemical to the Liberties of America." The legislators agreed—but not the citizens of Shrewsbury, who said they did not want a Tory living in their town.

A mob broke into Lewis's home and, on a sixteen-mile march to Northborough, took Jolley before a militia officer and his men. He was told to sign a statement saying that he would be shot through the heart if he left his brother's farm. Allen argued that he was being asked to sign his own death warrant. After some discussion the militiamen agreed to change the punishment to a lashing of five hundred strokes "on the naked Back." That done, the mob marched him back to his brother's farm. Sometime later, probably in 1779, he slipped away to New York and took sail to England. He died there in 1782.[52]

Crean Brush was still in jail in Boston on November 5, 1777, when his wife spent the day visiting him. When the turnkey told her it was time to go, her clothes left the cell—but she was not in them. As the very unobservant turnkey later told his story, he was greatly surprised the next morning, when, after he passed Brush's breakfast through an opening in the door, no hand appeared to take the food. He looked through the peephole and saw Mrs. Brush. She did not tell where Brush had gone. But his hunters established that Brush rode off to New York on a horse that had been hidden by Mrs. Brush. In New York he joined countless other Loyalists looking for British assistance in settling grievances. Brush wanted recovery of his disputed property, especially land in the New Hampshire Grants, which would become the state of Vermont. Depressed and penniless, in May 1778 he put a pistol bullet through his head.[53]

While militiamen and members of Committees of Safety throughout Massachusetts and every other colony were grappling with the question of justice for Loyalists like Jolley Allen and Crean Brush,

George Washington himself was forced to handle his own questions of justice. His issue involved a Loyalist spy posing as a staunch Patriot.

The case began in Newport, Rhode Island, where a Patriot was shown an encrypted letter that was to have been delivered to James Wallace aboard the HMS *Rose*, on patrol offshore. Captain Wallace was well known as an intelligence intermediary who accepted and then passed on messages from Loyalists. Patriots had intercepted at least one, a letter sent to him by Col. Thomas Gilbert, leader of the Tory militia in Freetown.

The encrypted letter had been given to a Newport baker by a former girlfriend who had asked him to get it to Wallace. Patriot leaders had the baker escorted to Cambridge, where Washington was shown the letter. The woman was found and brought to him. Washington began questioning her and "for a long time she was proof against every threat and persuasion to discover the author." Then, suddenly, she broke and gave up the name of the man who wrote the letter—her secret lover, Dr. Benjamin Church, a longtime Patriot whom Washington had just appointed surgeon general of the army and the director of the army's first hospital.[54]

Washington gave the letter to the Reverend Samuel West, a Continental Army chaplain and an amateur cryptanalyst familiar with encryption techniques used by merchants who did not trust the royal mail system. Washington also asked for a separate decryption by Elbridge Gerry, a member of the Massachusetts Provincial Congress and the Committee of Safety. The two codebreakers produced identical translations of the message, which was addressed "To Major Cane in Boston, On His Magisty's Sarvice." Cane had been on General Gage's staff.

"I hope this will reach you; three attempts have I made without success," the letter began. Church told of his visit to Philadelphia, where "I mingled freely & frequently with the members of the Continental Congress." He gave standard agent reports on the number and location of cannons and soldiers. And he provided some political observations: The Congressmen "were united, determined in opposition, and appeared assured of success" and "A view to independence gr[ows] more & more general." Revealing his fear of being caught, Church concluded: "Make use of every precaution or I perish."[55]

Church, who for years had been trusted by leading Rebels, probably had spied for the British since the Sons of Liberty first gathered secretly at Boston's Green Dragon Tavern. Paul Revere later recalled that during the blockade of Boston, Church had once gone into the city, ostensibly to obtain medicine. He claimed to have been captured by British soldiers and released after interrogation by Gage himself— a cover story for a meeting with his spymaster.[56]

Hanging was the usual fate of spies, but Congress had not passed any laws about treason or espionage. Washington, as commander in chief of the army, decided he had the power to call and head a court-martial. In a short trial Church insisted he was innocent, but Washington found him guilty of "holding criminal correspondence with the enemy." Congress ordered Church put in a Connecticut prison. Later, claiming illness, he got himself transferred to Massachusetts. Finally, around 1778, he was allowed to sail to the British West Indies. His ship was lost at sea.[57]

In the vacuum left by the withdrawal of the British, Congress recommended that the colonies establish new governments. Former royal legislatures became Rebel legislatures with varying views of their duties and their power, including how to deal with Tories. The Continental Congress was concerned with creating and regulating a Continental Army rather than a continental nation. So there was no national government. But, as summer came to Philadelphia, Congress inched closer to taking a first step toward nationhood by beginning to contemplate, in secret sessions, what Church's spying had revealed to British authorities: "a view to independence."[58]

In March, Congress sent Silas Deane, a young, ambitious Connecticut delegate, to France. Deane was under cover as "Timothy Jones," a purchasing agent of "goods for the Indian trade" (meaning British India). Deane's covert instructions came from Ben Franklin, a member of the Congressional Committee of Secret Correspondence. Deane was told to negotiate secretly with representatives of the French government and learn what France would do if the colonies were "forced to form themselves into an independent state."[59] Deane was given a special ink to write invisible reports between the lines of what would appear to be an innocuous letter when it was almost inevitably opened by British spies or ship captains in their hire.[60]

Deane's French counterpart was Pierre-Augustin Caron de Beaumarchais, a writer,* a friend of King Louis XVI, and a spy for the Comte de Vergennes, the French Foreign Minister. After conferring with Deane, Beaumarchais formed a front company, Roderigue Hortalez et Compagnie, which would buy munitions and other war supplies from the French government and ship them to America, accepting payment in tobacco, rice, cotton, and indigo. The company was financed by France, Spain, and a group of French merchants.[61]

France, still stinging from its defeat in the French and Indian War, wanted to help an independent America as much as it wanted to hurt England. Supplying arms to America meant that Redcoats would be shot by French bullets from French guns—and French soldiers would not have to endanger themselves by pulling the trigger.

The idea of independence from Britain was now in the revolutionary air. Back in the pre–Stamp Act days, there had been no reason for the word "Loyalist." Everyone involved was a British subject. They were engaged in a political row that got out of hand: The Boston Massacre. The Boston Tea Party. Lexington. Concord. Bunker Hill. Even after Bunker Hill optimists could see a way back. After all, the Continental Congress had passed the Olive Branch Petition to the "Most Gracious Sovereign." The king's ministers and certain members of Parliament were to blame for the lengthy crisis. Peace, if royally proclaimed, could somehow prevail.

A substantial number of subjects fled to Halifax because the old

_____

* Among his writings were plays that would later become operas: Mozart's *The Marriage of Figaro* and Rossini's *The Barber of Seville*.

ideas of dissident and disaffected had given way to hatred and violence, to death by musket and bayonet. Some who sailed to safety believed that they would be returning after the Redcoats extinguished
the rebellion. Most of those who sailed on to England believed they
would never return. That dreadful place had become *America*; it no
longer seemed to be a part of the British Empire.

Then, in July, in the wake of those transports to Halifax, came the
Declaration of Independence.

To many Loyalists and Tories, the Declaration was a surprise,
seemingly coming out of nowhere. But it had come, most of all, from
the pen of Tom Paine, an immigrant from England who attacked not
the king's ministers but the king, writing that "a thirst for absolute
power is the natural disease of monarchy. . . . To the evil of monarchy
we have added that of hereditary succession; and as the first is a deg-
radation and lessening of ourselves, so the second, claimed as a matter
of right, is an insult and an imposition on posterity."[62]

On July 18, 1776, British officers who were prisoners on parole in
Boston were invited by their American captors to a ceremony at the
Towne House, the handsome brick structure where the royal governor once reigned. "As we passed through the town," one of the officers later wrote, "we found it thronged; all were in their holiday
suits; every eye beamed with delight, and every tongue was in rapid
motion. The streets adjoining the Council Chamber were lined with
detachments of infantry tolerably equipped, while in front of the jail
artillery was drawn up, the gunners with lighted matches. The crowd
opened a lane for us, and the troops gave us, as we mounted the steps,
the salute due to officers of our rank. "[63]

As the clock on the Towne House struck one, Sheriff Stephen
Greenleaf, standing on the balcony below, began reading the Declaration of Independence in his weak voice: "When, in the Course
of human events, it becomes necessary for one people to dissolve the
political bands which have connected them with another . . ." Then,
sentence by sentence, the words were boomed forth by Col. Thomas

Crafts, Jr.—Son of Liberty, leader of the mob that hanged a placeman in effigy from the Liberty Tree, dumper of tea in Boston Harbor, artillery officer at Bunker Hill. A "shout, begun in the hall, passed to the streets, which rang with loud huzzas, the slow and measured boom of cannon, and the rattle of musketry. . . . large quantities of liquor were distributed among the mob; and when night closed in, darkness was dispelled by a general illumination."[64]

The British officers who listened to the Declaration being read were in a new world. When the time came for their exchange, the transaction would take place between representatives of two nations. Once their British Army had been fighting for the king to quench a rebellion, had killed British subjects, and had been killed by British subjects. Royal Navy ships had burned towns occupied by British subjects. Now the troops and the warships would be fighting citizens of a new nation that had vilified the king in its Declaration of Independence.

To Loyalists, the Declaration contained line after line of libel—hundreds of words maligning George III, making him, not his ministers, responsible for rebellion: "He has affected to render the Military independent of and superior to the Civil Power. He has combined with others to subject us to a jurisdiction foreign to our constitution, and unacknowledged by our laws; giving his Assent to their Acts of pretended Legislation. . . . He has plundered our seas, ravaged our Coasts, burnt our towns, and destroyed the Lives of our People."

And a new accusation: "He is at this time transporting large Armies of foreign Mercenaries to compleat the works of death, desolation, and tyranny. . . ."

The words "at this time" appeared in the Declaration because members of the Congress, meeting in Philadelphia, had just received intelligence from Europe that representatives of King George had gone to Hesse-Kassel and other German states to bargain for the hiring of nearly thirty thousand troops.[65] (Because most of the troops were from Hesse-Kassel, they all came to be called Hessians.)

Contracts for the German troops varied, but essentially they called for King George to bear the expense of a hired soldier killed in battle (with three wounded equaling one dead) while German rulers replaced at their own cost any soldier who deserted or died of sickness.[66] King Frederick the Great of Prussia so loathed the deal that he said he would impose the customary "cattle tax" on all the hired troops who passed through Prussia "because though human beings they had been sold as beasts."[67]

France could not openly give full support to colonies of Britain. But if those colonies broke off from Britain and claimed the status of a nation, then France could become an ally. To prevent that from happening, Britain had a more crucial reason to crush the Rebels. The Declaration of Independence meant that what had been armed response to a rebellion—a civil war between British subjects—had become a war between Britain and a new nation whose existence threatened Britain itself. As for the loyal subjects, their existence was also threatened, and so they must fight. And fitting Loyalist military forces into this new war presented a new challenge to the generals and the statesmen.

BEATING THE SOUTHERN DRUMS

Nova Scotia and North Carolina, 1776

The southern people . . . have got to the highest pitch of raving madness. The best friends of Great Britain are in the back parts of the Carolinas and Georgia.

Lieutenant Governor John Moultrie of East Florida[1]

Nearly a month after the evacuation of Boston, Alexander Mac-Donald, a Loyalist recruiter, wrote to his wife, Susey, from Halifax, to assure her that the fight to quell the rebellion would go on. He knew that a new phase of the war had begun: the mustering of Loyalist military units. They would fight for King George III against fellow Americans who had traitorously turned against their monarch.

MacDonald had left Susey and their children on Staten Island in the fall of 1774 when he set forth to raise soldiers for the Crown. Now he was in Halifax, an eager ally of the British Army, which, he wrote, had "embarked on board their transports with all their baggage stores and artillery and every thing that was worth the Carrying without the Least hindrance or Mollestation, or the Loss of one Single Man. . . ."[2] What MacDonald did not say was that Susey soon would see those Redcoats, for they would land on Staten Island to begin fighting to turn New York City and Long Island into Loyalist strongholds.

No one knew how many Tories remained in Boston after the evacuation. The city had lost many Loyalist leaders—men like Ned Winslow, Brigadier Ruggles, and Colonel Gilbert—who had been the architects of a Tory militancy. Ruggles's three sons also went to Canada, but his wife and their four daughters stayed behind.[3]

In Massachusetts and beyond, "Friends of the Government," as they once were known, now preferred the label "Loyalist," an assertion that they staunchly supported British rule. But few openly proclaimed their allegiance to the Crown. Most of them kept covert faith in the ultimate triumph of the British Army.

Those who chose the boldest role were the Loyalists who volunteered to shoulder arms in the numerous military units that began to appear in the colonies soon after the evacuation of Boston. MacDonald was one of the recruiters of those militant Loyalists, who would be seen by some British strategists as the key to squashing the rebellion. In that view the rebellion was a civil war, and the Loyalists had to be mobilized to fight the Rebels.*

MacDonald, born in Scotland, had been a lieutenant in a Highland Scots regiment eventually named the 77th Regiment of Foot, a proud unit that went to war in Highland kilt and sporran, with dirk and broadsword. He had served with the 77th in the French and Indian War and later in the French West Indies and Havana. Rather than return to England or Scotland, MacDonald and many other veterans decided to stay in America. He retained his commission as a senior lieutenant and was put on half pay, a status analogous to a ready reservist today. Some of the ex-soldiers chose to settle on the Canada–New York frontier, in the Mohawk Valley above Albany. There Sir William Johnson, Britain's superintendent of Indian affairs, reigned like a feudal lord over an immense realm and welcomed newcomers, who added to his legion of tenants.[4]

MacDonald instead decided to live on Staten Island, where Susey had inherited property. She had been born into a branch of the Living-

---

* A full description of the units of the Provincial Corps, the Loyalist equivalent of the Continental Army, can be found at ToriesFightingForTheKing.com.

stons, one of the two powerful New York families then engaged in a political struggle. The leaders of the Patriots were Livingstons, the colony's richest landed gentry. The De Lanceys, whose wealth came from trade, led the Tories. The families' well-known religious affiliations added to the "hott & pepper on both Sides," as a Livingston referred to the rivalry. The De Lanceys were Anglicans and the Livingstons were Presbyterians, like most outspoken opponents of British policy toward the colonies.[5]

As the New York political rivalry escalated toward rebellion, MacDonald's Patriot in-laws urged him to join their side. But having dabbled in trade and become friendly with Tory merchants, MacDonald remained pro-British and retained his army ties, winning a promotion to captain in 1772. As a soldier and a Highlander he could not merely support the Loyalists. He wanted to raise a regiment of Highland immigrants recruited to fight alongside the British Army.

MacDonald started outlining his idea in letters to veterans in the Mohawk Valley and to an old army friend, Maj. John Small, who was on the staff of General Gage. MacDonald also wrote to his cousin, Allan MacDonald in North Carolina, a leader of that colony's growing population of Highland immigrants. MacDonald called his recruiting drive "a beating," evoking the traditional image of a military conscription party seeking militiamen by marching through the town with a drummer steadily beating a summons.

A recruit had to be at least seventeen years old (drummers could be as young as ten), appear healthy, possess all limbs, be no shorter than five feet three inches, have no ruptures or be troubled by fits, and have at least two teeth that met (so he could bite through the paper that wrapped the gunpowder and the ball of a musket cartridge).[6] Men recruited in North Carolina were promised two hundred acres of land, to be confiscated from Rebels, and a twenty-year tax exemption. A married man received an extra forty acres for his wife and each child.[7]

When he had recruited more than one hundred men, MacDonald journeyed to Boston and formally presented his prospective muster roll to Gage. To his surprise MacDonald learned not only that a similar suggestion had come from Lt. Col. Allan Maclean but also that the Royal Highland Emigrant Regiment already existed and was commanded by Maclean. Gage had divided the regiment into two battalions, one to be commanded by Maclean and the other by Major Small. MacDonald was made a captain in Small's battalion.[8]

Expecting regimental command, MacDonald seethed. While Maclean went off with his battalion to defend Quebec and win glory, MacDonald and Small spent most of their time in Nova Scotia training recruits and begging for supplies. In a typical appeal MacDonald wrote: "Here I am with above 300 Men and Officers, the Most of the Men almost Naked and all of them bare footed. . . . P.S. We have formed a Mess and a poor one it is without a drop of any kind of wine or Spirits, only Spruce beer. . . . I beg you would send us a pipe of Good Madeira also a hogshead of good port wine."[9]

Eventually MacDonald's men of the Mohawk Valley were joined by volunteers from Boston, New York City, North Carolina, Nova Scotia, and other parts of Canada. To get more recruits Small and MacDonald received permission to intercept immigrant ships on the high seas, go aboard, and recruit the male passengers. The New York–bound immigrant ship *Glasgow*, for instance, was redirected to Boston. Somehow—coercion seems probable—Small enlisted and diverted to Nova Scotia one hundred men who had expected to settle in North Carolina.[10] The women and children who accompanied them were issued army rations and put in the care of the garrison at Fort Edward in Windsor, Nova Scotia.[11]

After Maclean's victory in Quebec, dozens of Continental Army prisoners of war enlisted in his battalion, though they were neither Highlanders nor emigrants. Some of them, enlisting only to end their prisoner status, deserted and made their way back to America. Deserters who were captured were given the choice of severe and potentially lethal flogging or lifelong exile as Redcoats deployed to protect slave dealers and other British subjects in western Africa.[12]

* * *

Andrew MacDonald's kinfolk included Flora MacDonald, who had saved Bonnie Prince Charlie after his defeat on the moors of Culloden. The prince, pursued by British soldiers to the Outer Hebrides, made a daring escape later celebrated in Scottish poetry and song. Flora disguised him as an Irish maidservant and took him from her island to the Isle of Skye on his way to safety in France. Arrested by the British, she was imprisoned in the Tower of London and released under a general amnesty in 1747.[13] Three years later she married Allan Mac-Donald, another member of Clan MacDonald and supporter of the Jacobite cause.

Life in the Highlands, which encompassed Scotland's Western Isles and northern mainland, had become a struggle for survival. No longer sustained by the land, bankrupted by ever-rising rents, denied their culture and even their native language by harsh British edicts after the Battle of Culloden, Highlanders headed for America. They were seeking what had become a dream, conjured up by letters from kin and friends.

Samuel Johnson and James Boswell, traveling on the Scottish mainland and the islands of the Highland realm in the fall of 1773, were entertained on the Isle of Skye by a group of couples doing a complex dance. "Each of the couples, after the common *involutions* and *evolutions*," Johnson wrote, "successively whirls round in a circle, till all are in motion; and the dance seems intended to shew how emigration catches, till a whole neighbourhood is set afloat." The islanders called the dance "America."[14]

The migration of thousands of Highlanders—the exact number is unknown—had begun about 1732, surging in the late 1760s and peaking in 1774 and 1775.[15] Among the emigrants were Flora and Allan MacDonald, who sailed across the Atlantic in 1774, just in time to find themselves again choosing sides in a conflict between Rebels and the king. This time they chose the king.

Many Highlanders made the same choice, even though they could recite a long litany of mistreatment by the British: They had cleared the Highlands of farmers to make way for sheep, a better investment. They had, through blandishments and power dispensing, co-opted the clan chiefs and shattered the clans. They had banished nearly one thousand Highland men, women, and children to America, where they were sold as indentured servants.[16] The banishment was a variation on "transportation," the euphemistic term for a judicial order that exiled a convicted felon to America. By 1776 about fifty thousand men and women, many of them guilty only of minor crimes, had been sent to America in transportation from England and Scotland.[17] Still, fighting for the Crown had become a Highland tradition. The Highlanders drawn to America had sailed through a complex history that tugged many of them toward the Crown rather than toward the Rebels. Their political differences of the recent past were eclipsed by the new reality that Highlanders discovered in America.

New England was far away, geographically and politically, from the part of North Carolina where Flora and Allan MacDonald settled. An estimated twelve thousand former Highlanders lived in the colony, most of them occupying the upper Cape Fear River in Cumberland County. Overwhelmingly they were Loyalists.[18]

For the Highlanders who migrated at the beginning of the Revolution, the choice of the Loyalist side was pragmatic: The Rebels were losing. By fighting for the British the Highlanders assured their position in this promising colony of the powerful British Empire.[19] The king rewarded army service with land grants; the Patriots offered nothing.

Highlanders had been welcomed into the British Army since the first Jacobite rebellion, in 1715, when warriors from loyal clans were stationed in the Highlands as peacekeepers between clans. In 1739 King George II authorized the formation of a regiment consisting of "natives of that country and none other." The regiment was known as the Black Watch—"black" for the dark colors of the soldiers' tartans and "watch" for the regiment's original mission to keep watch over the Highlands.[20] Impressed by the Highlanders' soldierly qualities, British Army commanders continued to recruit men from the land that Robert Burns would call "The birth-place of Valour, the country of Worth."

Additional Highland regiments were raised in 1756, 1757, and 1777.[21] Officers and enlisted men who fought in the French and Indian War received land grants as large as twenty thousand acres in Canada and northern New York. When recruitment of the Royal Highland Emigrants began, many of those veterans joined, resuming army life.[22]

The recruitment of American Highlanders was so well known to the people of Scotland that a Scotswoman traveling to America in 1774 noted in her journal the military significance of the Highlander immigration. Aware of looming trouble in America and seeing hardy America-bound Highlanders aboard her ship, she wrote, "Should levys be again necessary, the recruiting drum may long be at a loss to procure such soldiers as are now aboard this Vessel, lost to their country for ever, brave fellows, who tho' now flying from their friends, would never have fled from their foes."[23] She was right. The lack of recruits for the British Army at the beginning of the Revolution had been the principal reason that Britain hired Hessians.

In America, Highlanders usually retained their Gaelic language and Highland dress. And they dwelled apart from another, distinctly different immigrant group, the Scotch-Irish, who overwhelmingly joined the ranks of the Patriots. As for Highlanders, Thomas Jefferson often referred to them as "Scottish Tories."[24] In an early draft of the Declaration of Independence, Jefferson wrote a phrase denouncing the British for "permitting their chief magistrate to send over not only soldiers of our own blood, but Scotch and other foreign mercenaries." The phrase was struck out at the urging of John Witherspoon, a Scots-born delegate from New Jersey. He was a Presbyterian minister, an eloquent champion of independence, and a signer of a Declaration that, thanks to him, was not blemished by a libel on Scots.[25]

Jefferson's distrust of the Scots stemmed from common knowledge that in the many parts of the southern colonies a "Scot" and a "Tory" were almost always one and the same. In and around Norfolk, Loyalists were often referred to as members of "the Scotch Party," and, when the Queen's Loyal Virginia Regiment was raised, most of its men were Scots merchants and their employees. Virginia's tartan-clad royal governor Dunmore had served as a page to Bonnie Prince Charlie.[26] His fellow Scots were merchants and placemen whose influence extended from Alexandria, Virginia, to St. Augustine, the principal city of the East Florida Colony.[27] Dunmore spoke the warrior language of many a Tory Scot when he said, "I once fought for the Virginians. By God, I would let them see that I could fight against them."[28]

"BROADSWORDS AND KING GEORGE!"

> *To sign, or not to sign? That is the question,*
> *Whether 'twere better for an honest man*
> *To sign, and so be safe; or to . . . fly . . .*
> *And, by that flight, t' escape*
> *Feathers and tar . . .*
>
> —*"The Pausing American Loyalist"*[1]

Royal governors in Virginia, North Carolina, and South Carolina had been urging the mobilizing and arming of Loyalists since the first hints of rebellion. The governors believed that the path to victory coursed through the southern colonies, where there were enough friends of the king to help the British Army quickly put down the revolt. Ultimately twenty-nine Loyalist military units, mostly of regimental size, would fight the Continental Army or Rebel militias in the southern colonies.[2]

North Carolina governor Josiah Martin, a former lieutenant colonel in the British Army, envisioned a Loyalist army of twenty thousand men, enlisted around a core of three thousand Highlanders, along with tough backcountry settlers known as the Regulators. The Regulators had protested a corrupt system by attacking dishonest tax collectors and judges and rebelling against the royal rule of William Tryon, Martin's predecessor.

In May 1771 Tryon, also a former lieutenant colonel, had led a royal militia army against the Regulators in a battle near today's Alamance, killing about twenty, hanging seven without trial, and laying waste their farms.[3] The Regulators, who had seen their comrades' bodies ripped by royal grapeshot at Alamance, were unusual candidates for a Loyalist force. But their allegiance to the king transcended all else, and many of them joined Loyalist regiments.

As for the Highlanders, Governor Martin counted on their loyalty to the king—and his expectation that they would barter their warrior skills for generous land grants. Martin had been providing those grants to men who had only to swear "their readiness to lay down their lives in the defence of his Majesty's government."[4] Allan MacDonald and Flora bestowed their prestige on Martin's plan for a Loyalist force. Allan would be second in command; the commanding officer would be Donald MacDonald, a veteran of Culloden who had been commissioned a brigadier general by Martin.

Martin was working on his plan when Janet Schaw, a sharp-eyed woman from Scotland, appeared and became an unexpected witness to the stirring of rebellion in North Carolina. After a stormy crossing in a small ship, she and her younger brother Alexander arrived at a plantation, near Wilmington, which was owned by their older brother Robert. He, like all North Carolinians, was in the midst of choosing sides. He had served in the militia when it was royal and had fought under Governor Tryon in the Battle of Alamance. When the militia came under control of the Rebels, he remained an officer and was commissioned a colonel. But revolutionary events were moving so swiftly that, coincidental with his sister's arrival, he had to decide to support Governor Martin or go with the Rebels.

Janet Schaw almost immediately discovered the complexity of the choice her brother Robert faced, for the Rebel–Loyalist fissure reached deep into southern society. A neighboring hostess was warned by the Rebels' local Committee of Safety that she had to cancel a forthcoming ball because of a congressional edict against "Balls and Dancing."

Then, surprisingly, the committee relented, Janet got her invitation and, "dressed out in all my British airs with a high head and a hoop," she trudged to the ball "thro the unpaved streets in embroidered shoes by the light of a lanthorn carried by a black wench half naked."[5]

The Committee of Safety may have changed its mind because of the influence of a visitor to the plantation, Robert Howe. He "has the worst character you ever heard thro the whole province," Schaw wrote. "He is however very like a Gentleman." A prominent Patriot related to Robert Schaw by marriage, Howe, too, had served under Tryon.[6] While he was chatting and flirting with Janet Schaw, he had something else on his mind: a plot to kidnap Governor Martin.[7]

Howe was with Janet Schaw one day when, by her account, he stopped a Rebel mob from tarring and feathering a hapless Tory. She, Howe, and others were watching a mustering of Rebel militiamen—"2000 men in their shirts and trousers," she wrote, "preceded by a very ill beat-drum and a fiddler, who was also in his shirt with a long sword and a cue at his hair, who played with all his might. They made indeed a most unmartial appearance." But, she shrewdly observed, "the worst figure there can shoot from behind a bush and kill even a General Wolfe."[8]*

She, Robert Schaw, and others on the plantation were invited by Martin and his wife to go to the governor's palace in New Bern on June 4 to celebrate the king's birthday. But they learned that "the Govr's house had been attacked, himself obliged to get down to the man of war, and send off his wife, sister and children in a little vessel, with directions to land them in the first safe port. What renders these circumstances the more affecting is that poor Mrs. Martin is big with child, and naturally of a very delicate constitution."[9]

Martin escaped the kidnappers, who had been managed by Howe, and reached a fort on Cape Fear. His family sailed to New York and then journeyed to Martin's father-in-law's Long Island home, where

---

* A reference to the slaying of Gen. James Wolfe, commander of the British expedition that took Quebec in 1759, assuring British victory in the French and Indian War.

Martin's daughter Augusta was born. When Rebels were about to take and burn the fort, Martin fled to the HMS *Cruizer*, a British warship anchored in the Cape Fear River. Among the people who accompanied him were Janet and Alexander Schaw. While they were aboard the *Cruizer*, Martin asked Alexander to sail to England with dispatches that would inform Lord Dartmouth about the rebellion in North Carolina.

Janet Schaw's journal contains an emotional description of the Rebel martial law that pertained over much of North Carolina in the wake of Martin's flight:

> *An officer or committeeman enters a plantation with his posse. The Alternative is proposed, Agree to join us, and your persons and properties are safe; you have a shilling sterling a day; your duty is no more than once a month appearing under Arms at Wilmingtown, which will prove only a merry-making, where you will have as much grog as you can drink. But if you refuse, we are directly to cut up your corn, shoot your pigs, burn your houses, seize your Negroes and perhaps tar and feather yourself.*[10]

When Robert Schaw refused to take an oath supporting the Rebels, his punishment was exile and the confiscation of his land.[11] But Thomas Macknight, an extremely wealthy Scottish merchant in North Carolina, suffered the kind of punishments described in Janet Schaw's journal. Although Rebels had once labeled him "always friendly to the cause of American liberty," when he showed a tolerance for Tory friends in New Bern, he was deemed "inimical." Rebels plundered his house, his merchandise, and his crops, abducted his slaves, and even tried to kill him, according to his recollections. In 1776 he exiled himself to England.[12]

Janet cut short her stay and boarded a ship that was returning to Britain after bringing more Scot immigrants to North Carolina.[13] She bade a bitter farewell to an "unhappy land, for which my heart bleeds in pity. Little does it signify to you, who are the conquered or who the victorious; you are devoted to ruin, whoever succeeds."[14]

* * *

Among the documents that Alexander carried to England was a letter that Martin had written Lord Dartmouth on June 30, boasting that he could "reduce to order and obedience every colony to the southward of Pennsylvania." All he needed were ten thousand muskets and ammunition, along with some artillery "and a supply of money as might be necessary for the support of such a force."[15]

Dartmouth had been getting similarly optimistic reports from Crown officials in other colonies, especially from Dunmore in Virginia. Focusing on Martin's proposal, Dartmouth saw the possibility of a combined offensive that would stamp out the rebellion in North Carolina and so terrify the Rebels in the other southern colonies that they would surrender. Dartmouth's successor, Lord George Germain, enthusiastically endorsed the use of Loyalist soldiers allied with a British Army expeditionary force that would land at Wilmington, a port about fifteen miles up the Cape Fear River. The conquest of Wilmington, Dartmouth believed, would quickly lead to the surrender of the bigger port of Charleston, South Carolina, some 170 miles south. Thus began Britain's first venture into a strategy aimed at severing the southern colonies from those in the north.

Coincidentally, while Governor Martin was making his plans, Lt. Col. Donald MacDonald and Capt. Donald McLeod of the Royal Highland Emigrants Regiment arrived in New Bern. They had been dispatched as recruiters by General Gage. When picked up for questioning by the Rebels' Committee of Safety, they said they were former British officers who had been wounded at Bunker Hill and had come to North Carolina with the idea of possibly settling there. The committee released the officers with a warning to refrain from aiding the Tories.[16]

Martin issued a proclamation calling on all loyal North Carolinians to support the king. And he asked several prominent North Carolina Loyalists—many of them Scots—to raise a Loyalist force called the North Carolina Provincials. Between twelve and fourteen hundred volunteered. In Martin's vision local Highlanders and Regulators would be his principal warriors.[17] Martin also expected to mobilize Loyalist volunteers from other backcountry immigrants: the Scotch-

Irish. But unlike the Highlanders, the Scotch-Irish were not dependably loyal, and most of them had, at least, Patriot leanings.

The Scotch-Irish were descendants of transplanted Scots who in the early 1600s had been sent to Northern Ireland by King James I. He had tried to solve his problems with the Irish and the Scots by handing over to the Scots a vast Irish territory that came to be known as the Ulster Plantation. King James specifically barred Highlanders from Ulster because his plan was to introduce the ways of the British to the Irish. They, in his view, needed civilizing.[18]

Early in the eighteenth century the people of Ulster began sailing to America in a flight from what a Pennsylvania newspaper called "Poverty, Wretchedness, Misery and Want."[19] By the 1750s the immigrants were being called the Scotch-Irish, a shorthand acknowledgment of Scotch descent and Northern Ireland origin. The newcomers usually called themselves Scotch, avoiding any suggestion that they were partially Irish.[20]

Between 1717 and 1775, in America's first great surge of immigration, more than two hundred thousand Scotch-Irish arrived in the colonies.[21] How many others were lost at sea in the long perilous voyages will never be known. The ships were built to carry cargo, not passengers. Immigrants were stowed in the dank hold by shipowners who saw them as profitable ballast. On one ship starving survivors ate the bodies of the dead.[22] In another "about 100 of them dyed," said a Pennsylvania newspaper in 1728, reporting on the explosive growth and frequent tragedies of immigration. "Children from 1 to 7 years rarely survive the voyage," a survivor wrote. "I witnessed misery in no less than 32 children in our ship, all of whom were thrown into the sea. The parents grieve all the more since their children find no resting-place in the earth, but are devoured by the monsters of the sea."[23]

Most of the early Scotch-Irish immigrants landed at Philadelphia and found land along Pennsylvania's western frontier.[24] Descendants of the first settlers and new waves of immigrants moved westward along the Great Wagon Road, which had evolved from a network of old Indian trading trails stretching from Pennsylvania to Georgia east of the Appalachians. The road took newcomers, their livestock, and their wagons to the backwoods of Virginia and the Carolinas. In North Carolina the immigrants settled along the piedmont, which extended from the fall lines of seaboard rivers westward to the foot of the Appalachians. Besides being far from the riches and power of the seacoast, the people of the piedmont were also outside the economic sphere of the coastal ruling class.[25]

The immigrants built their dirt-floor log cabins near springs or streams. There were no village greens because the settlers had no time to form villages. They had to clear the land and then plant their first crops. They raised corn, wheat, flax, and cotton as well as sheep on farms whose average size was about 175 acres.[26] Most clothing was made from cotton cloth woven at home.

The Scotch-Irish were usually Presbyterians, but among them there were also evangelical Baptists. The Anglican Church, rebuffed by the toiling class back in Ulster, had not attracted the Scotch-Irish in America. To bring them the Anglican faith in the new land, missionaries were sent from Britain to the Carolina backcountry. Among the missionaries was the Reverend Charles Woodmason, a British-born Anglican clergyman who traveled the region in the 1760s.

Woodmason was shocked by the people's "low, lazy, sluttish, heathenish, hellish Life." He had difficulty averting his eyes from the "Young Women," who had "a most uncommon Practice, which I cannot break them of. They draw their Shift as tight as possible to the Body, and pin it close, to shew the roundness of their Breasts, and slender Waists (for they are generally fairly shaped) and draw their Petticoat close to their Hips to shew the fineness of their Limbs." When he performed a church service, he wrote, most of the congregation "went to Revelling Drinking Singing Dancing and Whoring— and most of the Company were drunk before I quitted the Spot."[27]

Woodmason was seeing a new American breed: people who had migrated not from England but from Ulster, people to whom Scotland was but a folk memory, a place few of them had even seen. And, as Presbyterians, they had turned away from the hierarchical structure of the Anglican Church in favor of the democracy of the meetinghouse. As a North Carolina minister—and Patriot—explained Presbyterian beliefs: God "had long ago implanted into man's nature a capacity for civic responsibility. God had taught men to consider themselves His stewards, had given them talents and opportunities, and expected them to make the most of those endowments."[28]

Many Loyalists believed that the Revolution itself had emerged from a conflict between Presbyterians and Anglicans: "Presbyterianism is really at the Bottom of this whole Conspiracy, has supplied it with Vigour, and will never rest till something is decided upon it," a representative of Lord Dartmouth wrote in 1776.[29] A disinterested witness in the form of a Hessian captain had noted this in a letter home: "Call this war, dearest friend, by whatever name you may, only call it not an American Rebellion; it is nothing more nor less than Irish-Scotch Presbyterian Rebellion."[30]

When word of Governor Martin's Loyalist plan reached Patriot leaders, they sent pro-Patriot propagandists, including five Highlanders, into the Scotch-Irish backwoods to meet with "the gentlemen who have lately arrived from the highlands in Scotland" and "to advise and urge them to unite with the other Inhabitants of America in defence of those rights which they derive from God and the Constitution."[31] The delegation did not attempt to recruit militiamen but chose only to gain sympathy for the Rebel cause.[32]

Governor Martin, meanwhile, received word that British strategists had enhanced his plan. Already at sea was a convoy bringing seven Redcoat regiments from the British Isles under the command of Lord Charles Cornwallis. Another large force was coming from Boston under Maj. Gen. Sir Henry Clinton, who said his mission was "to support Loyalists and restore the authority of the King's government in the four southern provinces."[33]

The Loyalists were to march to the coast down the southwestern side of the Cape Fear River and at Wilmington rendezvous with the

British troops coming by sea. Joined up, the combined force would then begin a campaign to fight the Rebels for control of North Carolina. That triumph was to be the inspiration for a general rising of the Loyalists and the return of the southern colonies to the Crown.

Highlanders, expecting that the seaborne British would soon arrive, began assembling at the hub of their power, Cross Creek (now Fayetteville), about eighty miles north of Wilmington. They hoped to raise as many as five thousand men, including a large complement of Regulators. But Regulators were scarce on February 15, 1776, when about fourteen hundred men gathered at Cross Creek. And only 520 of them carried muskets. Leaders canvassed people in the area and found muskets to arm about 130 more. Raiders also took powder that had been stored by Patriots and got additional powder, along with provisions, from Loyalist merchants.[34] About sixty people slipped out of the Cross Creek area to an encampment seven miles away, where some eleven hundred Patriot militiamen had mobilized. The Cross Creek refugees warned Col. James Moore, the Patriot commander, that the Loyalist troops were ready to march.

On February 19, Donald MacDonald—promoted to brigadier general by Martin—sent a messenger to Moore under a flag of truce. The message warned Moore that if he and his men did not accept the authority of the king by noon the next day, they would be attacked as enemies. Moore replied that he would continue "the defense of the liberties of mankind." Dozens of men unexpectedly turned away from battle as MacDonald led his troops, most of them Highlanders, not toward Moore but toward the coast.[35]

Moore sent some men to block one likely route and then marched off, hoping to pursue the Loyalists and force a battle. MacDonald continued at a slow pace, scouting for possible ambushes and strengthening bridges for his wagon train. Detachments of Patriots maneuvered along the Cape Fear River, seeking to close on MacDonald.

Finally, on February 26, MacDonald learned that about one thousand Patriots were ahead, at a bridge crossing swampy Moore's Creek, which flowed into a tributary's confluence with the Cape Fear River. MacDonald, an ailing old soldier, sensed disaster, called a council of war, and urged caution. The younger, bolder Donald McLeod, who was second in command, ordered an attack at dawn. Capt. John Campbell was to lead about eighty men brandishing the Highlanders' storied weapon, the broadsword.

About one a.m. the next day, scouts on the western side of the creek reached the bridge, a span of about fifty feet, and saw that the Patriots had removed its planks and greased the horizontal log supports, called stringers, with what smelled like tallow and soap. Entrenched on the eastern side, the Patriots guarded the road the Loyalists needed to take to Wilmington. Undaunted by the scouts' report, McLeod readied the charge.

The steady beating of drums and keening of war pipes broke the silence of the cool dawn. Then came three cheers from all the men, the signal for the charge. Shouting a rallying cry—"King George and Broadswords!"—the eighty bravest Highlanders, led by McLeod on one stringer and Campbell on the other, began slowly crossing the slippery logs, using their broadswords like spiked canes.[36]

The Patriots held the fire of their muskets and two cannons until McLeod and Campbell reached the eastern side of the creek. Suddenly cannons and muskets fired. The cannons were loaded with swan shot—a canvas bag that burst on firing, spewing twenty or more lead pellets.[37] Riddled by bullets and swan shot, McLeod and Campbell fell, both mortally wounded. McLeod, waving his broadsword, tried to rise, then died in another burst of bullets. One by one, two by two, other Highlanders fell, some to die by bullets, others by drowning. On the other side of the creek, Highlanders returned fire. Regulators and other Loyalists fled.

Sharp Patriot fire killed at least thirty Loyalists; the exact toll is not known because the bodies of some who died in the creek were not recovered. Two Patriots were wounded, and one died of his wounds. General MacDonald and about 850 others were taken prisoner, among them Flora MacDonald's husband, Allan.

Many of the escaping Loyalists went into hiding, some of them for the rest of the war. At least one managed to reach Florida and join another Loyalist military organization, the East Florida Rangers.[38]

An unknown number of Loyalists who were bystanders quietly disappeared, later sailing to safety in Nova Scotia and New York City.[39] The most significant effect of the victory came after exuberant Patriots reported it to their delegates at the Continental Congress and urged that the colonies sever ties with Britain. Thus North Carolina became the first colony to vote for independence.[40]

General Clinton had sailed out of Boston on January 10, expecting to rendezvous with Martin's men and the Cornwallis fleet in mid-February. But, because of bureaucratic delays and the usual mishaps of eighteenth-century ocean voyages, Clinton and his two hundred infantrymen did not arrive off North Carolina until March 12. Cornwallis was not there. His fleet would not begin to appear off Cape Fear until April 18.[41] Deciding that he might as well put his men to good use, Clinton broke camp, reloaded the troops aboard his ships, and sailed south to seize Charleston (then called Charles Town).

When Clinton set sail, three royal governors were trying to run their colonies from cramped quarters aboard Royal Navy warships bobbing at anchor off rebellious shores: Martin on the HMS *Cruizer* in the Cape Fear River, Dunmore from HMS *Fowey* on the James River in Virginia, and Lord William Campbell, on HMS *Tamar* off Charleston, in South Carolina.

Unlike the royal governors in New England, Dunmore and Martin reacted to rebellion in their colonies by mobilizing their Loyalists and using them as troops in a civil war aimed at regaining royal control. Campbell did not mobilize, but he had frequently urged British officials to invade his colony and link up with South Carolina's multitude of Loyalists. "Charles Town," Campbell wrote Lord Dartmouth, "is the fountainhead from which all the violence flows. Stop that and the rebellion in this part of the continent will, I trust, soon be at an end."[42]

Beyond the fountainhead of rebellion, both Patriots and Tories sought to win over the backcountry. In July 1775 the Patriots' Council of Safety in Charleston sent spokesmen into the "interior parts" of

the colony to urge people to join the Patriots' cause "in order to preserve themselves and their children from slavery," taking no apparent
notice that they were in a colony full of slaves. The spokesmen were
dispatched to counteract "the arts, frauds, and misrepresentations"[43]
of Moses Kirkland, a prominent South Carolina landowner who had
switched sides to become a peripatetic Tory operative. He had helped
Dunmore in Virginia, would serve with the British Army in Savannah, and would recruit Indians to fight Patriots in Florida.[44]

Tory and Rebel leaders in the backcountry at first dueled with
words and petitions, some changing sides. Finally, the local version
of civil war erupted at a trading post called Ninety Six, near today's
Clemson. After Rebels seized and imprisoned a Tory leader, about
fifteen hundred of his followers surged into Ninety Six, seeking revenge, surrounding a fort manned by some five hundred Rebel militiamen. In three days of fighting, one man was killed. His was the
first blood of a Patriot shed in South Carolina. The battle ended with
a truce, but conflict continued elsewhere.[45]

Later in 1775 more than five thousand Patriot militiamen scoured
the backcountry of Loyalists and captured their leaders, including one
who hid in the hollow of a sycamore. In a swirling storm—ending
what became known as the Snow Campaign—the Patriots made
Loyalists sign pledges to lay down their arms or have their property
confiscated.[46] Thirty-three who were jailed in Charleston finally promised "to Endeavor to Settle Peace to Your Satisfaction." But an added
note—"there is Different Circumstances Amongst us"—reflected the
reality that although peace had seemingly come to the backcountry,
conflicts would continue to simmer.[47]

The ferocity of the rebellion in South Carolina appalled Governor
Campbell. Thomas Jeremiah, a "free negro" well known as a fisher-
man and pilot called Jerry, was heard to say that if British warships
sailed into Charleston, he would guide them. Patriots arrested him
and, in a mockery of a trial, speedily convicted him of plotting insurrection against the Rebels. He was sentenced to be hanged and his
body publicly burned. Lord Campbell had the power to intervene—
"my blood ran cold when I read on what grounds they had doomed a

fellow creature to death," he said in a letter to Dartmouth. But Patriots warned Campbell that if he granted a pardon, "they would hang him at my door." The sentence was carried out; he was one of several South Carolina slaves executed after conviction on similar charges.[48]

For General Washington, the Revolution shifted southward after he learned of the movement of British warships and troops to Charleston. He sent Maj. Gen. Charles Lee to Charleston as an adviser to local Patriots. Lee, a former British officer who was contemptuous of the Continental Army, examined Charleston's defenses and called the city's key defense, the Sullivan's Island fort, a "slaughter pen." Lee told the commander, Col. William Moultrie, he should abandon it. Moultrie, backed by his Patriots, politely refused, though Lee seemed to have a point. The fort was made out of palmetto logs and sand, and when Clinton's fleet appeared off the harbor, the fort was only partially complete.[49]

In late June 1776 the British landed on unfortified Long Island (now the Isle of Palms), east of Sullivan's Island. Clinton planned to have his men wade across an inlet to Sullivan's Island at low tide, after the British fleet leveled the palmetto fort with an intense bombardment. But the inlet turned out to be too deep for wading, and the palmetto logs were so spongy that they absorbed cannonballs. The fort did not fall.

Lord Campbell, a Royal Navy veteran, had spent four months of his governorship on a British warship. When the battle began he was aboard the fifty-gun *Bristol*, the fleet flagship. He volunteered to take command of its gun deck. Patriot fire concentrated on the *Bristol*, the cannon hits producing arrowlike splinters that caused painful, often lethal wounds. A lucky cannon shot severed the ship's cable, which controlled its swing at anchor. Accurate Patriot fire raked its hull and rigging fore and aft.

The *Bristol*'s captain, John Morris, struck several times, stayed on the quarterdeck until his right arm was shot off. He died in a few days. Campbell was thrown to the deck by the blast of a cannonball. He never recovered from his injury, and died two years later in England.[50] The captain of HMS *Experiment* lost his left arm but survived. Forty men of the *Bristol* were killed; the *Experiment* lost twenty-three men. Of the more than 120 who were wounded, most would not survive.[51] Seventeen Americans in the fort were killed and twenty wounded.

The battered British fleet withdrew to the outer harbor, and after waiting for favorable winds, withdrew in mid-July. It then sailed north to join the British attack on New York.

In Virginia, Dunmore had been much more successful than Martin in leading armed attacks on Rebels. He had made his first move against rebellion on the night of April 20, 1775, a month after Patrick Henry declared, "Give me liberty or give me death!" A party of Royal Marines, carrying out the governor's order, slipped into the Virginia capital of Williamsburg, entered the powder magazine, disabled muskets that were stored there, and carried off fifteen half barrels of gunpowder to the warship HMS *Fowey*. The next morning, drums boomed on the streets of Williamsburg, rallying protesters, including members of the House of Burgesses.[52]

The mayor and other local officials called on Dunmore and told him he had to return the powder to assure residents that the militia would have powder if their slaves rose in insurrection. Dunmore did return the powder, but he raged: "The whole country can easily be made a solitude, and by the living God, if any insult is offered to me, or to those who have obeyed my orders, I will declare freedom to the slaves, and lay the town in ashes." Dunmore then fled to York, about twelve miles from the capital, and boarded the *Fowey*. [53]

Dunmore made good on part of his threat on November 7, 1775, when he proclaimed freedom for all slaves or indentured servants belonging to Rebels, as long as they "are able and willing to bear arms" and join "His Majesty's Troops."[54]

Dunmore's proclamation stunned Virginia, where there were nearly as many slaves as white persons. Among the Virginians who would lose slaves was Thomas Jefferson. When British forces invaded the state in 1781, twenty-three of his slaves ran away, or, as one entry in his farm book says, "fled to the enemy."[55] Dunmore's move induced nightmares of armed slaves rising in insurrection against their masters. Fear spread to South Carolina, where there were two slaves for each white.[56] The proclamation had a profound effect on the war, transforming countless slaveholders into Rebels and drawing thousands of slaves to the Loyalist side.

In reaction to Dunmore's proclamation, the Continental Army had begun enlisting free blacks, who for a time had been banned from the army. More than one hundred black Americans fought at Bunker (Breed's) Hill. But, when George Washington arrived to take command, he expelled blacks, accepting a resolution of the Massachusetts Committee of Public Safety, which said, "you are not to enlist . . . any stroller, negro, or vagabond" into the regiments of the "Massachusetts Bay Forces."[57] After Dunmore's proclamation Washington partially rescinded his order, allowing free blacks, but not slaves, into the army. Rhode Island, offering freedom to slaves who enlisted with the consent of their owners, later raised what became known as the Black Regiment, which included Indians.[58]

Within a month after Dunmore issued his proclamation, more than five hundred slaves left their masters and became black Loyalists. About three hundred joined Dunmore's Ethiopian Regiment and donned uniforms emblazoned with "Liberty to Slaves" across the chest.[59] Dunmore gathered a force of British soldiers, members of the Ethiopian Regiment, and white Virginia Loyalists, to launch an attack at Great Bridge, a shipping point for nearby Norfolk. Virginia Patriots, with allies from North Carolina, stopped the invaders with a musket barrage that killed or wounded 102 of Dunmore's men.[60] The only Patriot casualty was an officer who suffered a slight hand wound. Thirty-two members of the Ethiopian Regiment were captured and shipped off to be sold in the Caribbean.[61]

On January 1, 1776, three ships in Dunmore's impromptu navy shelled Norfolk, setting the city afire and destroying "a Work of great

Value and publick Utility, with a large stock of Rum and Molasses."[62] After a short, defiant stand on a Chesapeake Bay island, Dunmore sailed away to New York, taking with him survivors of the Ethiopian Regiment.[63]

For General Cornwallis, General Howe, and General Washington, the war would now shift to New York, where more Loyalists awaited a call to arms.

WAR IN THE LOYAL PROVINCE

NEW YORK, APRIL 1776–DECEMBER 1776

This mob . . . searched the whole town in pursuit of tories . . .
they placed them upon sharp rails with one leg on each side; each
rail was carried upon the shoulders of two tall men, with a man
on each side to keep the poor wretch straight and fixed in his seat.
In this manner were numbers of these poor people . . . paraded
through the most public and conspicuous streets in the town.
— *A Loyalist watches Patriots occupy New York City[1]*

When General Washington arrived in New York City from Boston in April 1776, the city was under the control of Patriots. But Royal Navy warships swayed at anchor in the harbor, and the colony was clinging to its Tory reputation: It was "the Loyal Province" and New York City was "Torytown."[2] The men aboard those ships, including Royal Governor William Tryon, were eating food sold to them and delivered to them by enterprising New York merchants. Tories were routinely commuting between ship and shore.

Washington, outraged, learned that Tryon, seven months before, had won an agreement from New York Patriots, who allowed provisions to be delivered to HMS *Asia*, a sixty-four-gun warship intimidating the city. By the time Washington brought the war to New York, Tryon had followed the example of other royal governors by going to sea. Fearing kidnapping, he had fled first to a Royal Navy warship and then to the British supply ship HMS *Duchess of Gordon* in New York harbor.[3] He continued to get water and provisions, along with visits from spies and royal placemen. He was running the Tory underground while the Patriots' provincial congress was the government.

Washington assumed that Tryon was supervising a network of spies and passing their information on to British military officials. Long after the war General Gates's papers yielded a letter from Tryon containing detailed information about Continental Army plans. What Washington did not immediately discover was that his own fate—a decision to either kidnap or kill him—was being discussed by plotters who boarded Tryon's ship as easily as they would have stepped onto a harbor ferry.[4]

Washington issued a proclamation that denounced the "sundry base and wicked persons" who profited from dealing with the British ships, declaring that "if any person or persons shall hereafter presume to have, hold, or to continue to carry on such intercourse, or any kind of correspondence whatsoever, or furnish and supply the said ships of war, and other vessels in such service with provisions and necessaries of any kind, that he or they, so offending, will be deemed and considered as an enemy or enemies."[5]

The commander in chief was combating the double standard of the New York Patriots: harshly treating Loyalists in general while exempting certain merchants from anti-Loyalist policies. The unwieldy, jittery provincial congress had 119 members; at least 19 were Tories, enough of a division to slow down the Patriots' ability to govern.[6]

Washington had seen for himself the Tory-Patriot split in New York when he passed through the city in June 1775 on his way to Cambridge to take command of the Continental Army. On the same day Governor Tryon, who had been on leave in England, happened to return to New York. In the afternoon, when Washington stepped off the boat that carried him to New York from New Jersey, members of the provincial congress and leaders of both Tory and Patriot factions greeted him with huzzahs and flowery speeches.

A few hours later a boat from Tryon's ship deposited the governor in Manhattan, where he received huzzahs from many of the same people who had welcomed Washington. The two generals accompanying Washington—Charles Lee and Philip Schuyler—joined in the welcome, shaking Tryon's hand and wishing him long tenure in his royal post.

The next day Washington, Lee, and their entourage went on their way to war while Schuyler remained behind to arrange for the forwarding of ammunition and other Rebel supplies to Albany for use in the imminent invasion of Canada. Schuyler was lodging with his cousin, whose house was across the street from Tryon's. Donning his splendid new blue-and-buff uniform of the Continental Army, Schuyler decided to call on Tryon. The governor, not wishing to continue the choreography of make-believe, coolly rejected the visit, saying that he did not know anyone named *General* Schuyler.[7]

"Keep a watchful eye upon Governor Tryon," Washington had told Schuyler, adding that he should also "watch the movements of the Indian Agent and prevent, as far as you can, the effect of his influence to our prejudice with the Indians."[8] This was a reference to Col. Guy Johnson, the powerful British viceroy to the Indians who, even as Washington wrote his orders, was recruiting a Loyalist force along New York's northern frontier.

Royal governors had shifted in 1771, Tryon of North Carolina replacing Lord Dunmore as governor of New York, and Dunmore moving to Virginia. Dunmore did not take the switch gracefully, for he believed that his noble title trumped Tryon's mere military background and familial links with the British aristocracy.

Tryon liked New York, especially its Loyal Province status. He had been tough enough to shoot and hang dissident constituents in North Carolina, and he was ready to do the same if rebellion came to his new colonial post. Tryon had served as an officer in the elite Foot Guards and had been wounded in France during what the French and British called the Seven Years' War. His sister was lady-in-waiting to Queen

Charlotte, wife of King George. His wealthy wife was the daughter of a merchant who, because of his ties to the East India Company, had served as governor of the Bombay Colony.[9]

Before Tryon took over as governor, back in the Stamp Act days, New York Sons of Liberty had followed the lead of compatriots in Massachusetts. As soon as news of the Lexington and Concord battles reached New York City, the Sons had stoked the rebellion. They broke into the armory at City Hall, carrying off more than five hundred muskets, along with hundreds of bayonets and boxes of ball cartridges. They responded to a recommendation from the Continental Congress by convening a provincial congress. Acting as New York's real government, the provincial congress moved aggressively at first, taking control of militias, manufacturing gunpowder, and even hiring gunsmiths to produce gun barrels, bayonets, and musket ramrods.[10]

But revolution in the New York Colony was slowed down by a large Tory presence. Adjoining New York City was another Loyalist stronghold: Queen's County (which then encompassed what are now Queens, Brooklyn, Long Island's Nassau County, and part of Suffolk County). In September 1775, two Patriot militias, ordered to enter the county to disarm the Tories, called off the mission as too dangerous. There were also large and militant Tory populations in Poughkeepsie, in Albany, and on the frontier of the New York–Canada border, where Col. Guy Johnson was ruling over an ever-growing and well-armed Loyalist realm.[11] By one estimate, half of the colony's two hundred thousand people were Loyalists, most of whom kept their allegiance to themselves.[12]

In the region dominated by the great harbor of New York City, Tories had long wielded enormous influence, especially through the machinations of the De Lancey family. De Lancey Toryism evolved from political squabbling with Whigs to the commissioning of family members in Loyalist regiments that would bear the De Lancey name and carry it onto battlefields from New York to Yorktown. Oliver De Lancey, a member of the Governor's Council, met with Tryon aboard the *Duchess of Gordon* as part of a floating Loyalist government, not only preparing to resume power after a British victory but also mobilizing Tories to fight the Rebels in battle.[13]

The New York Patriots who had forced Tryon to leave dry land faced a complex task. In March 1776, for example, the Continental Congress recommended the disarming of all "who are notoriously disaffected to the cause of *America*, or who have not associated, and refuse to associate to defend by arms these United Colonies." The New York Committee of Safety reacted by passing along its own recommendation to its Patriots: Ask for an oath to defend New York against the British "until the present unhappy controversy between the two countries shall be settled." As for the disarming, the Committee of Safety mildly urged "all possible prudence and moderation."[14]

Now came Washington and his ten thousand troops, introducing New Yorkers—both Patriots and Tories—to the reality of war in their colony.

In anticipation of a British assault, Washington's troops built fortifications on the Battery at the southern tip of Manhattan, on Brooklyn Heights, and on the shores of Long Island. The defenses did not reassure New Yorkers. Nor did the Continental reinforcements, a motley force streaming in from Pennsylvania, Maryland, Virginia, and New Jersey. New Yorkers saw menace on all sides. The Continental Army's invasion of Canada had failed, leaving a route for a British attack from the north. The geography of New York Harbor offered another threat: Washington did not have the ships needed to defend a city whose every shore lay open to the world's mightiest navy.

Patriots beyond New York bristled at the lenience being shown British sympathizers in a city girding for war. Frustration inspired the taunting and torturing of Tories, especially by newly arrived troops from other colonies. "Here in town very unhappy and shocking scenes were exhibited," Gustavus Shewkirk, a Loyalist who was a Moravian minister, wrote in his diary. "On Munday night some men called Tories were carried and hauled about through the streets, with candles forced to be held by them, or pushed in their faces, and their heads burned; but on Wednesday, in the open day, the scene was by far worse; several, and among them gentlemen, were carried on rails;

some stripped naked and dreadfully abused. Some of the generals . . .
had enough to do to quell the riot, and make the mob disperse."[15]

In October 1775 the Continental Congress had made a drastic recommendation for Patriot governments like the New York provincial congress: "arrest and secure every person who, going at large,
might, in their opinion endanger the safety of the colony or liberties
of America."[16] Shortly later, Washington gave his endorsement to the
Tory roundup in a letter to Jonathan Trumbull, governor of Connecticut, urging him to seize dangerous Tories in his state.[17] Isaac ("King")
Sears, a volatile New York Patriot and somewhat of a provocateur,
took it upon himself to carry out the arrest-and-secure order.

Sears went to Connecticut, rounded up about one hundred mounted
Patriots, and led his posse across the border to arrest three leading
New York Tories. One of them was the Reverend Samuel Seabury,
rector of Saint Peter's Anglican Church in Westchester (now The
Bronx). He was known to be the anonymous author of *Letters of a
Westchester Farmer*, in which he had written, "[I]f I must be enslaved,
let it be by a KING at least, and not by a parcel of upstart, lawless
committee-men. If I must be devoured, let me be devoured by the
jaws of a lion, and not gnawed to death by rats and vermin!" He had
joined with other New York Loyalists in a vow to support the king
"at the hazard of our lives and properties."[18] To Sears, the Reverend
Seabury was a major catch.

Seabury later complained that the Patriot raiders had pawed through
his desk, thrust a bayonet through his daughter's cap, and cut up the
quilt she was making. Some of the raiders hauled Seabury and their
two other captives off to prison in Connecticut. The rest had a new
target: the printing shop of James Rivington, at the foot of Wall Street
in Manhattan. They destroyed his press, and carried off his type. He
called himself "Printer to the King's Most Excellent Majesty" and published *Rivington's New-York Gazetteer*, a pro-Tory newspaper that bore
the royal arms on its masthead.[19]

Rather than cheer for the raid and the silencing of the *Gazetteer*,
the New York Provincial Congress indignantly protested to the Connecticut governor, Jonathan Trumbull. Deciding to keep the peace

between his state and New York, Trumbull ordered the release of the three prisoners.[20]

While moving against public Tories like Seabury, neither the provincial congress nor Washington realized the size and effectiveness of the Tory underground that Tryon was running from the *Duchess.* Tryon directed an organization so well connected with Tory supporters ashore that a New York shoemaker could run a pick-up-and-deliver service repairing the shoes of Royal Navy sailors. Agents stationed on the *Duchess* routinely carried messages to and from prisoners held in Rebel jails.[21]

The Continental Army did not yet have a formal intelligence-gathering organization, but talkative prisoners can sometimes substitute for counterintelligence agents. On the evening of June 22 Lt. Col. David Mason wrote a terse report to his commanding officer, Maj. Gen. Henry Knox. "I just Recd Intelligence from a Gent in the City," Mason said. He had been told that Lara Fraga, a Continental Army private jailed on charges of attempted counterfeiting, had important information. Fraga claimed that a number of men "have inlisted in the minesterall Troops," as the British Army was sometimes called. Fraga, who was in an artillery unit commanded by Capt. Alexander Hamilton, said he would point out the men whom he knew had gone over to the British.[22]

That jail-cell tip led to the discovery of a widespread plot whose mastermind was Tryon. His principal assistant was David Matthews, the Tory mayor of New York, described by a Tory writer as "a person low in estimation as a lawyer, profligate, abandoned, and dissipated, indigent, extravagant, and luxurious, over head and ears in debt, with a large family as extravagant and voluptuous as himself."[23] When Continental officers and Patriot civilians began to unravel the plot, they found that Tryon's key agent on board the *Duchess* was a "Mulotto Coloured Negro dressed in blue Cloathes."[24]

Few knew that the man in blue was John Thompson, a freeborn black servant of Edmund Fanning, Tryon's private secretary and close friend. Fanning had been a key aide to Tryon in North Carolina. As a corrupt judge, he was the type of royal magistrate who inspired the rise of the Regulators.[25] Tryon depended upon Thompson to pass out money and instructions to Matthews. The cash was used to bribe Continental soldiers and to buy weapons that were to be distributed to Tories when they rose against fellow Americans.

When the battle for New York began, the bribed soldiers were to change sides and fight for the British as guerrillas. They would blow up magazines, disrupt Continental Army strategic plans, and seize a battery of artillery—probably Hamilton's—to turn the cannons against their comrades. A murky phase of the plot involved the kidnapping or assassination of officers, including Washington. Many aspects of the plot were kept secret at the time, presumably to prevent local Patriots from realizing how vulnerable Washington was and how susceptible Continental soldiers were to bribery.[26]

At Washington's request the provincial congress's newly created Committee to Detect Conspiracies sent a detachment of troops to arrest Matthews and confiscate his papers. John Jay, the future chief justice of the Supreme Court, turned a Wall Street tavern into a hearing room and questioned witnesses rounded up by Continental soldiers. Isaac Ketchum, a jailed counterfeiter, testified that two soldiers, Sgt. Thomas Hickey and Pvt. Michael Lynch, had attempted to recruit him for the British. Like Fraga, they were in jail on suspicion of counterfeiting. As members of the Life Guard unit that protected Washington, they were uniquely placed for a kidnapping or assassination attempt.

Hickey, an Irish-born deserter from the British Army, was quickly charged with "exciting and joining in a mutiny and sedition, and of treacherously corresponding with, inlisting among and receiving pay from the enemies of the United American Colonies." Apparently for security reasons, no mention was made of a plot to kidnap Washington or the plan to subvert as many as seven hundred American soldiers. Jay's hearing implicated other soldiers and many Tory civilians. Only Hickey was court-martialed.

Four Continental Army brigades marched to a field near the Bow-ery on the morning of June 28 and formed ranks around a newly erected gallows. Thousands of civilians gathered behind the troops. Eighty soldiers, twenty from each brigade, escorted Hickey to the gallows. "He appeared unaffected and obstinate to the last," a witness wrote, "except that when the Chaplains took him by the hand under the Gallows and bad him adieu, a torrent of tears flowed over his face; but with an indignant scornful air he wiped 'em with his hand from his face, and assumed the *confident look*."[27]

Word of the plot spread through the ranks of the Patriots, unleashing an anti-Tory campaign that focused on Queen's County, especially the areas of Long Island where British landings were anticipated. Action against Queen's had been simmering since 1775 when most Queen's voters refused to send delegates to the New York Provincial Convention.

King Spears had tried to force Long Islanders to make their commitment known by having them sign an oath, swearing by the "Almighty and Tremendous God" not to "convey any intelligence" to the British and to do nothing "to intimidate or dissuade other men, from embarking in the cause of their country and liberty."[28] Nothing much came of Spears's campaign, but it did launch a war of oaths that both the Patriots and the Loyalists would wage alongside the shooting war. When territory was conquered by one side or the other, residents could expect a proclamation demanding that they swear fidelity to the latest conqueror.

The Continental Congress rejected a military proposal to round up all the Tories and exile them to a place where they could not harm the Patriot war effort. The idea was unreasonable because of the size of the Tory population. The roundup proposal was modified to a recommendation that known Tories be disarmed. In one swoop on Long Island, Continental troops gathered about two hundred muskets and ammunition.[29]

Washington, anticipating a British invasion, later sent men to Long Island with orders to root out Tory leaders. But the hunt did little to tame the Tories. Brig. Gen. Nathanael Greene, commander of Continental forces on Long Island, joined other ranking officers in a report that said: "With regard to the disaffected inhabitants who have lately been apprehended, we think that the method at present adopted by the County Committee, of discharging them on their giving bonds as a security for their good behavior, is very improper and ineffectual."[30]

A more detailed description of the Tory hunt came from a contemporary Tory historian, who wrote that "all fled and hid themselves in swamps, in woods, in barns, in holes, in hollow trees, in corn-fields, and among the marshes. Numbers took refuge in the pine barrens in Suffolk, while others in small boats kept sailing about the Sound, landing in the night, sleeping in the woods, and taking to the water again in the morning."[31] One of the hunted who escaped would have been a prize catch: John Harris Cruger, who was married to Ann De Lancey, the daughter of Oliver De Lancey. Cruger later found his way to General Howe, who commissioned him as a lieutenant colonel in one of the major Loyalist military units, the De Lancey Brigade.[32]

On July 4 the Continental Congress in Philadelphia voted, by a unanimous tally of the delegates of twelve colonies, to adopt the Declaration of Independence. The New York delegates, unsure of their colony's sentiments, did not vote to accept the Declaration until five days later. Coincidentally, on that same day Washington ordered his brigade officers to pick up copies of the Declaration at the office of the adjutant general and have it read in full to their men. Civilians gathered around the troops, listening to the words that indicted the king and severed the colonies from his royal rule.[33]

One sentence in the Declaration would soon come to life across the Narrows: "He is at this time transporting large armies of foreign mercenaries to complete the works of death, desolation and tyranny." Anyone looking toward Staten Island could see a forest of Royal Navy masts. More ships were coming, and among them would be transports carrying to America thousands of foreign mercenaries—Germans, drawn from many principalities but all known as Hessians.

"He has plundered our seas, ravaged our coasts, burned our towns, and destroyed the lives of our people." As New Yorkers heard the words of the Declaration, many of them pictured a golden King George III, enshrined on a golden horse and wearing a golden laurel wreath as an incarnation of Marcus Aurelius. When the last words were spoken, a mixed mob of soldiers and civilians headed for Bowling Green, at the foot of Broadway, where the golden statue stood.

They slung ropes around the statue—four thousand pounds of gilded lead—and pulled it off its pedestal. They cut the fallen king and his horse into pieces and chopped off the king's head, putting it aside with the intention of impaling it later on a pole. Most of the pieces were loaded onto a wagon and taken to Litchfield, Connecticut, where Patriots cast from the royal lead more than 42,000 bullets. Before the head could be exhibited, however, Tories stole it and buried it to hide it from the desecrating Rebels.[34]

Anglican clergymen agreed that praying openly for the king was too dangerous to them and their congregations. So they shut their churches rather than continue services without the prayers for the king. "To have prayed for him had been rash to the last degree—the inevitable consequence had been a demolition of the churches, and the destruction of all who frequented them," one of the clergymen wrote. Even closing churches "was attended with great hazard; for it was declaring [in] the strongest manner our disapprobation of independency, and that under the eye of Washington and his army." Later, when the Continental Army began its retreat from New York, the provincial council authorized Washington to remove and melt down all bells from Anglican churches because the council did not want "the fortune of war . . . to . . . deprive this State . . . resource for supplying our want of cannon."[35]

Ever since late June the British troops had been streaming from their transports and setting up camps on Staten Island. There was no armed resistance. Local Tories, as Patriots expected, welcomed the occupation. Many who had previously posed as Rebels were among "the first to join the British as soon as they appeared in force," Elias Boudinot, a young New Jersey lawyer and a future president of the Continental Congress, noted in his journal.[36]

Desperate Tories from New York and Connecticut, telling of friends and relatives who had been jailed and mistreated by the Rebels, arrived on Staten Island in search of refuge. Flight to the Loyal Province would continue throughout the Revolution. And Tory civilians were not the only refugees. Noting the influx of Continental Army deserters, a Staten Island man wrote a friend in London: "Some of their Rifle-men have joined our army, and many more are watching a convenient opportunity to come over."[37]

Washington's force by then numbered some twenty-three thousand men. But disease and desertion kept whittling down the number of able-bodied troops to about nineteen thousand, scattered about fortifications along New York and New Jersey shores.[38] While Washington maintained control over stretches of New York and Connecticut territory, New York's islands—Manhattan itself, Staten Island, Long Island—were so well guarded by the Royal Navy that they became not only British Army strongpoints but also places where Tories could find refuge and raise regiments.

One day in August, when British warships headed up the harbor toward the mouth of the Hudson River, Patriot shore batteries began firing. The British answered with a swift bombardment, damaging a few houses. "This affair caused a great fright in the city," Gustavus Shewkirk wrote in his diary. "The smoke of the firing drew over like a cloud, and the air was filled with the smell of powder."[39]

Now well aware of Tryon's spy and sabotage network, Washington ordered the posting of troops along the harbor shores and sent out patrols in whaleboats—narrow vessels about thirty-six feet long, with pointed bows and sterns, sometimes armed with small cannons. He also ordered his men to seize or destroy every small boat and canoe they saw. Still, numerous Tories managed to reach the British warships. Among them was Oliver De Lancey, who, like Cruger, made his way to join Howe on Staten Island, where he would become Brigadier General De Lancey, commanding officer of his own Tory brigade.[40] Another New Yorker, charged with conspiracy by the Rebels, swore that he had not been recruiting for the Loyalists but "had advised and persuaded men to enlist in the Continental service." He was released—and then became a commander of the De Lancey Brigade.

General Howe had assembled more than thirty thousand men: his own, brought from Halifax; the Redcoats and Hessians brought across the Atlantic by his brother, Admiral Sir Richard Howe, and the troops who had returned after their failed invasion of Charleston, under the command of Maj. Gen. Sir Henry Clinton. On the morning of August 22 Howe landed fifteen thousand men from all of these troops on Long Island, beginning an offensive that drove the defenders off Long Island and won him a knighthood.

Many Long Island Tories who had sought refuge on Staten Island joined the invasion, landing with the British and Hessian troops. They wore red rags tied to their hats to distinguish themselves from Rebels. Some "red rag men" became informers, pointing out real or suspected Patriots and targeting their homes for looting.[41] The red rag men were also prime recruits for the Loyalist regiments being raised in areas now occupied by British troops. A major recruiter was Brigadier Ruggles, who began enlisting Loyalists on Staten Island and Long Island soon after he arrived in New York from Halifax. Some of his three hundred men built camps, cut wood for the British Army's cooking and heating fires, or foraged for hay and food supplies. Others served on privateer ships prowling Long Island Sound.[42]

General Howe did not include the newly raised Loyalist regiments in his invasion force when he landed about fifteen thousand men at Gravesend Bay on Long Island on August 22. But Loyalists proved invaluable to him, for Tory spies found a weak spot in Washington's defense line and quickly passed the intelligence to the British. Howe flanked the Continentals and began to drive them back in the first act of what would be the evacuation of Long Island and a retreat across the East River to Manhattan.

A storm held back Royal Navy warships as Washington's men crossed the river in small, wind-buffeted boats.[43] In his epic nighttime retreat, Washington succeeded in loading all of his 9,500 men into boats, canoes, and sloops—"every kind of water craft . . . that could be kept afloat"[44]—bound for Manhattan.[45]

Of Washington's men 1,012 were killed or wounded on Long Island, compared with British losses of 392.[46] Washington lost more men as soon as his army reached Manhattan and deserters began to disappear. "Great numbers of them have gone off, in some instances almost by whole Regiments, by half ones and by Companies at a Time," Washington wrote to the president of Congress.[47]

As four thousand British troops made an amphibious landing at Kips Bay (near the foot of today's East Thirty-third Street), Washington's broken army retreated north, pursued by Howe at a leisurely pace. Accompanying Howe as both a civil and military aide was Governor Tryon, finally ashore from the *Duchess of Gordon*. When Washington reached the dense woodland of Harlem Heights,* he dug in, creating a triple defense line of trenches and earthworks to hold off the British and gain time.[48]

Washington had wanted to burn down New York City, which he saw as a nest of Tories, but Congress had refused to give him permission. On the night of September 20 flames had swept the heart of the city, blackening a mile-long path and destroying nearly five hundred homes. Enraged Loyalists and British soldiers attacked suspected arsonists, hanging at least two and tossing another into the flames. Some Loyalists blamed vengeful Patriots. The modern verdict is that the conflagration began with an accidental fire in a dockside house.[49]

Loyalists celebrated the liberation of New York City, which would be the capital of Tory America throughout the war. As British troops marched down Broadway, a two-way exodus was under way: Patriots out of the city and Loyalists into it. "Joy and gladness seemed to appear in all countenances," the Reverend Shewkirk wrote in his diary, "and persons who had been strangers one to the other formerly, were now very sociable together, and friendly." People were painting the letters *GR*, for *George Rex*, on Rebels' houses so that, as Shewkirk gloated, "all the houses of those who have had a part and a share in the Rebellion were marked as forfeited."[50]

---

* Now called Morningside Heights, near Columbia University.

Loyalist military units patrolled the streets of New York City while Howe and his staff planned strategy. He and his successors governed New York as an occupied, rather than liberated, Loyalist city—"in fact, a Garrison," a British official said.[51] Soldiers turned houses into barracks and Presbyterian churches into stables. High-ranking British officers appropriated the finest homes, vacated by the Rebels. The poorest civilians, many of them black Loyalists, moved into the burned section, putting tents over the foundations of gutted buildings, creating a large neighborhood called "Canvas Town."[52]

People endured food shortages and inflation: Flour that cost twenty shillings a barrel in 1775 would cost seventy shillings by 1781; a barrel of pork doubled in price. Smugglers and privateers of both sides so disrupted commerce that the only coffee and sugar in the city came from the holds of blockade-running Rebel ships captured by British privateers. Firewood became so scarce that residents walked on the ice of the frozen harbor to tear apart derelict ships.[53]

Howe enjoyed New York, where he presided over a fine table, spent evenings with fellow gamblers, and had the company of the lovely wife of Joshua Loring, a Massachusetts Loyalist. Elizabeth and Joshua Loring had gone from Boston to Halifax with the British Army. By the time Howe took his army to New York, she was Howe's mistress. Loring himself certainly benefited from his wife's relationship with Howe. The grateful general made Loring the commissioner of prisoners of war, a post that promised a fortune in bribes from sellers of provisions.[54]

Howe drew other Americans into his military command. Edward Winslow, whose Loyalist life had begun with the demise of the Old Colony Club in Plymouth, had fled Boston with Howe, offered his services in Halifax, and now in New York was Lieutenant Colonel Winslow, muster master general of all Loyalist forces in North America. Serving with him as provincial commander in chief was Governor Tryon, an old soldier who eagerly changed roles from civilian to martial and began planning military expeditions.[55]

Winslow looked prosperous and resplendent in his uniform—"a blue coat, scarlet cape, and a scarlet lining, with plain white buttons."

But he was in great need of extra cash. After learning that Rebels had given his aged father the choice of joining the Continental Army or going to jail, Winslow had paid a fine to release his father and then provided the money to bring both his parents and his two spinster sisters to New York, where they lived with him, his wife, and their three young children.[56] Loring and his brother-in-law (one of the socially prominent Lloyds of Lloyd Neck, Long Island) had founded a company for selling rum and wine to Howe's officers. To get some extra income, Winslow, who traveled often to Loyalist outposts garrisoned by thirsty officers, became a partner in the enterprise.[57]

Among the Tories sojourning in New York was Lord Dunmore, back in the colony he had once governed. He leased a house on Broadway for the winter and would later sail to England.[58] Dunmore took some members of his Ethiopian Regiment with him to New York. The unit, decimated by smallpox and other diseases during its service in Virginia, was disbanded in New York, but many of the black Loyalists continued to serve in guerrilla units in New York and New Jersey.[59]

A new military unit awaited the regiment's veterans: the Black Pioneers, founded by General Clinton during his aborted 1776 invasion of North Carolina. After Clinton's fleet had gone up the Cape Fear River, the British controlled the area around Wilmington for two months. During that time, about seventy slaves, responding to Dunmore's freedom proclamation, approached the ships to enlist. One of Clinton's officers, Capt. George Martin of the Royal Marines, organized them into a Loyalist military organization called the Black Pioneers. Clinton admired the ex-slaves. He took only one Loyalist unit—the Black Pioneers—with him when he led the invasion and occupation of Newport, Rhode Island, in the fall of 1776.[60]

In the British Army, a pioneer was a soldier-engineer who cleared grounds for camps, dug latrines (called necessaries), and built bridges and fortifications. Separate detachments served in various regiments so that the black soldiers' activities usually did not appear in regimental histories. The scant records on the Black Pioneers indicate that they served as spies, couriers, and guides. They also wrangled livestock, mended uniforms, and played in army bands.[61]

So much Loyalist recruiting was going on among blacks that a Connecticut slaveholder noted the phenomenon in a *Connecticut Courant* advertisement:

TO BE SOLD

This country born NEGRO WENCH, 24 years old, now pregnant, and bids fair to make more recruits for Lord Dunmore. Any person, or family, where that is esteemed no fault, can make an easy bargain with her present mistress.

Enquire of the PRINTERS[62]

Washington chose as his Harlem Heights headquarters the country mansion of Roger Morris, a former British Army officer. Morris had abandoned the home when he and his wife, along with other prominent New York Tories, fled to England before the arrival of Continental troops.[63] The home, on a hill with a strategic view of the Harlem and Hudson rivers, was chosen with a soldier's eye for terrain.[64] The house also held poignant memories for Washington.

In 1756, on the eve of the French and Indian War, when Washington was a young officer in the Virginia militia, he twice stopped in New York to visit Beverley Robinson, a friend he had known back home. Robinson had moved to New York from Virginia and prospered by becoming a merchant partner of Oliver De Lancey and the husband of Susannah Philipse, heiress of a great Hudson River estate. When Washington dropped by, he met Susannah's beautiful sister Mary.

At the time romantic stories circulated that he was smitten, had seen her again and later proposed, only to be turned down. In 1758 Mary married Morris, who had known Washington during the French and Indian War.[65] Now, eighteen years later, Washington was the temporary master of their house; Mary and Roger Morris were among the

Loyalist refugees in England; and Robinson was an enemy, a Loyalist leader spying for Howe and raising a Loyalist regiment.[66]

Washington lacked the kind of spy network that served Howe. Hoping to get some actionable intelligence on Howe's future strategy, Washington sent a large raiding party across Long Island Sound to pick up Tories for interrogation. A British naval patrol spotted the raiders, attempted to capture their boat, and thwarted the operation. Washington then turned to a veteran of the Battle of Bunker Hill, Lt. Col. Thomas Knowlton, the commander of the Ranger regiment, a new, elite reconnaissance unit. Washington asked Knowlton to find a volunteer for a spy mission.[67]

Knowlton's officers refused to volunteer for a task that they considered dishonorable. One of them said that he was willing to die in battle, but he refused to be caught as a spy and then be "hung up like a dog."[68] But a newly arrived twenty-one-year-old captain named Nathan Hale was troubled by the officers' response. Hale turned to Capt. William Hull, a Yale classmate, for friendly advice. Hull tried to talk him out of volunteering. But Hale believed that he could not avoid a call to service. He told Knowlton that he would accept the mission.

A sloop sailing out of Norwalk, Connecticut, deposited Hale ashore at Huntington, Long Island, on the night of September 15–16. He gave himself the cover of a Connecticut schoolteacher who, like so many real Loyalists, was fleeing Rebel persecution. He wore a brown suit, a round broad-brimmed hat, and had removed the silver buckles from his shoes so he looked more like a poor refugee. As teaching credentials he carried his Yale diploma. He did not seem to realize that "Nathan Hale" on the diploma might alert any curious British officer who knew that Maj. Samuel Hale was General Howe's deputy commissioner of prisoners. Samuel, a Loyalist from New Hampshire, was Nathan's first cousin.[69]

Hale apparently spent about five days as a spy before he was caught. The only British document containing information about the end of his mission is an orderly book kept by a British officer. There is one sentence, dated September 22: "A spy from the Enemy (by his own full Confession) Apprehended Last night, was this day Executed at 11 o'Clock in front of the Artilery Park."[70]

Most accounts of Nathan Hale's execution assume that Samuel Hale had recognized his cousin and betrayed him. But a dramatically different narrative surfaced in 2000 when the Library of Congress received a manuscript history of the Revolution, written during or shortly after the end of the war. The author was Consider Tiffany, a cantankerous, well-educated Tory under house arrest in the town of Barkhamstead in northwestern Connecticut. The manuscript had been handed down by one Tiffany generation to the next until a family member donated it to the library. Because the manuscript was written by a Tory, the library said, it presented "a point of view virtually absent from the literature of the American Revolution."[71]

According to Tiffany, Lt. Col. Robert Rogers, commander of Rogers' Rangers, was looking for American spies on Long Island when he spotted Hale as a suspect. Rogers, posing as a spy for the Patriots, "invited Captain Hale to dine with him," Tiffany says, continuing:

> Capt Hale repaired to the place agreed on, where he met his pretended friend, with three or four men of the same stamp, and after being refreshed, began the same conversation as hath been already mentioned. But in the height of their conversation, a company of soldiers surrounded the house, and by orders from the commander, seized Capt Hale in an instant. But denying his name, and the business he came upon, he was ordered to New York. But before he was carried far, several persons knew him and called him by name.

Those several persons were probably some of the many Connecticut Tories who had fled to New York.

Details of Hale's death and an enduring story about his last words— "I only regret that I have but one life to lose for my country"—came in an odd way. On the evening of September 22, Capt. John Montrésor, Howe's chief engineer, approached Washington's lines under a flag of truce. Montrésor—creator of the map that Arnold had used in the invasion of Canada—had spent more time in America than most British officers. He had been wounded in the French and Indian War and later lived for a time in America. He owned an East River property (Montrésor Island, now Randall's Island) and had married a stunning New York woman; they both had their portraits done by John Singleton Copley.[72] Perhaps it was Montrésor's connections with America, along with his basic decency, that inspired him to pass on the tragic news about Hale.

Montrésor's journals show him to be a faithful subject of the king and an officer disgruntled by the army's failure to promote him. He had been in many battles, had built many fortifications—and had helped New York Tories take and hide the head of the fallen statue of King George III. As soon as the British began occupying New York, he had the relic dug up and arranged for it to be delivered to Lord Townshend, whose notorious Townshend Acts were the roots of the Revolution. Montrésor said he sent the head to Townshend "in order to convince them at home of the Infamous Disposition of the Ungrateful people of this distressed Country."[73]

Under the flag of truce Montrésor said that General Howe had ordered him to present General Washington with a formal protest about the use of an inhumane weapon: musket balls with pieces of nails inserted into them. Many had been found when the British searched abandoned American quarters in New York City. Washington officially apologized for the "wicked and infamous weapon," and Montrésor's mission was over.

But he lingered so that he could, unofficially, report Hale's execution, which he said he had witnessed (though the hanging is not mentioned in his journals). Captain Hull, who had tried to discourage Hale from the mission, was at headquarters. He heard the British officer's news, introduced himself, and got the details of Hale's final hours: that he had been sentenced to death without a trial; that he had not been allowed a Bible or a visit from a clergyman; that he had uttered those storied last words. After the war Hull reported what he had been told. Without Montrésor, we might never have known about Hale's noble death.[74]

Tiffany's manuscript, bringing together Rogers and Hale, adds a new piece to the puzzle of Hale's capture. The account also illumines Rogers's role in the New York campaign to create Loyalist regiments.

Rogers, born in 1731 in northeastern Massachusetts to Scotch-Irish parents, had gone from a farm in New Hampshire to fame in the French and Indian War. He became the commander of the Rogers' Rangers, warriors who fought in the wilderness, scouting, ambushing, and introducing the British Army to guerrilla warfare. Rangers wore green jackets and moccasins, and stuck tomahawks in their belts as advertisements that Indians were not the only scalpers.[75] After the war Rogers returned to New Hampshire but soon appeared in England, where he publicized himself in bestselling journals, got into debt, and, after a stint in debtor's prison, returned to America in 1775.

Soon after his arrival the Committee of Safety in Philadelphia arrested and briefly jailed him. The committee released him after giving him a parole that said he "solemnly promised and engaged on the honor of a gentleman and soldier, that he would not bear arms against the American United Colonies in any manner whatsoever, during the American contest with Great Britain." He then wrote to Washington, saying, "I love America; it is my native country and that of my family, and I intend to spend the evening of my days in it." Washington, suspecting that Rogers was a spy, rejected his request for a permit to visit American military encampments. Rogers then slipped away, passed through American lines, and offered his services to the British Army, which commissioned him as a lieutenant colonel.[76]

In August 1776, making his headquarters on Staten Island, Rogers began raising the Queen's American Rangers, a Loyalist regiment of about four hundred officers and men, most of them from Westchester County and Long Island.[77] In a printed circular he promised recruits "their proportion of all Rebel-lands," a promise he had no authority to make.[78] But he had discovered what would later become a standard recruitment pledge. British officials came to realize that Loyalists would more speedily take up arms if they were promised a grant of land as a reward. And, as one officer put it, land grants would "at once detach from the Rebels, the common Irish and other Europeans who make the Strength of their armys."[79]

Tryon, though still officially governor, was spending most of his time trying to find the five thousand armed Loyalists he had promised Howe. Tryon rallied several recruiters, including Rogers, De Lancey, Robinson, and other leading New York Loyalists. Fanning, Tryon's faithful secretary, became a colonel in command of a battalion in Brigadier General De Lancey's brigade.[80]

Encounters between Patriots and Loyalists often ended swiftly and fatally. One of Rogers's Westchester recruits, Capt. William Lounsbury, was accused by the Committee of Safety of leading a group of Tories who spiked Continental cannons in Westchester County. Rebels tracked Lounsbury down and told him to surrender. The Tories with him fled but he stood his ground. When the Patriots threatened him with bayonets, he tried to defend himself with a club and was fatally stabbed. In his pockets were found a commission signed by Rogers and a muster roll of the men he had enlisted, all of them potential targets of Rebel retribution.[81]

By December 1776 seven hundred Rogers' Rangers were raiding Patriot outposts in Westchester. And Colonel Fanning had a commission to raise two more Ranger battalions.[82] Recruitment was so successful that the Committee of Safety appealed to Washington for help. "Nothing can be more alarming than the present situation of our State," the letter said. "We are daily getting the most autheritick intelligence of bodies of men inlisted and armed, with orders to assist the enemy. We much fear that those cooperating with the enemy will seize such passes as will cut off all communication between the Army and us, and prevent your supplies. We do not trust any more of the Militia out of this County."[83]

Nor, certainly, could the Patriots trust the great Hudson River Valley landowners and their tenants—especially the tenants. In 1775,

New York Patriots circulated a petition, called Articles of Association, supporting the actions of the Continental Congress. Robert R. Livingston, Jr., owner of an immense Hudson River estate, noted that "many of our Tenants here refused to sign . . . and resolved to stand by the King." One tenant, Livingston said, vowed that if he were armed and put in a Rebel militia, "the first person he would shoot would be his captain."[84] In those days the tenants had hoped for British victory and postwar land-grant rewards to loyal subjects. Soon, as the war produced defeat after defeat for the Continental Army, their envisioned rewards seemed just over the horizon. Emboldened Tory tenants roamed the valley in bands. Livingston's mother, Margaret Beekman Livingston, wrote, "Some say their number is 4000. . . . They . . . have three boxes of gun powder that has been sent to them by some as bad as themselves."[85]

Frederick Philipse III, who called himself "Lord of the Manor of Philipsburg," presided over an estate that ran along the eastern bank of the Hudson River for about twenty-four miles, from the Croton River to Yonkers and much of the rest of Westchester County almost as far as south New York City. In 1776 he refused to sign an oath of allegiance to the Patriot cause and rallied many of his tenants and his neighbors to the Tory side. A detachment of armed Patriots arrested him and sent him to house confinement in Connecticut. He appealed to the New York Committee of Safety, asserting that he had "done nothing (upon the Strictest Examination) Inimical to the Liberty's of My Country." The Patriots relented and allowed Philipse to return to his manor in Yonkers. But he broke his parole and took his wife and children to New York City. Even though his estate lay within the theoretical control of the Patriots, he continued to receive rents on some of his properties.[86]

By the end of 1776 recruiters had sworn in about eighteen hundred Loyalist soldiers, most of them from Staten Island, Long Island, and Westchester County.[87] Colonel Fanning later became commander of the King's American Regiment, one of the war's most active

Loyalist units. For the chaplain of his regiment, Fanning chose the Reverend Samuel Seabury, who had been the victim of a Rebel kidnapping.[88]

Long Island was a magnet for Loyalist families, especially those who could easily sail there from Connecticut coastal towns across from Long Island's northern shore. Many of these displaced Loyalists clustered around Lloyd Neck or Eaton's Neck, towns that jutted into Long Island Sound, opposite Stamford. Eaton's Neck harbored a large Tory community that included 118 refugees from Connecticut. At first they lived "free from the tumultuous bustling world." But their freedom gave way to martial law, and they were ordered to aid the British Army by gathering wood or moving military supplies. They endured a constantly growing feeling of exile—and fear, both of their Rebel foes and their British Army guardians.[89]

A local civil war soon broke out between the Connecticut refugees and the "whaleboat men"—Connecticut Patriots who sailed across the Sound on raids. They struck the Loyalist communities, plundering homes and kidnapping people for ransom or to swap for prisoners taken by the British or by Loyalists conducting their own raids.

Every Loyalist family was a potential target for looting or abduction.[90] Supposedly British soldiers, quartered on Long Island, protected them. But, as a contemporary Tory later wrote, the soldiers

> Robbed, plundered, and pillaged the inhabitants of their cattle, hogs, sheep, poultry, and in short of every thing they could lay their hands upon. It was no uncommon thing of an afternoon, to see a farmer driving a flock of turkeys, geese, ducks, or dunghill fowls, and locking them up in his cellar for security during the night. . . . It was no uncommon thing for a farmer, his wife, and children, to sleep in one room, while his sheep were bleating in the room adjoining, his hogs grunting in the kitchen, and the cocks crowing, hens cackling, ducks quacking, and geese hissing, in the cellar. Horned cattle were for safety locked up every night in barns, stables, and outhouses. This robbing was done by people sent to America to protect loyalists against the persecution and depredations of Rebels. To complain was needless; the officers shared in the plunder.[91]

* * *

The depredation did not go entirely unpunished. Records of De Lancey's battalions show, for example, that two soldiers found guilty of robbery were sentenced to one thousand lashes each (remitted to five hundred by General Clinton). And two other soldiers, found guilty of robbery, murder, and rape, were hanged.[92]

While Nathan Hale was sailing across the Sound, Colonel Knowlton was killed leading an attack on a British outpost near Harlem Heights. So Knowlton died not knowing what had happened to Hale. The deaths of Knowlton and Hale profoundly changed Washington's attitude about the gathering of intelligence. Instead of relying entirely on military officers, he would use civilians, primarily Patriots who could spy while posing as Tories.[93] Eventually, Washington would oversee an elaborate and productive spy network centered on New York City.

Fearing that Howe would attack him from the rear, Washington decided to withdraw from Manhattan, leaving a force behind at Fort Washington, near the northern end of the island, to block Howe's advance. Connecticut's Governor Trumbull had set up a supply depot at White Plains, a Westchester town about twenty-five miles north of Manhattan. The town became Washington's goal.[94] He also ordered a detachment of Virginia and Delaware Continentals, guided by local Patriots, to the village of Mamaroneck, about seventeen miles south of White Plains.[95]

For some time Tories had been "getting the upper hand" in Mamaroneck, which lay within the De Lancey fiefdom.[96] As a Patriot told the Provincial Congress in November 1775, Tories walked around the village in small, armed groups and warned Rebels that soon "there would be bad times," including kidnappings. The threat seems to have been inspired by the furor over the killing of William Lounsbury, the Tory leader.[97]

On October 21 the Continentals approached Mamaroneck at night, hoping for a surprise attack on the armed Loyalist unit known to be bivouacked there. Due to the local guides' negligence or treachery, the Loyalists discovered the attackers and, in a hand-to-hand clash, two Continentals were killed and fifteen were wounded; about twenty Loyalists were killed or wounded. The Continentals withdrew, taking thirty-six Loyalist prisoners with them.[98]

The skirmish was overshadowed a week later by the Battle of White Plains, which ended in another defeat and another retreat for Washington. But the clash by night at Mamaroneck introduced Patriots to the mixed fighting skills of Rogers's regiment: Some men fought savagely when Rogers directly rallied them, shouting, "Steady, boys, steady! Fire! Fire!" Others, reeling back, submitted to capture. Rogers himself managed to escape with his survivors.[99]

Rogers, exploiting his fame in the French and Indian War as the founder of Rogers' Rangers, promised better pay and better food as he recruited in New York, Maine, and Canada. He even bragged about the skills of the tailor who made the Rangers' uniforms. Gen. Frederick Haldimand, royal governor of Quebec, complained about Rogers, saying, "He has totally given himself up to the lowest debauchery and unworthy motives of obtaining money to gratify it."[100] A few months after the Mamaroneck battle, an inspector general ordered to examine Loyalist units found the Rangers below standard, a verdict that led to Rogers's forced retirement. But under a new name, the Queen's American Rangers, and a new commanding officer, Lt. Col. John G. Simcoe, the unit went on to fight in battles throughout the war.[101]

Fort Washington, on the highest point of Manhattan Island, was thought to be impregnable. But Howe captured the fort, on November 16, 1776, because a Continental officer went over to the enemy. He slipped out of the fort and sent Howe plans showing the fort's vulnerable points. When Howe attacked, the defenders fought gallantly but were overwhelmed. About 150 were killed and wounded and three thousand taken prisoner. British Regulars and Hessians lost about five hundred men, including wounded.

The fall of the fort meant that Patriot authority over the Island of Manhattan had ended for the rest of the war. The traitorous officer, in a petition for a reward after the war, said he had "Sacrificed all I was Worth in the World to the Service of my King & Country."[102] His words reflected the sentiments of many Americans. In December 1776, when Governor Tryon reviewed the royal militia in Queen's County, 820 men were mustered. And when Tryon circulated a statement supporting the Crown, nearly three thousand men signed it.[103] For George Washington, the rise of Tory militancy in the Loyal Province was a foretaste of what awaited him when, following the fall of Fort Washington, he crossed the river and entered New Jersey.

11

TERROR ON THE NEUTRAL GROUND

New York and New Jersey, November 1776–Spring 1777

*Why is it that the enemy have left the New England provinces,
and made these middle ones the seat of war? The answer is easy:
New England is not infested with Tories, and we are. I have been
tender in raising the cry against these men. . . . The period is now
arrived, in which either they or we must change our sentiments,
or one or both must fall.*

—Tom Paine, The Crisis[1]

General Howe sent Lord Cornwallis and 4,500 men in pursuit of
Washington, hoping to trap him in New Jersey and end the war.
Cornwallis first had to take Fort Lee, on the western side of the Hud-
son. Braced for a British invasion, Continental patrols in New Jer-
sey kept watch along the river. But the Continental lookouts ignored
Closter Dock Landing, a patch of shore with a steep, narrow path
leading up the sheer Palisades, about six miles north of Fort Lee. The
Continentals did not believe that an army could climb that steep trail.

Cornwallis and his men did. They landed on the scrap of land from
boats that shuttled between the riverbanks. Then, in a driving rain,
they began their ascent. They had been led to that unguarded landing by Maj. John Aldington of Bergen County, commander of the
Guides & Pioneers, a Loyalist unit raised for reconnaissance missions.[2]

Aldington had a personal stake, for he owned a brewery that the Continentals had taken over to use as a storehouse. But Aldington, called "a zealous Loyalist" by Cornwallis, was not merely interested in recovering his property. Through much of the war he would stay in command of his unit, which was later attached to Beverley Robinson's Loyal American Regiment.[3] With him were two other Tories serving as guides—William Bayard, who ran a Hudson River ferry, and John Ackerson, who owned the Closter dock. In peacetime, farmers used the dock to ship produce to New York City; during the war the site was a popular connection for Tories slipping in and out of New York.[4]

The Guides & Pioneers, who served in several places during the war, rarely mustered more than 150 men at a time. Each man had invaluable local knowledge about terrain and people. Some of the Guides had commissions as officers but did not command the Guides or any other unit. They were spies who hoped that if they were caught the commissions in their pockets would give them the status of prisoners of war and save them from the gallows.[5]

The fall of Fort Washington so jeopardized the weaker Fort Lee that General Washington ordered Maj. Gen. Nathanael Greene to abandon the fort and lead the two-thousand-man garrison toward Hackensack, where Washington waited. Greene was unaware of Cornwallis's bold nighttime climb until morning, when word reached him, probably from a British deserter.[6] He hastened the evacuation and led his men southwestward as Cornwallis began his march on the fort.

When the British reached the fort, they found breakfast teakettles boiling, fifty cannons unspiked, and three hundred tents still standing. Some soldiers—sick or drunk—had been left behind by their fleeing comrades. Food and other stores were scattered about the fort, and knapsacks littered the road that led to Washington's headquarters. He was six miles south in Hackensack, staying in the home of Peter Zabriskie, a Patriot whose mansion on the village green was surrounded by the homes and farms of Loyalists.[7]

About two thousand Continentals were still on the eastern side of

the Hudson because their commander, the arrogant, self-aggrandizing
Maj. Gen. Charles Lee, refused to obey orders that Washington had politely proffered. "There are times when we must commit treason against
the laws of the State for the salvation of the State," Lee explained. "The
present crisis demands this brave, virtuous kind of treason."[8]

General Lee finally crossed the river and moved slowly southward
along a road about twenty miles west of the British Army. Instead of
camping with his men, Lee chose to stay in a tavern about three miles
away. Tories routinely kept British officers aware of the movement of
Lee and his men.

Before his military career took him to colonial America and into the
Continental Army, Lee had been in Portugal as commander of the
British Army's 16th Light Dragoons. Now, fourteen years later, local
Tories gave men of the 16th the location of Lee's vulnerable quarters.
Cornet Banastre Tarleton, a cavalry officer of the lowest commission
(comparable to ensign in the British infantry), began his own illustrious
career in America that day. He would become the commander of the
Loyalist British Legion—later called Tarleton's Legion. At least one of
his Tory guides that day would become an officer in the legion, whose
horsemen, later in the war, would spread terror wherever they rode.[9]

After British cavalrymen quickly routed or killed Lee's sentinels,
they shot up the tavern and threatened to burn it. Lee surrendered
and was taken away, tied hand and foot to a Tory guide's horse and
reputedly still in his nightshirt. As a deserter from the British Army,
he faced execution. But he would live in a comfortable captivity that
would raise questions about his views of treason.[10]

Lee's host, General Howe, treated Lee as a fellow general and gave
him relatively free range in New York City. Lee was confined in the
council chamber in the City Hall and provided with firewood and
candles. He was allowed to order dinner for six with "what liquor he
wanted, and of what kind he pleased," wrote an eyewitness, Tory historian Thomas Jones. "He had the privilege of asking any five friends
he thought proper to dine with him each day. This was all furnished

at the expense of the [British] nation." Robert Hull, who ran a popular tavern on lower Broadway, "waited on him by General Howe's orders, with a bill of fare every morning, and Lee ordered his own dinner and his own liquors. It was cooked at Hull's, and always on the table at the time appointed. His servant had free access to him at all times."[11]

Lee was talkative during his sojourn as Howe's guest. Suspicions arose that he had helped himself by helping the British. But not until the nineteenth century did stunning evidence of his treason appear among Howe's papers: a sixteen-hundred-word "Plan," in which Lee stoked the British faith in a Loyalist rising and gave Howe a strategy that would "bring matters to a conclusion." As second in command of the Continental Army, Lee had great authority in British eyes.

To "unhinge or dissolve . . . the whole system or machine of resistance, or in other terms, Congress Government," Lee wrote, the British should look for aid from Loyalists. "If the Province of Maryland or the greater part of it is reduced or submits, and the People of Virginia are prevented or intimidated from marching aid to the Pensylvania Army, the whole machine is dissolved and a period put to the War."[12] Lee's plan centered on Howe's taking a major force into Maryland via Chesapeake Bay, where he would find pro-Loyalist residents. Howe's subsequent maneuvers seem to have been at least partially based on Lee's advice.

In New Jersey, Washington was racing time as well as Cornwallis, for the enlistments of many Continentals would soon expire. With the addition of Greene's men, Washington had a force of about three thousand troops. But they were "much broken and dispirited men."[13] A Patriot saw the soldiers who had fled from Fort Lee entering Hackensack: "The night was dark, cold and rainy, but I had a fair view of Greene's troops from the light of the windows as they passed on our side of the street. They marched two abreast, looked ragged, some without a shoe to their feet and most of them wrapped in their blankets."[14]

Deciding to march without Lee's men, Washington bade Zabriskie farewell. According to family tradition, Zabriskie asked the general where he was heading. And, the story goes, Washington leaned down from his saddle and whispered, "Can you keep a secret?" Zabriskie assured him that he could, and Washington said, "I can, too."[15]

The story underlined the Patriots' distrust of New Jersey people. "A large part of the Jerseys," Washington bitterly observed, "have given every proof of disaffection that a people can do."[16] As Washington led his troops out of Hackensack and headed toward Newark, the eyes of countless Tories watched. Some of them belonged to members of the big Zabriskie family.

As soon as Washington's troops left Hackensack, young men from local homes and outlying areas began to appear around the village green. They were Loyalists who had secretly enlisted in the Fourth Battalion of New Jersey Volunteers, the state's first Loyalist regiment, which had been raised soon after the British landed on Staten Island.[17] The regiment was one of New Jersey's three major Loyalist units.

Now the men of the Fourth Battalion emerged to receive their weapons and the green uniforms that would give them—and many armed Loyalists to come—the nickname Greencoats. Lt. Col. Abraham Van Buskirk, a former lukewarm Patriot who commanded the Fourth Battalion, appeared with his officers. The overwhelming majority of enlisted men were of Scotch-Irish stock; the officers were all scions of old Dutch families, known as the Tory Dutch.[18]

The members of the battalion could parade around in their new uniforms without fear because Hackensack had become, literally overnight, a Tory village. Worrisome Patriots, especially farmers, tried to earn a living while wondering where and when Tory raiders would strike. Sometimes men of the Fourth Battalion staged small-scale raids, picking up some cattle here, a few horses there. Or there might be a major foraging expedition, when several hundred British troops and members of the Fourth Battalion would put Patriots in jail and then plunder their homes and farms.

Reporting on one of those raids, a Patriot militia officer wrote, "I keep out large patrolling parties every night in that neighborhood for the protection of the inhabitants, but the enemy have so good intelligence of our thoughts and every motion that it is beyond my power to give protection."[19] Patriots retaliated by ambushing small foraging parties. "Not a stick of wood, a spear of grass, or a kernel of corn, could the troops in New Jersey procure without fighting for it," a Tory wrote. ". . . Every foraging party was attacked in some way or another." Each side was losing men in sudden skimishes on this strange, unexpected battlefield called the Neutral Ground.[20]

As Washington retreated into New Jersey, taking the war westward, the allegiance of Loyalists also began to turn. In parts of New Jersey, as Washington would learn, Tories were in the majority and in control. Back in New York there were so many Loyalists in some areas that they pinned down Patriot militiamen who might otherwise be aiding Washington in New Jersey. One Westchester unit, the King's American Regiment, expanded its horizon by going to sea and becoming privateers. They made Lloyd's Neck, the Loyalist refugee community on Long Island, their home port.[21]

Washington assumed that all his movements were being reported to the British by Loyalist spies. He knew that Loyalists stole powder from magazines and drove horses and cattle through the American lines to be sold to the British. From presses operated by the British Army came counterfeit versions of "Continentals," the paper money that helped to finance the Revolution. British operatives passed great wads of the bogus bills to Loyalists, who circulated them in a campaign that depreciated the value of the money.[22] And the phrase "not worth a Continental" entered the American language.

In his writings Washington once alluded to "half tories," saying they might be useful as spies, for, in his low view of Tories, he believed that they were amoral and would just as well work for the Americans as for the British. That idea of half loyalty permeated the areas where the Continental Army had fought and retreated, fought and retreated, passing through defeats from New York into New Jersey.[23] Each defeat produced more Loyalists not only in the area of the defeat but in the territory beyond. New Jersey's government was under Patriot control, but the state's population included thousands of Tories.

In Washington's army fleeing across New Jersey was Tom Paine, whose *Common Sense* had stirred the Rebels and thrust them toward independence. Now, in a dark December of defeats, in "times that try men's souls," when the "summer soldier and the sunshine patriot . . . shrink from the service of their country," he looked around and began to envision what he soon would write in *The Crisis*. He saw an infestation of Tories, and he realized that the time had come when Rebels and Tories would repeatedly fight each other, no matter whether the British Army was present.

There were truly and clearly two Americas, one governed by the British military operating from New York and the other a group of colonies in rebellion but not quite governed. "The Declaration of Independence," wrote Thomas Jones, the Loyalist historian, ". . . was the first act that put an end to the courts of law, to the laws of the land, and to the administration of justice under the British crown. . . . The revolt was now complete. . . . A usurped kind of Government took place; a medley of military law, convention ordinances, Congress recommendations and committee resolutions."[24] Every American now had a choice: to remain a subject of King George III and thus a traitor to a new regime called the United States of America or to support the rebellion and become a traitor to the Crown.

Loyalists by the thousands signed loyalty oaths administered by Tryon, who traveled to territory occupied by British troops. Few people refused to swear allegiance to the Crown.[25] And those who chose the king had another way to show their choice; in taverns and meeting halls throughout New York City, Tryon's recruiters signed up wealthy and well-connected young men for commissions in Loyalist regiments. For enlistments in the ranks there were many—among them farmhands, men without jobs, and ambitious sons turning away from their Rebel kin. The Loyalist recruits were issued weapons and uniforms, usually designed by their regimental commanders.

Eventually New York would send more men into Loyalist regiments than into the Continental Army.[26] Tory Dutch farmers in New Jersey found a lucrative market in the British garrison in New York. The area closest to Manhattan, around Leonia and Englewood, contained Dutch farms whose farmers who could speak English, and this enclave became known as the English Neighborhood, a place where no Patriot could openly live. A New Jersey merchant fleet carried produce and goods to Tory New York. Vessels ranged from schooners to Dutch *pettiaugers*, small ships without keels; they could sail in shallows but had leeboards that could be lowered when a keel was needed.[27]

James Rivington, the Tory publisher driven out of America by Rebels who smashed his press, had returned from England to a New York City much more to his liking. He began publishing *Rivington's New-York Gazetteer*, which produced such recruiting advertisements as this one:

ALL ASPIRING HEROES

have now an opportunity of distinguishing
themselves by joining

THE QUEEN'S
RANGERS HUZZARS,

Commanded by
Lieutenant-Colonel Simcoe*

Any spirited young man will receive every encouragement, be immediately mounted on an elegant horse, and furnished with clothing, and accoutrement, &c to the amount of forty guineas, by applying to Cornet Spencer, at his quarters No. 133 Water Street, or his rendezvous, Hewett's Tavern, near the Coffee-house....

Whoever brings a recruit shall instantly receive two guineas.

Vivant Rex et Regina.[28]

---

* John G. Simcoe, a successor to Lt. Col. Robert Rogers, added the Huzzars (Hussars)—a light cavalry troop—to the Rangers, named in honor of Queen Charlotte, the wife of King George III.

As soon as the British Army took root on Long Island, scores of young Connecticut men sailed across the Sound to enlist. Many described themselves as "Churchmen," Anglicans who equated service for the king with their religious beliefs. The names of men from Fairfield, Stratford, Stamford, Norwalk, New Haven, Newtown, Waterbury, Middletown, and Redding appeared on the musters of the Queen's Rangers, the King's American Regiment, and the Prince of Wales's American Volunteers.[29] By July 1777 the King's American Regiment alone had enlisted twenty-six sergeants, nine drummers and fifers, and 415 rank and file, the military label for noncommissioned troops.[30]

The Prince of Wales' American Volunteers brigade was the creation of Montfort Browne, a prisoner of war. Governor General of His Majesty's Bahamas, Browne had been captured in March 1776 on the island of New Province when Cmdr. Esek Hopkins, commander in chief of the fledgling Continental Navy, in its first amphibious operation, landed 280 marines from whaleboats.

Among the island residents were several Loyalists, described by Browne as "licentious, poor, haughty and insolent."[31] No one offered resistance when the Continental sailors cleared the island's fort of military stores—including eighty-eighty cannons and a ton of gunpowder—and took Browne prisoner.

Browne was placed under house arrest in Middletown, Connecticut, which was considered far enough inland to keep the governor from mischief. But Browne managed to raise a Tory regiment by smuggling out invitations to friends and friends of friends, much as he might have arranged a dinner party on New Province. He dispatched two young Connecticut Loyalists, who easily passed through the porous Continental lines, to New York City, where they conferred with General Howe. He arranged the exchange of Browne for Continental Army general William Alexander (who claimed the disputed title of Lord Stirling despite his Patriot fervor). He had been captured during the Battle of Brooklyn.[32]

Browne set up his headquarters in Flushing, Long Island, and began issuing warrants to recruiters, claiming that he had been commissioned by "His Majesty's Commissioners for Restoring Peace and Tranquility to the Deluded Subjects in America."[33] In a letter to Muster Master General Winslow, Browne said that he had "about eighty or ninety men" who had agreed to serve without pay. Soon, he said, he would get many more, thanks to a nautical recruiter who was sailing along the Connecticut coast in an armed sloop and picking up Browne's enlistees.[34]

In New Jersey, Cortlandt Skinner, a member of one of the state's oldest and wealthiest families, raised six battalions of New Jersey Volunteers. Patriots, unaware that Skimmer had long spied for the British,[35] had offered him command of Rebel forces in New Jersey. He chose instead to accept a brigadier general commission from General Howe.[36] Skinner originally envisioned a force of three thousand men. But four battalions of four hundred men each were raised, and they fought in battles from New Jersey to Virginia.[37] A merchant and shipowner who had been a member of New Jersey's Provincial Congress was commissioned a major and recruited two hundred men for one of the battalions.[38]

Officers and men were outfitted in green uniforms—a frequent color selection for Loyalist units—and became known as Skinner's Greens.[39] Skinner, besides commanding the Volunteers, would continue to spy as British troops marched across New Jersey in pursuit of Washington. He once wrote as his own testimonial that "there was scarcely any Material Information of the Encampment of the Rebel Army which I did not obtain the first Intelligence of."[40]

Loyalists in Bergen County, adjacent to Staten Island, provided the British in New York City not only with food but also with spies and recruits, many of whom Tryon secretly signed up while he was aboard the *Duchess of Gordon*. Those shipboard enlistees were told to return home and tell no one about their enlistments until British troops arrived in New Jersey. This was an unprecedented move, going beyond usual British military doctrine by setting up advance Loyalist units in places that the British Army had yet to invade.[41]

Volunteers signing up for Loyalist regiments were given their equipment and were paid in gold-backed British money, not in the ever-declining currency printed by the Continental Congress. Some Loyalist recruiters promised a prospective soldier a five-guinea signing bonus, rather than the advertised forty guineas, but added the lure of two hundred acres of land, with an extra one hundred acres for his wife and fifty acres for each child. A Patriot enlisting in a Rebel militia typically had to provide his own musket and bayonet, a sword or tomahawk, cartridge box and belt, twenty-three rounds of cartridges, twelve flints, a knapsack, a pound of gunpowder, and three pounds of bullets in reserve. All this was an expensive outlay for a poor farmer.[42]

Stephen De Lancey, of the powerful Tory clan, was commissioned a lieutenant colonel of the Volunteers' 1st Battalion. His commissioning had been delayed by a brief encounter with Patriots. On the evening of June 4, 1776, he and several others had been celebrating King George's birthday at a Tory tavern in Albany, New York. The Committee of Correspondence denounced the birthday party as an "indecent meeting" and ordered De Lancey and five others deported to Connecticut, a frequent venue for the safekeeping of Tories, whose crimes ranged from not accepting Continental currency to taking up arms against the Revolution. Patriots usually charged them for their rooms and meals.[43] When De Lancey was released in December, he went to New York and received his commission.[44]

Each battalion had, in addition to its complement of commissioned officers, a surgeon and a chaplain, all drawn from New York and New Jersey. The most distinguished of the chaplains was the Reverend Charles Inglis, assistant minister at Trinity Church, New York City's most esteemed Anglican congregation. Inglis was a passionate Loyalist who, after the occupation of the city, became an eloquent propagandist.[45] Responding to Tom Paine's *Common Sense*, for example, Inglis wrote, "The Americans are properly Britons. They have the manners, habits, and ideas of Britons. . . . Limited monarchy is the form of government . . . which is best adapted to the genius and temper of Britons; although here and there among us a crack-brained zealot for democracy or absolute monarchy, may be sometimes found."[46]

* * *

Loyalists who enlisted or were commissioned in areas under Patriot control had to make their way to safe grounds in New York City or Long Island. These traveling new soldiers of the "Provincial Corps," as the British Army collectively called the Loyalist units, were treated by Patriots sometimes as spies and sometimes as armed foes. One of the largest packs of recruited Loyalists, numbering about fifty, was assembled in northern New York's Ulster County. At Wallkill, about eighty-five miles north of New York City, militiamen spotted them, and in a brief firefight three Patriots were wounded and the Loyalists got away.

An alarm spread through the countryside. The recruits, supported by local Loyalists, hid out in the woods or in the cellars or barns of sympathizers by day and slogged through creeks and along old trails by night. They had not gone far before a militia patrol found them and captured about thirty. Eleven were accused of "levying war against the United States of America" and five of "aiding and assisting [and] giving Comfort" to the enemy. They were brought before a court-martial ordered by Gen. George Clinton, a former member of the Continental Congress and soon to become governor of New York.

The court-martial, after listening to Patriots who told of encounters with the armed Tories, resolved that "an immediate Example was necessary and requisite to deter intestine Enemys from continuing Treasonable Practices against the State." Fourteen men "were adjudged to suffer the Pains and Penalties of Death by being hanged by the neck until they are dead." But after hearing petitions and statements the provisional state government (the Convention of the Representatives of the State of New York) ruled that only the leader and his assistant were to be executed. The others received various sentences, ranging from immediate parole to confinement until the end of the war.[47]

One of the black veterans brought to New York by Dunmore was a former slave known as Titus. He had run away from his Quaker master in Colt's Neck, New Jersey, and managed to reach Virginia, where he enlisted in Dunmore's Ethiopian Regiment. Under the nom de guerre "Colonel Tye," Titus became a leader of guerrilla warriors who fought on the Neutral Ground.[48]

When the New York Provincial Congress in October 1775 described widespread Tory recruiting as a "conspiracy from Haverstraw [New York] to Hackensack [New Jersey]," the phrase roughly encompassed the Neutral Ground. It stretched along both banks of the Hudson River from above the New York–New Jersey border south to Sandy Hook. In Westchester County, the term referred to the land between the British-held Bronx and American-held Peekskill. The label was ironic, for on this so-called Neutral Ground both sides would fight, not to gain territory but to forage for food and firewood, to demand loyalty oaths, to kill each other in skirmishes—and to spy.[49]

The label was made famous in the postwar generation by James Fenimore Cooper. His Revolutionary War novel, *The Spy*, was published in 1821 with the subtitle *A Tale of the Neutral Ground*. Cooper, who had married into the still-powerful De Lancey family in 1811, wrote the book while living in Scarsdale, a Westchester County town that had been part of the Neutral Ground. Cooper's hero, Harvey Birch, was based, at least partially, on Enoch Crosby, a true spy of the Neutral Ground. Crosby, masquerading as a Tory recruiter, secretly worked for John Jay, chairman of the New York State Committee and Commission for Detecting and Defeating Conspiracies. Jay had the power to send Tories to the notorious "fleet prison," a string of former privateer ships anchored off Kingston, New York.[50]

Cooper never publicly linked the real Crosby and the fictional Birch. But he did say that Jay had told him spy stories, and presumably Cooper learned about Loyalist activities from his wife and in-laws. In his petition for a federal pension, Crosby told his own story, which began in the summer of 1776, when his eight-month enlistment in a Connecticut regiment ended and he found himself in the Neutral Ground. On his way to join another regiment, he met a stranger who took him to be a Tory. Realizing that the stranger "intended to go to the British," Crosby instantly decided to string the Tory along. The talkative stranger told Crosby where and by whom a Loyalist regiment was secretly being raised.

Crosby took his information to a member of the Westchester County Committee of Safety. After being vetted by the committee, Crosby became an agent. Among the Tories he found were some thirty men recruited by Lt. Col. Beverley Robinson, Jr., son of the commander of the Loyal American Regiment.[51] Crosby's true status was withheld from the principal hunter of Tory spies in the area, Capt. Micah Townsend of Westchester, commander of Townsend's Rangers. Townsend did capture Crosby as a genuine spy. And Crosby genuinely escaped, risking his life to preserve his secret identity. The escape helped to establish his reputation among Tories.

Crosby's real and staged escapes, under various names and in various Neutral Ground places, did not raise suspicions. In the Neutral Ground war, many real Tories were captured and did get away from inept Rebel guards. But the ruse could not last forever, and, after nine months as a secret agent, Crosby enlisted in a new regiment and served as a sergeant on regular service.[52]

While Crosby was hunting Tories in the Neutral Ground, a small but brutal civil war was being fought by the Dutch Tories and the Dutch Rebels of New Jersey. British forces, Loyalist volunteers, and Patriot militiamen launched hit-and-run raids, keeping residents jittery. No one ever knew when a Tory or a Rebel might shatter the night. Bergen County was particularly split because of a schism in the Dutch Reformed Church, with some congregations supporting American-trained clergy and the Rebels, while others, backing the more conservative faction, favored the Loyalists. Many spoke only Dutch. When local broadsides on the Revolution were published, they were often in English and Dutch.[53]

An American officer, writing about the "good people of Bergen County," said they "lay greatly exposed to both internal and external enemies, and the internal enemies have a free recourse to New York, the center and head of all British activity in America." Tories, allying with Quakers, won elective offices. Patriot governor William Livingston complained, "I have seen tories members of Congress, judges upon tribunals, tories representing in our Legislative councils, tory members of our Assemblies."[54]

In New York's Orange County, which bordered on Bergen County, a militia leader reported that "matters are come to such a height that they who are friends of the American Cause, must (for their own safety) be cautious how they speak in public" and that some of those "who have been active in favour of our Cause, will soon (if any opportunity offers) be carried down to New York." Patriots who were carried down to New York City faced imprisonment in a prison ship or confinement in the Sugar House, a sugar refinery turned into a dank stone prison where many Patriots died.[55]

Several British prison ships lay at anchor in a small Brooklyn bay. The ships were the dreaded dungeons of thousands of captured Continental Army soldiers, Rebel militiamen, and Patriot civilians. Loring, the commissioner of prisoners, showed little interest in his captives, except as a source of income from contractors' kickbacks. Prisoners were jammed into holds, where so many died of disease or starvation that each day guards would open the hatches and yell down, "Turn out your dead!" Bodies were either tossed into the sea or buried ashore in shallow graves. Estimates of the total death toll ran as high as 11,500.[56]

The Sugar House, although less notorious than the prison ship, was as horrifying. "Cold and famine were now our destiny," a survivor wrote. "Not a pane of glass, nor even a board to a single window in the house, and no fire but once in three days to cook our small allowance of provision. There was a scene that truly tried body and soul. Old shoes were bought and eaten with as much relish as a pig or a turkey; a beef bone of four or five ounces, after it was picked clean, was sold by the British guard for as many coppers."[57]

Tory historian Thomas Jones, who lived in New York and knew Loring, wrote that he "was determined to make the most of his commission and, by appropriating to his own use nearly two thirds of the rations allowed the prisoners, he actually starved to death about three hundred of the poor wretches before an exchange took place." Noting General Howe's fondness for Mrs. Loring, Jones wrote, "Joshua made no objections. He fingered the cash: the General enjoyed Madam."[58]

* * *

As Cornwallis pursued Washington across New Jersey, a guerrilla war was declared. In a special order Howe empowered his troops, including Loyalist forces, to treat their retreating foes as outlaws: "Small stragling parties, not dressed like Soldiers and without Officers, not being admissable in War, who presume to Molest or fire upon Soldiers, or peaceable Inhabitants of the Country, will be immediately hanged without Tryal as Assassins."[59]

Vicious little actions, hardly noticed in the chronicles of the Washington-Cornwallis saga, cost uncounted lives. Kidnappings were common. One had a famous Patriot as its victim: Richard Stockton, a member of Congress and a signer of the Declaration of Independence.

When the British were about to capture Princeton, Stockton fled with his family to a friend's house in Monmouth County, New Jersey. Local Loyalists kidnapped him and handed him over to the British. A descendant of an old and distinguished Quaker family and a member of Princeton's first graduating class, Stockton was a prize catch. He was jailed as a criminal and cruelly treated until he broke under duress and signed an oath of allegiance to the king—an act that disavowed his signature on the Declaration.

Sometime around mid-March 1777, Stockton was released without any public explanation and returned to his magnificent Princeton home, Morven. The mansion—Cornwallis's headquarters during his occupation of Princeton—had been ruined. Hessians were blamed for making firewood of fine furniture, drinking their way through the wine cellar, bayoneting family portraits, and stealing Stockton's horses and livestock. Tories had led the looters to hastily buried silver plate and other treasures.

Stockton had returned from captivity under a cloud because congressmen possessed unpublicized knowledge that he had foresworn the Declaration. Later an ailing Stockton tried to recant his oath to the king by signing oaths of adjuration and allegiance prescribed by the New Jersey legislature as a way to redeem tainted citizens.[60]

* * *

Tryon joined in the guerrilla war by organizing a troop of Westchester County horsemen, under the command of Col. James De Lancey, for raiding in the Neutral Ground. The cavalry soon became known as De Lancey's Cow-boys, a term that started out meaning cattle and horse rustlers and soon became a name for raiders.[61] A typical raid, as reported in the Tory *New York Gazetteer*: "Last Sunday Colonel James DeLancey, with sixty of his Westchester Light Horse went from Kingsbridge to the White Plains, where they took from the Rebels, 44 barrels of flour, and two ox teams, near 100 head of black cattle, and 300 fat sheep and hogs."[62]

Patriots savagely retaliated against De Lancey's Cow-boys by attacking Oliver De Lancey's country home at Bloomingdale, about seven miles up the Hudson from New York City. As the victims later recounted, strange noises awakened De Lancey's teenage daughter Charlotte and her friend Elizabeth Floyd, daughter of a Long Island Loyalist. They ran to a window, opened it, and shouted, "Who is there?" From below a gruff voice answered, "Put in your heads, you bitches!"

Men entered the house from front and rear and, prodding the teenagers with muskets, ordered them out of the house "as fast as you can." The elder Mrs. De Lancey got out on her own, but, able only to hobble, she hid in a dog kennel under the stoop. The girls fled into a swampy wood (today's Central Park) and spent the night "sitting upon their feet to keep a little warmth in them" and watching the house burn to the ground.[63]

Stephen De Lancey's horsemen also foraged on Long Island, another site of clashes between mounted Tories and amphibious Rebels. In one of the biggest whaleboat attacks Col. Return Jonathan Meigs led 170 men across the Sound to Sag Harbor, pounced on the foragers, and killed six of them. After burning the Tory boats and forage, Meigs took ninety prisoners back to Connecticut. The entire operation lasted twenty-five hours. Congress rewarded Meigs with a commendation and a sword.[64] The British soldiers had just been paid and were "pretty well boozey," wrote a Rebel raider. "Some had nothing but his shirt on, some a pair of trowsers others perhaps 1 stocking and one shoe."[65]

Governor Tryon, acting as General Tryon, waged what he called "desolation warfare." He sent the Cow-boys and Emmerich's Chasseurs, another mounted Loyalist unit, on terror raids, torching the homes of leading Patriots, whom he believed he disparaged by calling them mere "committeemen." During one of the horsemen's most terrifying raids, they burned down parts of Tarrytown, a Hudson River community split between Loyalists and Patriots.[66]

The participation of Emmerich's Chasseurs in the raids added European cavalrymen to the Loyalist guerrilla forces. The unit was raised by Andreas Emmerich, a native of Germany, where he began his military career. He emigrated to England, then to America, where he was commissioned a lieutenant colonel and raised a corps of light troops named after himself. About half of the officers were Europeans, who did not get along with their American counterparts.

Mutinous American officers asked General Clinton to court-martial Emmerich for "employing Soldiers, Negroes, & [Tory] Refugees, to robb and plunder" civilians, taking a cut of the loot from the looters. The officers also accused him of "imprisoning, Whipping and cruelly beating the inhabitants without cause or trial," selling British Army horses, and stealing army funds. Clinton sidestepped the court-martial by transferring the men into other regiments. Emmerich somehow managed to keep his colonelcy.[67]

For a time Major General Putnam had his headquarters at Peekskill, New York. Putnam tried to rein in Tory raiders by sending Colonel Meigs down the Hudson to attack pillagers. Rumors circulated that, in retaliation, Tryon planned to kidnap Putnam. Coincidentally Patriots reported a sudden surge of Tory spies in the area. One of them, a Loyalist lieutenant named Nathan Palmer, managed to infiltrate the headquarters encampment. Soldiers discovered him, and Putnam brought him before a court-martial to be tried as a spy. Tryon tried to intercede, threatening Putnam personally if he did not release Palmer.

Putnam replied:

> *Sir—Nathan Palmer, a lieutenant in your king's service, was taken in*
> *my camp as a spy. He was tried as a spy; he was condemned as a spy, and*
> *you may rest assured, sir, he shall be hanged as a spy.*
> *I have the honor to be, &c.*
> *Israel Putnam.*
>
> *P. S. Afternoon. He is hanged.*[68]

Casual malice like Putnam's laid bare the fact that gentlemanly warfare had disappeared in the Neutral Ground. Joseph Galloway, a leading Tory, charged that marauding and even rape was officially tolerated by the British and the Loyalists. Galloway said that "indiscriminate and excessive plunder" was witnessed by "thousands within the British lines." In a "solemn inquiry," backed by affidavits, he said, "it appears, that no less than twenty-three [rapes] were committed in one neighborhood in New Jersey; some of them on married women, in presence of their helpless husbands, and others on daughters, while the unhappy parents, with unavailing tears and cries, could only deplore the savage brutality."[69] Similarly, in New York City, citizens and officers accused Hessians, Redcoats, and Loyalists of robbing houses, raping women, and murdering civilians. Thomas Jones wrote that even murderers sentenced to death were set free. The crimes stirred futile demands for an end of martial law and the return of civil courts.[70]

Tryon's desolation warfare shocked many British officers and outraged Patriots. Justifying his tactics, he said, "Much as I abhor every principle of inhumanity or ungenerous conduct, I should, were I in more authority, burn every Committee Man's house within my reach, as I deem, those Agents the wretched instruments, of the continued calamities of this Country."[71]

The Neutral Ground expanded, if not geographically, then psychologically, so that the label was applied to whatever territory that armed Loyalists were trying to control. The ebb and flow of Loyalists, Continentals, and Patriot militias created a climate of lawlessness. At times anarchy ruled much of the Pine Barrens of southern New Jersey and the Hudson Highlands along both banks of the river.[72]

New dens for refugees appeared as Washington, retreating across New Jersey, left in his wake more of the Neutral Ground and more weathervane Tories who changed allegiance with the winds of victory. When General Howe issued a proclamation offering pardon and protection to anyone who within sixty days should swear allegiance to the Crown, three thousand persons did so.[73] People sensed the death of the Continental Army. They were seeing what a British officer had seen a few days before the proclamation: "The fact is their army is broken all to pieces . . . it is well nigh over with them."[74]

In December 1776 Washington reached the Pennsylvania side of the Delaware River, directly across from Trenton. He sent men thirty miles up the icy river and thirty miles down to collect all the boats they could take back to camp and then destroy the rest. Washington especially wanted Durham boats. Flat-bottomed and forty to sixty feet long, they hauled freight to and from Philadelphia. Each boat, Washington figured, could hold as many as forty armed men.[75]

Amidst his planning there was anguish. "I think the game is pretty near up," he wrote to his brother John, "owing, in a great measure, to the insidious Arts of the Enemy, and disaffection of the Colonies . . . but principally to the accursed policy of short Inlistments, and placing too great a dependence on the Militia, the Evil consequences of which were foretold 15 Months ago with a spirit almost Prophetick."[76] Enlistments for many of his troops would end on January 1, 1777.

In the first hours of December 26, through wind-whipped snow and hail, Washington launched his complex plan to cross the Delaware and take Trenton by overwhelming the city's Hessian guard. The morning was still dark when the Americans struck, startling the sleeping Hessians. In a short, furious battle the Americans killed twenty-two Hessians, wounded eighty-four, and took 918 prisoners. Four Americans were wounded and two were killed.[77] On January

3, 1777, Washington outmaneuvered the British and led an audacious attack on Princeton, thirteen miles north of Trenton. His exhausted troops—including many volunteers who fought even though their enlistments had expired—improbably won another battle.

The Hessians were shocked at how quickly Trentonians, who had hailed their occupiers and sold them provisions, became newly hatched Rebels. "When the Hessians entered Trenton and occupied the region, the inhabitants swore their allegiance to the King of Britain," a Lutheran minister wrote. "But as soon as the American troops attacked on Christmas, the inhabitants shot at the Hessians from their houses. In fact, even a woman fired out of a window and mortally wounded a Captain."[78]

From Princeton, Washington led his army to winter quarters in the heights of Morristown, New Jersey. The Continentals' camp, a two-day march from New York City, lay behind the Watchung Mountains, which formed a natural barrier. Washington sent out orders to militiamen in the Neutral Ground to wage guerrilla war and "harass their troops to death."[79] In another letter to his brother, Washington wrote, "Our Scouts, and the Enemy's Foraging Parties, have frequent skirmishes; in which they always sustain the greatest loss in killed and Wounded, owing to our Superior skill in Fire arms."[80]

While Howe enjoyed another pleasant winter in New York and Washington watched his army shrink to barely eight hundred men, the recruitment of Loyalist soldiers continued from New England to Georgia.[81] Howe had given Col. Edmund Fanning, Governor Tryon's confidant, a warrant to raise the King's American Regiment.[82] Fanning then bestowed officers' commissions on recruiters who set out in search of volunteers in New York and New England.

One of Fanning's recruiters was Capt. John McAlpine, who, while seeking enlistees in the upper Hudson Valley, was captured by Rebels and imprisoned in Poughkeepsie. He was awaiting trial and a probable sentence of execution when nineteen of his recruits attacked the jail and freed him. Aided by Loyalists along the way, the Tory rescuers and McAlpine spent twenty-six days eluding Rebels until they reached the safety of regimental headquarters on Long Island.[83]

Most of the regiment's men came from Long Island and the lower Hudson Valley, where some fathers and sons enlisted together, as did a number of brothers. Recruiters were less successful in New England. Recruiting in Connecticut was Capt. Moses Dunbar, a thirty-one-year-old native of Wallingford. Like so many other Loyalists from that state, Dunbar had fled to Long Island after the British conquest of New York.[84] Dunbar twice slipped back into Connecticut, first to get married and then to bring his pregnant wife back to Long Island. On the second trip, in his words, "I accepted a captain's warrant for the King's service in Colonel Fanning's regiment."

Tory hunters caught Captain Dunbar and charged him with recruiting for the British. He later was indicted for waging war against Connecticut, under a new state treason law that carried the death penalty. He was denounced by his Patriot father, who was said to have declared when his son was arrested that he would furnish the hemp to make a rope for him.[85]

Dunbar was tried and hanged in Hartford before a "prodigious concourse of people," including his wife, who was forced to attend.[86] The Tory press reported the hanging of six other recruiters in Connecticut.[87]

Around this time, a Loyalist spy identified only as "The Woman" reported "a very considerable Store of Provisions & cloathing and 80 pieces of Cannon" in Danbury, Connecticut, near the New York border, about twenty-five miles from the coast.[88] General Howe authorized Tryon to lead a fleet of twenty-six ships to the mouth of the Saugatuck River, at Norwalk, where he landed more than eighteen hundred men, including both Regulars and the Prince of Wales's American Volunteers Brigade. Tryon's style of pitiless warfare did not appeal to Howe, but this time he gave Tryon a sizable force because pilfering or destroying the Rebels' supplies made military sense.

Tryon met little opposition as his men marched to Danbury, where local Loyalists guided them to supply caches. The troops invaded and pillaged Patriots' homes, careful to spare the Anglican church and

Loyalists' homes. Great piles of tents and provisions and thousands of shoes went up in flames. His mission accomplished, Tryon led his troops back toward the coast. But he discovered that his way was blocked by militiamen who had been hastily mobilized and led by Brig. Gens. David Wooster and Benedict Arnold.

Arnold was home in New Haven, brooding about the Congress's failure to promote him to major general, when at three o'clock one morning a messenger came with the report that a large force of Regulars and Tories had landed near Westport and were bivouacked in Weston, obviously on their way to Danbury. Arnold galloped off and was soon joined by sixty-six-year-old David Wooster, commander of the Connecticut militia. They quickly gathered one hundred mounted militiamen. Meanwhile Brig. Gen. Gold Selleck Silliman, commander of the Fairfield militia, had rounded up about five hundred more.

The next day Tryon left a burning Danbury behind him as he led his troops back to the shore and the waiting transports. He stopped at Ridgefield, where he supervised the torching of a Presbyterian church and several Patriots' homes. Local Loyalists were said to have been spared because they had painted black and white stripes on their chimneys.[89] Militiamen trailed Tryon's force, starting a short, fierce battle in which Wooster was fatally shot. A Continental soldier saw Arnold ride "to our front line in the full force of the Enemy's fire, of Musquetry & Grape shot." His horse, riddled by nine shots, fell, trapping Arnold, who had been lamed in Quebec when a musket ball struck his left leg. As he struggled to free himself, a Redcoat demanded surrender at bayonet point. Arnold fatally shot him with a pistol and managed to escape.[90]

Fighting continued on the third day, when Arnold, Silliman, and scores of additional militiamen arrived, firing at the invaders from houses, stone walls, and barricades. Arnold had another horse shot from under him as Royal Marines from the fleet landed and turned the battle, opening the way for the invaders to make it to boats that took them to the ships. In the three days of the invasion, about two hundred men in Tryon's forces were killed or wounded, as were about sixty militiamen.[91]

Congress rewarded Arnold for his bravery by promoting him to major general. And Washington, responding to the start of a British offensive in northern New York, told Congress that "an active, spirited officer," a man who is "judicious and brave," was needed on the imperiled northern frontier. That man, Washington said, was Benedict Arnold. He was swiftly on his way to what would be an epic battle. His foes would include not only the British Army but two new enemy forces.[92]

A prophecy about those new enemies came in a curious way. Beyond the Neutral Ground, in the Highlands of northern New Jersey, the most notorious Tory raider was Claudius Smith. A young thief before the war, he became a guerrilla who profited from his loyalty to the king. The victim of his most infamous raid was Maj. Nathaniel Strong, a well-known militia officer who lived in Blooming Grove, a town on the Hudson. Smith had publicly vowed to kill Strong, along with several other prominent Patriots.

One night Smith and four Cow-boys plundered one house, and then, about midnight, invaded Strong's home. Smith told Strong that his life would be spared if he gave up his musket and surrendered. As Strong turned to set his gun against a wall, he was fatally shot in the back. That crime inspired Governor George Clinton of New York to offer a twelve-hundred-dollar reward for the capture of Smith and six hundred dollars for the capture of his outlaw sons.[93]

A wanted poster called Smith "a fierce looking man nearly 7 ft. tall, wearing a suit of rich broadcloth adorned with silver buttons," a "Notorious leader of a lawless band of men." Besides the murder of Strong, Smith was accused of stealing "money, pewter & silver plate, saddles, guns, oxen, cattle & horses from the American colonists and turning them over to the British in New York City." His leading confederate in the city was Mayor David Matthews, who had been implicated in the plot to kidnap or assassinate George Washington.[94]

After the wanted posters were tacked up, Smith fled to the protection of British-occupied Long Island, leaving his oldest son, William, in charge of the Cow-boy gang. Patriots stormed the gang's mountain hideout and fatally shot William and another Cow-boy. Several others, including Richard Smith, Claudius's youngest son, got away.

In a daring whaleboat raid, other avenging Patriots rowed to Long Island from Connecticut, captured Claudius Smith in his bed, bound and gagged him, and carried him off to the hidden boat. They took him, under heavy Patriot guard, to Goshen, New York, about twenty miles west of West Point. There he was chained to a prison cell floor while awaiting trial. Convicted not of murder but of robbery and breaking and entering, he was nevertheless sentenced to death, along with members of his gang.

He and two other Cow-boys were taken in a cart to a gallows surrounded by a large crowd. When the nooses were put around the necks of the three men, according to witnesses, Smith kicked off his shoes, with the observation that his mother, scolding him for his evil behavior, had often told him that he would die with his shoes on: By dying barefoot he would make her a liar.[95] The cart was then driven off, and the three men dangled from their nooses.

Soon after the execution Claudius's son Richard Smith went on a rampage, singling out Patriots for death. The first known victim was found with a note pinned to his body:

> A Warning to Rebels
>
> You are hereby forbid at your peril to hang no more friends to government as you did Claudius Smith. . . . We are determined to hang six for one, for the blood of the innocent cries aloud for vengeance. . . . There is particular companies of us that belongs to Col. Butler's army, Indians as well as white men and particularly numbers from New York that is resolved to be revenged on you for your cruelty and murders. . . . This is the first and we are determined to pursue it on your heads and leaders to the last till the whole of you is massacred.[96]

Smith's vengeful warning reminded the Patriots of the two new enemies who were waging the new war in the north: Colonel Butler's Loyalist Rangers and Britain's Loyalist Indians.

"INDIANS MUST BE EMPLOYED"

NEW YORK FRONTIER, SPRING 1775–OCTOBER 1777

Col Johnson's Conduct in Raising Fortifications round his house,
keeping a Number of Indians and other armed men constantly
about him, and stopping and searching Travellers upon the King's
Highway, and stopping our Communication with Albany is very
alarming to this County, and . . . confirms us in our Fears, that his
Design is to keep us in awe, and to oblige us to Submit to a State
of Slavery.

—*Tryon County Committee of Safety*[1]

B y the spring of 1777 the Revolution had spawned two wars. One was being fought primarily by the British Army and the Continental Army on battlefields in New Jersey and Pennsylvania. The other, along the northern frontier, was just beginning. It would be a war of terror and vengeance that pitted Indians and Tories against Patriot settlers, Loyalist neighbors against Patriot neighbors, Indians' tomahawk and torch against the muskets of farmers and militiamen.

The war in the borderland would be sudden bloody encounters in Indian villages or isolated Rebel settlements. On the frontier between Canada and western New York, soldiers of the Continental Army would fight wilderness battles not only against the British Army and Loyalist regiments but also against what the Declaration of Independence called "the merciless Indian Savages." Left unsaid was the fact that Patriots tried as hard as the British and the Loyalists to gain the support of Indians.

The roots of Indian-settler conflict were deep along the Canada–New York border, where many Patriots, remembering how the French had forged an alliance with Indians in a previous war, now feared that Britain would do the same. France had led the Indians into white man's warfare by transforming tribal alliances made for the fur trade into a military strategy that gave its name to the French and Indian War. To American colonists of the time, the Indians were the most fearsome foes. Ben Franklin's "Join or Die" drawing of a snake in eight pieces, often cited as a symbol for the Revolution, was a much earlier warning, urging the colonies to unite against the French and their Indian allies.[2]

But Britain had an ambiguous policy toward the Indians. Robert Dinwiddie, royal governor of Virginia, saw the wisdom of seeking Indian aid in 1753 when he sent young Maj. George Washington of the Virginia militia into the wilds of the Ohio Country. Washington met with Indian leaders and told them, "I was destined, brothers, by your brother, the governor, to call upon you, the sachems of the nations. . . . His Honor likewise desired me to apply to you for some of your young men to . . . be a safeguard against those French Indians who have taken up the hatchet against us."[3]

Both the British and the French used a scalp as proof that an Indian had slain a foe and deserved a bounty. The British superintendent of Indian affairs for the southern colonies once noted that "large publick Rewards for Scalps given by Provincial Laws to Indians, are attended with very pernicious Consequences to his Majesty's Service." Indians, the superintendent said, were killing each other merely to collect bounty scalps, producing the possibility of unnecessary intertribal warfare that would jeopardize alliances with tribes favorable to the British.[4]

Victory in the war presented Britain with more land, more Indians, and more problems over colonists' incursions into Indian land. In 1768 British officials met with leaders of the powerful Iroquois at

Fort Stanwix (near the Mohawk River and today's Rome, New York).
The Iroquois agreed, in a formal treaty, to give up claim to all of their
lands east and south of the Ohio River in return for a guarantee that
they could retain their land in western New York. Less powerful
tribes, including the Delaware, Mingo, and Shawnee, did not accept
the treaty.[5]

Shawnee and Mingo attacked white settlers, who struck back in isolated encounters. Then, in May 1774, settlers killed eleven Mingo near
today's Steubenville, Ohio. As Mingo and Shawnee sought revenge,
Lord Dunmore sent one thousand militiamen into what is now West
Virginia. After a daylong battle between militiamen and more than
one thousand Shawnee warriors, Dunmore and tribal leaders par-
leyed and the Shawnees essentially agreed to abide by the Fort Stanwix treaty.[6] Dunmore's War, as his critics dubbed it, made Tories of
many western Pennsylvania frontiersmen. They endorsed his attempt
to settle the boundary quarrel between Virginia and Pennsylvania by
seizing Fort Pitt, the predecessor of Pittsburgh, in January 1774.[7]

Soon after the Battles of Concord and Lexington, the Continental
Congress received reports that the British were plotting to make Loyalist allies of the Indians along the New York–Canada border. Congress attempted to counter that move by negotiating a treaty with the
Six Nations (Mohawk, Oneida, Tuscarora, Onondaga, Cayuga, and
Seneca). The confederacy, which dated to the time before the arrival
of Europeans in North America, included English-speaking Indian
leaders familiar with British ways and wary of Rebels.

The congressional delegation and tribal leaders met at Fort Pitt,
at the forks of the Ohio River, then under control of Virginia's Rebel
government. "This is a family quarrel between us and Old England,"
the Indian leaders were told. "You Indians are not concerned in it.
We don't wish you to take up the hatchet against the king's troops.
We desire you to remain at home, and not join either side, but keep
the hatchet buried deep."[8] The delegates returned to Philadelphia believing that the Six Nations would remain neutral. But the nations

split. Oneida and Tuscarora broke away to support the Patriots, while the other tribes agreed to become warriors for the British and their Loyalist allies.[9]

The Indians mystically imagined their confederacy as "the Long House," with the Seneca of the Susquehanna and Ohio valleys guarding the western entrance, and the Mohawk the eastern door, while the Onondaga watched over the council fire in the center. The Six Nations raised maize and beans and lived in large wooden houses encircled by palisades. These they called by a word that was translated as "castle." Not by chance the Mohawks' principal castle stood near a fortified mansion that was the nerve center for the rapidly evolving alliance between the British and the Mohawk. The mansion was Johnson Hall, built by a man who owned more land in the Mohawk Valley than anyone before or since: William Johnson, Britain's regional superintendent of Indian affairs.

Johnson was born a Roman Catholic near Dublin, in 1715, and, planning his future in a British world, became an Anglican around 1738. This was about the time he sailed to America to manage his uncle's frontier estate in New York's Mohawk Valley. The uncle, Admiral Sir Peter Warren, had grown rich by seizing French and Spanish vessels for prize money—and by shady maritime trading in wine, rum, and slaves. As a successful trader he was welcomed into the great De Lancey family, marrying Susannah De Lancey, daughter of Stephen De Lancey, an influential, well-connected fur trader.[10]

Warren became a major New York property owner, acquiring tracts of land in Manhattan and Westchester. His showcase home at 1 Broadway would become the residence of New York City's successive commanding generals of the Revolution, including Washington, Howe, and Clinton.[11] He also owned some fourteen thousand acres near the junction of the Mohawk and Schoharie rivers, about 180 miles north of Manhattan. Warren invited William Johnson, affectionately called Billy, to come to New York with some prospective tenants and manage the vast northern property. While developing the wilderness into prosperous farmland, Johnson also set up his own trading business with the Mohawk. Operating independently of his uncle, Johnson grew wealthy himself.[12]

Johnson won acclaim from colonial officials in the prelude to the French and Indian War by leading Mohawk warriors on raids into border territory claimed by France, often taking the scalps of French defenders. The Lords of Trade in London put Johnson on the New York Governor's Council, and the Mohawk made him a sachem, giving him a name that meant something like "Chief Much Business."[13] He successfully straddled the two worlds of councillor and sachem, though to some compatriots in London and New York City he seemed more Mohawk than Briton.

Johnson's uncle resented the way Johnson had struck out on his own, and, when his uncle died in 1752, Johnson did not inherit the estate.[14] But Johnson had no need of any more land. And he had prospered as a trader who dealt honestly with Indians and spoke their languages. Colonial officials appointed him "Colonel, agent, and sole superintendent of all affairs of the Six Nations and other Indians."[15]

Johnson often wore Indian dress, painted his face, joined in tribal dances,[16] and took several Indian mistresses. One was the beautiful Molly Brant, sister of Joseph Brant, a shrewd and powerful Six Nations leader. Molly, a high-ranking Mohawk, bore Johnson eight children, seven of whom survived.[17] Both Joseph and Molly had Mohawk names and adhered to Mohawk customs. Their English names came with their British education, which introduced them to the world that Johnson represented and that they cautiously accepted.[18]

During the French and Indian War, Johnson met John Butler, the son of a British Army officer who had been stationed at Fort Hunter on the Mohawk River. Butler, like Johnson, was an officer in command of Indian allies. He was with Johnson when his British and Indian forces captured Fort Niagara. Butler knew Indian languages and had spent much of his life on the frontier. He married into a Dutch family that traced its American roots to 1637 and was well known on the Mohawk.[19] A postwar associate of Sir William, Butler lived in an imposing house on a hill overlooking the Mohawk River.

He had inherited his five-thousand-acre estate, near today's Fonda, New York, from his father.[20]

Nearby was Johnson's domain, named John's Town (today's Johnstown, New York), in honor of his son, one of his three white children. They had been born to Catharine Weisenberg, a teenage German immigrant who had come to America as an indentured servant. She had run away from her master in New York City and became Johnson's wife without a wedding ceremony.[21] Commoners could climb socially and politically on the frontier, whose rough-hewn leaders lacked the credentials of those who ruled the British establishment in New England and New York City.

Sir William Johnson died in 1774 during a conference with representatives of the Six Nations. His son, John Johnson, then thirty-two years old, inherited the family's fiefdom and retained a noble title via a knighthood bestowed by King George III. Educated in Philadelphia, widely traveled in England, and married to an Albany heiress, he was far more polished than his father.[22] While making John Johnson his heir, Sir William, confidently exceeding his authority, had bequeathed the royal post of superintendent of Indian affairs to his ambitious, Irish-born nephew Guy Johnson. Sir William correctly assumed that the king would confirm the choice.

Guy had also become Sir William's son-in-law by marrying Johnson's daughter Mary. Like the Indians, Sir William believed in kinship as a way of holding power. At Sir William's funeral, an Indian delegation hailed Guy and told him, "Continue to give good advice to the young men as your father did: . . . Follow his footsteps; . . . as you know very well his ways and transactions with us."[23]

Guy, a militia colonel, did follow those footsteps—just in time for the Revolution. The lieutenant colonel of Guy's militia regiment was John Butler, who also served both Sir John and Guy as an Indian interpreter.[24] The three began rallying fellow Loyalists in the Mohawk Valley, hoping to mobilize a major frontier force against the Rebels. Working with the Johnsons, especially among German immigrants like himself, was Christian Daniel Claus, an Indian agent married to

Sir William's daughter Ann.[25] They plotted an armed stand at Johnson Hall, the mansion-fortress that dominated the Johnson realm.

Word that Johnson was creating a Loyalist stronghold quickly reached the Patriots' Committee of Safety for Tryon County, which encompassed an immense swath of New York west of Albany, including the Mohawk Valley and the Johnson fiefdom. "We are informed that Johnson Hall is fortifying . . . ," the committee reported to Patriots in Albany in May 1775. "Besides which, we are told that about 150 Highlanders (Roman Catholics) in and about Johnstown, are armed, and ready to march." The committee also told of a report that Indians, "whom we dread most . . . are to be made use of in keeping us in awe."[26]

Both sides were maneuvering in a fog of rumors. Guy Johnson, aware of the local Patriots' appeal for aid from Albany, believed that Rebels were plotting to kidnap him and attack his followers. He fled to Canada with his pregnant wife, Mary; his children, other family members; and about 170 supporters, including Joseph Brant and ninety other Indians. Along the way, Johnson won from more than fourteen hundred Indians a promise that they would "assist His Majesty's Troops in their operations."[27] At Oswego, on the southeastern shore of Lake Ontario, after an exhausting journey, Mary Johnson died, an early casualty of the frontier war.[28]

When the Revolution began, many New York Tories, including John Butler and his sons Thomas and Walter, had fled to Canada. Butler had left behind his wife and other children. The Rebels seized them, giving Butler a motive for vengeance. (The family would not be reunited until a prisoner exchange in 1780.) Johnson, like Butler, did not intend to languish in Canada. Johnson wanted to muster an all-Indian military unit that would fight the Rebels in frontier settlements in western New York and Pennsylvania. But Sir Guy Carleton, royal governor of Quebec refused, saying he wished to use Indians only as scouts or in defense.[29]

Guy Johnson and his two young daughters, along with Brant and several others, sailed to England to put their dependents safely into

the hands of Loyalists in London and to gain official endorsement
of the Indian mobilization that Carleton had rejected. Johnson's vision of the British-Loyalist-Indian alliance was captured in a painting he commissioned by Benjamin West, who had left America years
before and was the king's court painter. West created an allegorical
portrayal of British-Indian relations. West put moccasins on Johnson
and added to his British Army uniform a Mohawk cap, an Indian
blanket, and a wampum belt. Hovering behind Johnson is the shadowy figure of an Indian, who points to a peace pipe while Johnson
holds a musket. Deeper in the background is an Indian family near
a British Army tent.[30]

Brant became a star of London society. The king presented him
with a silver gorget* bearing the inscription *The Gift of a Friend to
Capt. Brant*—royal notice of Brant's military commission in Canada's
Loyalist corps. In his portrait, by George Romney, a leading portrait-
ist of English nobility, Brant wears the gorget and a plumed head-
dress. He looks grim and he carries a tomahawk—a portent of the
future, when Brant and the Butlers would lead Loyalists and Indians
in bloody battles.[31]

Carleton sent Butler to Fort Niagara with orders to keep the Six Nations off the warpath but loyal to Britain. He was inadvertently aided
by the Reverend Samuel Kirkland, a Christian missionary from Connecticut who preached neutrality to Six Nations Indians while trying
to preserve the allegiance of those who supported the Rebels.[32] Kirkland, who lived among the Oneidas and spoke their language, succeeded in keeping them on the side of the Patriots throughout the war.

Butler, anticipating the frontier civil war to come, began enlisting Indians as spies, not only along the Mohawk River but also as
far west as the Mississippi. Ever since the Revolution began, Loyalists
had been fleeing to Canada. Butler also aided them, knowing that the

---

* An ornament worn with officers' uniforms. It evolved from medieval armor that
  protected the throat.

able-bodied men among them would be potential recruits for future Tory militias.

In January 1776, responding to a new round of reports that Sir John Johnson was arming his Loyalist tenants and supporters, Washington sent a force of some four thousand men to Johnson Hall, where they confronted about six hundred armed Loyalists, most of them Highlander immigrants. The Continentals disarmed Johnson's private army and ordered him to post a sixteen-hundred-pound bond that backed his promise to essentially remain under house arrest.[33] In May, Sir John broke his parole, fleeing to Montreal with about 170 tenants and followers, guided by Mohawks. Lady Johnson, who was seven months pregnant, was left behind. Rebels took her to Albany as a hostage. She appealed in vain to Washington.[34]

Once in Canada, with Carleton's approval, Sir John soon began raising the King's Royal Regiment of New York. He drew on his own followers, along with Loyalist refugees who had abandoned their homes and farms in the Mohawk and Schoharie valleys. They would return armed and ready to die to get back their land, which the Rebels had seized.[35]

In the spring of 1777 John Butler received a letter from Carleton, recently the heroic defender of Quebec against American invaders. Carleton had changed his mind. He ordered Butler to recruit as many Indians as possible for an invasion of New York.[36] Johnson's raising of the King's Royal Regiment and Carleton's Indian mobilization order to Butler were preparatory moves for the launching of a three-prong invasion of New York. Lt. Gen. Sir John Burgoyne, who developed the strategic plan, believed that the invasion would inspire a Tory uprising and ascendancy in New York, making the entire colony as loyal as New York City. The invasion would also drive a wedge between the colonies, reflecting the belief of British officialdom that the firebrands in New England were of a different revolutionary breed from Americans elsewhere. Isolating New England would begin the process of snuffing out the Revolution.

Burgoyne did not include the Royal Highland Emigrants among the Loyalists in his force. His decision not to take the Highlanders along may have stemmed from the distrust that many British officers felt toward erstwhile supporters of Bonnie Prince Charlie. Lt. Col. Allan Maclean, commanding officer of the Emigrants, had been pardoned and was officially regarded as a trusted Loyalist officer. But Burgoyne decided that Maclean's men would better serve as garrison troops, left behind as a home guard.[37]

Burgoyne proposed leading an army of about seven thousand men—Regulars, Loyalists, Hessians, and Indians—from Canada via the Richelieu River–Lake Champlain waterway to Fort Ticonderoga, then down the Hudson River Valley to Albany. A smaller force of two hundred Regulars, three hundred Hessians, eight hundred Indians, and two hundred Tories—including Butler's recruits—was to advance overland from Lake Ontario to Lake Oneida, march eastward along the Mohawk Valley and destroy Fort Stanwix, the only Patriot obstacle in the valley. That unit, commanded by Lt. Col. Barry St. Leger, would then head for Albany to link up with Burgoyne. The third force would head out of New York City and up the Hudson to block any Rebel attempt to move against Burgoyne from the south.

When Burgoyne presented his plan in London, he won approval from British officials in London and from King George III, who said in an endorsement written in his own hand: "The outlines of the plan seem to be on a proper foundation. . . . Indians must be employed, and this measure must be avowedly directed, and Carleton must be in the strongest manner directed to furnish as many Canadians as possible."[38] The king's command led to Butler's enlisting both Indians and white Loyalists. But they were mostly Americans, not the Canadians the king had expected.

Henry Simmons, a New Yorker, for example, left his wife and children and headed north when he heard about Burgoyne. "The Sixteeth Day of August, 1777," Simmons wrote in his journal, "I left my house at Claverack and Sat out with a Compiny of Seven and twenty Men and officers to go to General Burguins armey Which was at the time at Fort Miller." Claverack, New York, was about seventy miles south of Fort Miller on the Hudson. They traveled for eleven days "through enemy country, likely at night along untrodden paths, fording streams, and hiding in thickets by day" until they caught up with "the flyeing arme" and enlisted in the King's Royal Regiment.[39]

A Mohawk Valley Tory, brother of a Rebel leader, raised a company of sixty-three Loyalists and led them to Canada. Another Tory from that area, a captain of Rebel militia, changed sides and left New York with several relatives he had enlisted. More recruits came from the efforts of Adam Crysler, a friend of Joseph Brant and a prominent landowner in Schoharie, a settlement near Butler's mansion. Three of his brothers were known Tories. A fourth brother claimed to be a Patriot. But, suspected of lying, he was arrested by Rebel raiders and taken to Albany, where he was reportedly hanged.[40]

Crysler wrote in his journal that he had recruited seventy white men and twenty Indians from that area. He was instructed by Brant to hold his recruits in the valley for what was expected to be a Tory uprising. Crysler knew that a civil war was simmering in the valley and that recruiters risked their lives. Earlier in 1777 a party of Rebels, searching for a Tory recruiter, "levied a tax upon his poultry yard and ate up his chickens." Then they tracked him down and found him with twenty recruits. The captured Tories were delivered to Albany "for safe keeping." The recruiter was hanged, a fate that became common for others with the same mission.[41]

Unlike short-time Rebel militiamen and Continentals who served "for one year, unless sooner discharged," Tories enlisted for the duration of the war. Rebels typically signed up for months at a time because they had farms or businesses to run. Tories typically had forfeited their homes, land, and way of life, for they knew that if they returned they would be prosecuted under state confiscation and treason laws. Their only hope for return was a British victory. Washington envied the ability of the British and Loyalists to field permanent forces while he often had only "the throngs of militia, which at certain periods have been, not in the field, but in their way to and from the field," creating "two sets of men to feed and pay; one coming to the army, and the other going from it."[42]

* * *

On his way southward, up Lake Champlain toward Fort Ticonderoga, Burgoyne enlisted about four hundred Indians in his force and addressed them through an interpreter: "The great King, our common father and the patron of all who seek and deserve his protection, has considered with satisfaction the general conduct of the Indian tribes from the beginning of the troubles in America." He was diplomatically endorsing the Six Nations' neutrality. But now, Burgoyne said, "You are free. Go forth in might and valor of your cause, strike at the common enemies of Great Britain and America, disturbers of public order, peace and happiness, destroyers of commerce, parricides of state."

He pointed to the German and British officers with him and continued, "The circle round you, the chiefs of his Majesty's European forces and of the Princes his allies, esteem you as brothers in the war. . . . Be it our task, from the dictates of our religion, the laws of our warfare and the principles and interest of our policy, to regulate your passions . . . to suspend the uplifted stroke."

He then introduced the Indians to a new etiquette of battle: "Aged men, women, children and prisoners must be held sacred from the knife or hatchet, even in the time of actual conflict. You shall receive compensation for the prisoners you take, but you shall be called to account for scalps. . . . You shall be allowed to take the scalps of the dead when killed by your fire and in fair opposition; but on no account, or pretence, or subtilty, or prevarication, are they to be taken from the wounded or even dying." They could, however, show "less reserve" toward "base, lurking assassins, incendiaries, ravagers and plunderers of the country, to whatever army they may belong."

The Indians cried *"Etow! Etow! Etow!"*—a way of showing affirmation—and the leading chief, at the end of a much shorter speech, declared: "With one common assent we promise a constant obedience to all you have ordered and all you shall order."[43] More Indians would join Burgoyne on his way to Albany.

Burgoyne also added to his force by asking John Peters, a Loyalist refugee in Montreal, to raise a regiment that Burgoyne named the Queen's Loyal Rangers. Peters, born in Connecticut, had parted ways with his Patriot father. Peters's uncle the Reverend Samuel Andrew Peters was such a ranting Loyalist that General Gage, as a matter of public safety, politely urged him to leave New York. He sailed to England in 1774.[44]

Burgoyne's new Loyalist regiment became a magnet for Canadians, New Yorkers, and Vermonters who saw the invasion as a chance to join the winning side. Many along his battle route would fight and kill neighbors, for on this frontier the civil war became local and personal. The recruits of the Queen's Loyal Rangers were inspired not only by loyalty to the Crown but also by their desire to drive the Rebels from this fertile land and convert it to a Loyalist domain. Within a month the Rangers had 262 men. One of them was Ensign John Peters, Jr., Lieutenant Colonel Peters's fifteen-year-old son.[45]

The Queen's Loyal Rangers were in the vanguard of Burgoyne's army when Fort Ticonderoga came into view. As the force maneuvered for battle, Burgoyne realized that if artillery could be hauled up a mountain overlooking the fort, he could pulverize Ticonderoga. After hacking a road through the forest and hauling dismantled cannons up the slope, artillerists and pioneers placed a battery on the summit and began firing.[46] With the cannons threatening devastation and the defenders outnumbered almost four to one, Maj. Gen. Arthur St. Clair, commander of Ticonderoga, ordered the fort abandoned. He and his men slipped away under cover of night. Burgoyne now had a base for his advance on Albany—and a rallying point for more Loyalist recruits.

Burgoyne's political adviser was a powerful local Tory, Col. Philip Skene, who had acquired from British officials the title of royal governor. His realm included Crown Point, Fort Ticonderoda, and land that would eventually become part of Vermont. Skene had made a fortune as surveyor of His Majesty's woods around Lake Champlain, as landlord over tenants working thousands of acres of land, and as the owner of sawmills and forges. A wounded veteran of the French and Indian War with a distinguished British Army career, he became the Crown's leading representative and landholder in the area, and he commanded a Loyalist militia. His son Andrew, a graduate of King's (later Columbia) College, was a British Army officer.

Skene had made a long detour before joining Burgoyne. He was in England during the American invasion of Canada. While returning to America, his ship, as was customary, "spoke" an England-bound ship by coming close enough so that the ships could shout news to each other. From the ship out of America came the shouted news of Lexington and Concord. Skene and British officers reacted by trying to take over the ship and sail it to British-occupied Boston. But the ship's crew and captain overpowered and confined them. When the ship reached Philadelphia, Patriots arrested Skene and transported him to Connecticut,[47] where he would "still harangue the people from the prison windows."[48] Skene was exchanged in time to join Burgoyne's invasion.[49]

Both Skene and his son welcomed Burgoyne as a liberator. When the elder Skene and Burgoyne conferred in Skene's mansion at the southern end of Lake Champlain, Skene convinced the general that he would be joined by Loyalists as fervent as the Skenes, all along his march to Albany.[50] Skene became Burgoyne's principal aide, checking on the loyalty of civilians, watching over the delivery of supplies, and running a spy network. At least two of his spies, Continental Army deserters, were caught lurking in a Rebel camp. They both carried incriminating passes, signed by Skene, saying they were "on his Majesties Service" and had taken "the oath of allegiance." They were tried by court-martial and condemned to death.[51]

Burgoyne pursued part of St. Clair's army, whose rear guard fought a delaying action that enabled most of the men to escape. Burgoyne, with Philip and Andrew Skene at his side, continued pursuit, veering from his original plan and taking a route that led him into a dense forest laced with swamps. The decision led to a logging duel between

Patriots and Burgoyne's men: He had to fell trees to build a corduroy (log) road for his long train of artillery and supplies, and the Patriots ahead of him felled trees to block his path and dam streams, while sniping at his men as they struggled southward through sweltering heat and clouds of mosquitoes. (Because the soldiers' labors added a new forest-to-lumber-mill road to Skene's domain, there was some suspicion that Skene had recommended the route so he could get a road built.) Moving at about a mile a day, Burgoyne took nearly a month to reach the Hudson, giving his quarry more time to prepare for battle.[52]

Burgoyne's instructions to Indians about gentlemanly warfare had been empty words to many Indian Loyalists. And British officials dealing with the Six Nations knew they could not control the way Indians traditionally fought. Guy Johnson managed to convey that reality in a letter to the Rebels that simultaneously denied British attempts to recruit Indian warriors—but warned that "if the Indians find . . . their Superintendent [Guy Johnson] insulted, they will take a dreadful revenge."[53]

The Rebels had certainly "insulted" the Johnsons and John Butler by forcing them to flee, and so the Indians were supposedly justified in becoming allies and fighting as they customarily did, which meant scalping.

The first sign of Indian warfare came on the afternoon of July 27, when soldiers at Fort Stanwix heard four gunshots. Soldiers and civilians had been told to fire their muskets only in self-defense or as a signal for help. At the sound of the shots several soldiers ran from the fort to the edge of a patch of woods about five hundred yards from the fort. They were too late.

In a meadow speckled with berry bushes, two girls lay scalped and tomahawked, one dead and the other dying. A third girl, shot in the shoulder, had escaped. She emerged from the woods and told what had happened. The three girls had been picking raspberries when four Indians suddenly appeared and fired. Four armed Rebel soldiers had walked past the spot shortly before the attack and apparently had been ignored by the Indians. They had come, said a bitter report on the killings, "not to fight, but to lie in wait to murder; and it is equally the same to them, if they can get a scalp, whether it is from a soldier or an innocent babe."[54]

On the day that the girls were murdered, Jane McCrea, the twenty-year-old daughter of a Presbyterian minister and his wife, was staying in the home of Mrs. Sarah McNeal, near the village of Fort Edward, about one hundred and twenty-five miles east of Fort Stanwix. Fort Edward had been abandoned by defenders, who were retreating before Burgoyne's advance.[55] It lay at the southern end of the Great Carrying Place, where rapids and falls churned the Hudson River, making farther passage impossible. From there canoes and boats had to be portaged to the headwaters of Lake Champlain.[56]

The McCrea family had been sundered by the Revolution. Presbyterians generally were on the Rebel side, and Jane's brother John was a Rebel. But Jane was a Loyalist and the fiancée of Lt. David Jones, a local Tory who had joined the Queen's Loyal Rangers. Jane had been staying with her brother but had moved to the home of Mrs. McNeal, also a Tory, in anticipation of David's arrival with Burgoyne's army.

What happened to Jane McCrea produced myth and rage. The story that swept down the valley and then through the colonies told of a beautiful young maiden attacked by two Indians serving with Burgoyne. The Indians had stormed into Mrs. McNeal's house and carried off the two women. Mrs. McNeal was taken to Burgoyne's camp. Shortly she saw an Indian holding a bloody scalp that she quickly identified as Jane McCrea's. Soldiers rode off and found McCrea's mutilated body.[57]

Burgoyne did not punish the scalper because he did not want to disturb the tense relationship between the whites and Indians in his army. His decision further outraged the people in his path. Patriots seized on the propaganda value of Indian tomahawk killings. Maj. Gen. Horatio Gates, who would soon be fighting Burgoyne, wrote to him, condemning him for the "miserable fate of Miss McCrea" and adding: "Upwards of one hundred men, women and children have perished at the hands of these ruffians."[58] The letter, published in New England newspapers, produced an increase in Continental Army volunteers.[59]

As Burgoyne continued on toward the Hudson, the second, smaller invasion army under St. Leger had progressed up the St. Lawrence from Montreal, then southward on Lake Ontario to Oswego, New York. Besides a detachment of about eight hundred Regulars and Loyalists, there were some eight hundred Indians in the force. There was also a separate unit: Brant's Volunteers, about one hundred Mohawk and some white Loyalists who painted and dressed themselves as Indians. Unrecognized as members of an official unit, and thus unpaid, they expected to live on plunder.[60]

Fort Stanwix, St. Leger's objective, was not the easy conquest he had expected.[61] Its walls of turf and timber had been strengthened. Reinforcements, arriving just before the invaders, raised the number of defenders to some 750, but less than half the size of the approaching force. In June the Continental Congress had adopted the Stars and Stripes as the national flag. Tradition says that the first American flag to fly against an enemy—pieces of a red flannel shirt, a white shirt, and the blue petticoat of a soldier's wife—was hoisted above Fort Stanwix as the invaders approached.[62]

On August 3 St. Leger's force surrounded the fort and demanded surrender. Col. Peter Gansevoort, commander of the fort, replied that he would "defend this fort and garrison to the last extremity."[63] Both sides settled in for a siege. St. Leger issued a proclamation that called on Loyalists to come forth in support of his soldiers—and for Patriots to be beware of his "messengers of justice and of wrath," who would be bringing them "devastation, famine, and every concomitant horror."

When reports of the invasion first reached Brig. Gen. Nicholas Herkimer, commander of the Patriots' Tryon County militia, he asked for a meeting with Joseph Brant, a former neighbor. Herkimer wanted to talk about the chance of maintaining peace between Indians and Patriots. Brant's headquarters was then Unadilla, at the confluence

of the Susquehanna and Unadilla rivers, about eighty miles west of
Schenectady. They met at nearby Sidney, site of a new white settlement. Each man came with an armed escort but agreed to be personally unarmed when they talked face-to-face. Brant held firmly to his
allegiance to the king, and Herkimer departed, knowing that Brant
was his enemy. Herkimer returned to Fort Dayton, about thirty miles
downriver from Fort Stanwix, and prepared for a long-term war in
the Mohawk River Valley.[64]

Herkimer was the grandson of a German immigrant, one of many
who settled in the valley. His English neighbors called him a Dutch-
man and mocked his thick accent. His written orders needed deci-
phering as much as his verbal ones. One is said to have begun: *Ser yu
will orter your bodellyen do merchs Immiedietlih.* (Sir: You will order
your battalion to march immediately.) Herkimer, warning that an invading army of "Christians and savages" was on its way, ordered every
able-bodied man from sixteen to sixty years of age to mobilize with
musket and ammunition. Men older than sixty were to gather women
and children together and prepare to defend them. Any man who
refused to take up arms would be imprisoned.[65]

Herkimer's militia quickly grew, and on August 4 he headed for
Fort Stanwix with about eight hundred men and boys. Three men,
apparently secret Tories, slipped away and hurried to the invaders'
camp to warn of Herkimer's advance.

Among those who marched with St. Leger was Herkimer's brother.
St. Leger ordered Brant and Butler, with a force of about four hundred Indians and Tories, to intercept the militiamen. At the Oneida
village of Oriska, about six miles east of the fort, Brant set up an ambush in a deep marshy ravine.

The militiamen brought with them about four hundred ox-drawn
carts containing supplies for the fort, stretching their column out for
nearly a mile. At the head of the soldiers were about sixty friendly
Oneida, acting as scouts, and Herkimer on his white horse. As the
scouts and advance guard reached the ravine, the road narrowed to
a corduroy causeway surrounded by a dense forest thick with un-

derbrush. Even the Oneida, skilled woodsmen, could not see the hidden foes.

As soon as Herkimer and his first group of men were deep into the ravine, the ambushers—Brant's Indians and an Indian force assembled by Lt. Col. John Butler—struck, killing many militiamen with musket fire and cutting down others with tomahawks. Herkimer, his left leg shattered and his horse killed, ordered men to place him against a tall beech tree with his wounded leg resting on his saddle. He took out his pipe, lit it, and kept shouting orders. Many Oneida had melted away into the forest. But some Oneida—men and women—did fight. One warrior, shot in the wrist, continued to fight using his tomahawk.[66]

Some militiamen panicked and ran off, pursued by the Tory Indians, who killed many. But most men stayed and rallied around Herkimer. At first each time a man fired from behind a tree, an Indian would run to the tree and tomahawk him before he could reload. Herkimer ordered two men to pair behind trees so that while one man reloaded the second could kill the approaching brave.

A sudden rain, soaking muskets' primers, stopped the battle for a short time. During the pause Herkimer had his men form a circle. Ripped by concentrated fire, the Indians began to pull back, beginning what would be a ragged retreat to their camp near Fort Stanwix. Then from St. Leger's camp came reinforcements—a detachment of the King's Royal Regiment. The Tories had turned their green uniforms inside out. For a moment the militiamen believed they were getting help from the fort. Then one of the militiamen recognized former neighbors, cried out a warning, and slew three of them with his espontoon, an officer's half pike. Other men of the valley fought to the death—American against American, kin against kin—until the daylong battle suddenly ended, probably because Butler and Brant expected that a rescue force was on its way from the fort.[67]

Herkimer would die in a few days, joining about 160 militiamen who fell in the ravine. The invaders lost about 150 Tories and Indians, including several chiefs. Further losses came when Rebel raiders from the fort attacked the besiegers' camp before the surviving ambushers returned.[68]

St. Leger, confronted by angry Indians who had fought but gotten no plunder, tried once more to end the siege with a threat that if Colonel Gansevoort did not surrender his fort, the Indians could not be stopped from "executing their threats to march down the country and destroy the settlement with its inhabitants. In this case, not only men but women and children will experience the sad effects of their vengeance."[69]

Gansevoort again rejected the call to surrender. Meanwhile an officer left the fort in the night, got past St. Leger's sentries, and set out for Fort Dayton, where Benedict Arnold was preparing to lead about nine hundred militiamen and Continentals to Fort Stanwix to end the siege. Arnold, well known as a Tory hunter in Connecticut, had discovered a group of Loyalists who were planning an uprising. Among the men he captured were Ensign Walter Butler, Col. John Butler's twenty-five-year-old son, and Hon Yost Schuyler, a deranged young man whose madness was seen as mystical by many Indians. He was related to both Herkimer and Major General Schuyler. Butler and Hon Yost were tried by court-martial as spies, convicted, and sentenced to death.

Some of Arnold's officers appealed Butler's sentence because they knew him as a schoolmate when he studied for the law in Albany. So he was sent off to imprisonment—and subsequent escape, a frequent boon for Loyalist prisoners kept amid Tories. After clemency pleas for Hon Yost from his mother and brother, Arnold said he would spare his life if he would use his mystical powers to convince St. Leger's Indians that a huge army was on its way to the fort. Hon Yost agreed, and to add to his story of a narrow escape from Arnold's soldiers, he shot holes in his clothes with a borrowed musket. Arnold sent Hon Yost off and held his brother as a hostage.

Hon Yost reached the Mohawk camp outside the fort and, when asked how many men were on their way, pointed to the leaves of the trees and said each leaf was a soldier. Hon Yost repeated his rambling story to St. Leger. Frightened Mohawks began grabbing up British supplies, food, and liquor for a hasty retreat. Loyalists and St. Leger's Regulars joined the Indians, fleeing before the phantom army until they reached Oswego and the safety of Canada. They left behind tents, food, ammunition—and St. Leger's own escritoire, the portable writing table that contained his private papers, which historians long afterward used for descriptions of the siege and the flight of his army.[70]

To the north, Burgoyne continued to attract local Tories as he passed into the newly created republic of Vermont. Some Tories joined informally, marching with Burgoyne but not fully committing to a long-term enlistment. Those who openly joined Burgoyne risked their own private battles, especially in areas where Rebels had fled before the army, producing a temporary Tory enclave. Nathan Tuttle, a minor official in the village of Rutland, was a Rebel who stayed. Known to hate and taunt Tories, he disappeared around the time that Burgoyne's army passed through Rutland. In 1786 a former Tory revealed that Tuttle had been bayoneted during an argument, his body weighted with stones and thrown into a creek.[71] We will never know how many other private grievances were settled in a similar manner during the war.

In Vermont, Burgoyne met unexpected Rebel resistance. Short of supplies and in need of horses for his footsore German cavalrymen, he ordered a force of about eight hundred men to forage in the village of Bennington. Burgoyne had learned that Bennington was a Patriot supply depot stocked with corn and cattle. And, still relying on Skene's advice, Burgoyne expected that the area was full of Loyalists who would rush to join his army. But the man in command of the foraging force, Lt. Col. Friedrich Baum, would not be able to speak to any Loyalists because he spoke only German.[72] (Skene translated for Baum; Burgoyne spoke to him in French.)

Awaiting Baum were about fifteen hundred militiamen from New Hampshire and Massachusetts. Their commander was Col. John Stark, a tough old soldier who had been captured by Indians as a young man and bore the scars of heroic service with Rogers' Rangers during the French and Indian War. He had fought in the Battles of Bunker Hill, Trenton, and Princeton.[73] As more militiamen arrived, Stark drew up a complicated plan. After a day of impasse and rain, Stark led the attack.

Some of his men—farmers lacking uniforms—had stuck pieces of white paper in their hats to identify themselves as Rebels. But they looked like Loyalists to Baum, who welcomed them as they drifted into his ranks. When the attack began, the infiltrators started picking off their unwitting foes. Skene saw more men who did not look like soldiers and shouted, "Are you for King George?" An answering fusillade killed his horse. He ran back to a German redoubt.[74]

As at Oriska, neighbor fought neighbor. Lieutenant Colonel Peters, commander of the Queen's Loyal Rangers, spotted Patriot captain Jeremiah Post, a childhood playmate and a cousin of Peters's wife. Just as Post fired and missed, Post rushed toward Peters, shouting, "Peters, you damn Tory, now I have got you!" He shoved his bayonet into Peters's side. Peters fired his gun, "Though his bayonet was in my body," Peters later wrote, "I felt regret at being obliged to destroy him."[75]

Peters joined the retreat, lucky to survive a battle that left more than two hundred of his Loyal Rangers dead, wounded, or captured.[76] Indians scattered and vanished. The Germans valiantly held their redoubt until Baum fell, mortally wounded by a musket ball. Some Germans fought their way into the woods and escaped; some surrendered. Reinforcements for both sides arrived, but as the long day ended, the surviving invaders fled the field. About thirty of Stark's men were killed and forty wounded. The victory inspired militiamen for miles around to head north to join the fight against Burgoyne.[77] "Wherever the king's forces point," Burgoyne later wrote, "militia to the amount of 3 or 4 thousand assemble in 24 hours . . . and the alarm over, they return to their farms."[78]

Burgoyne now faced a growing enemy army while between nine hundred and a thousand of his own men had been killed, wounded, captured, or were missing, many of them dying of wounds, unseen in dense forests.[79] Maj. Gen. Philip Schuyler, who lost the confidence of

Congress after his retreat from Fort Ticonderoga, was now replaced by Maj. Gen. Horatio Gates. Added to his army, besides the swelling ranks of militia volunteers, were Arnold's men, freed by the ending of the Fort Stanwix siege, and Col. Daniel Morgan's three hundred riflemen, whose reputation as marksmen preceded them.[80]

Morgan, captured during the battle for Quebec, had been released in a prisoner exchange in January 1777 and given command of an independent special rifle corps. Morgan picked the men of the unit, who came from western Pennsylvania, Maryland, and Virginia. They wore the rifleman's standard uniform—loose-fitting hunting shirts and leggings.[81]

Gates blocked Burgoyne's path to Albany by entrenching on a Hudson River bluff known as Bemis Heights. On the morning of September 19 Burgoyne's advance guard—mostly Indians and Tories disguised as Indians—probed forward through a dense forest. Morgan's men and some of Arnold's were ordered to clear Indians spotted in the woods. An officer remembered Arnold's turning to Morgan and saying, "Colonel Morgan, you and I have seen too many Redskins to be deceived by that garb of paint and feathers; they are asses in lions' skins, Canadians and Tories; let your riflemen cure them of their borrowed plumes."[82]

As riflemen perched in trees picked off officers and artillerymen, Arnold brought his men forward, generating a ferocious battle that neither Burgoyne nor Gates had expected. Arnold sent word to Gates that with reinforcements he could defeat Burgoyne. But Gates ordered Arnold back to the American redoubt. Gates had lost about three hundred men, compared with Burgoyne's six hundred, many of them officers. Burgoyne had won the day, but Gates's army still lay across the path of the invaders. And the Loyalists in the area had not risen to aid Burgoyne.

While Burgoyne contemplated his future moves, the third facet of his plan—a thrust up the Hudson—was carried out by Sir Henry Clinton. Leaving the defense of New York City to Regulars augmented by Loyalist regiments, Clinton took a force up the river and captured two Continental forts.[83] To punish local Rebels, hundreds of

British soldiers swarmed into Kingston, the interim capital of New York State, and burned much of the town, which the British commander called "a Nursery for almost every Villain in the Country." Another landing party put the torch to Clermont and Belvedere, the manorial homes of the Hudson Valley's best-known Patriots, the Livingstons.[84]

By October 7 Burgoyne's troops were on half rations. Desperate, he led what he called a reconnaissance in force to turn a Gates flank and burst through to Albany. Two miles away, at Gates's headquarters in the well-fortified American camp, Arnold learned of Burgoyne's maneuver and asked Gates to let him lead a counterattack. Gates, born in England and a haughty veteran of the British Army, despised Arnold, a fifth-generation New Englander and an ex-clerk in an apothecary shop. Gates refused to send Arnold into battle. Furious, Arnold mounted his horse and galloped to the sound of the guns, racing away from an officer Gates had sent to stop him.

Arnold sighted a British strongpoint and headed toward it, shouting, "Come on, brave boys, come on!" He led a charge across the line of fire, through cannon shells, musket balls, and grapeshot. His riddled horse fell, throwing Arnold. A wounded German shot him—in the same leg that had been hit at Quebec. "Don't hurt him!" Arnold ordered as his frenzied men sprang forward to bayonet the German.

On a litter made of a British tent and ridgepoles, he was carried off in the last hour of the battle that was the turning point of the war.

The next day Burgoyne withdrew, pursued by Gates's men closing in for the kill. Finally convinced that Clinton would not come to his rescue, Burgoyne surrendered on October 17. France, which had been clandestinely supplying arms to the Patriots for some time, would now openly begin the process that would lead to French recognition of the United States as an independent nation and a partner in a military alliance against Britain.[85]

* * *

Under the terms of the surrender's "Convention"—Burgoyne preferred that word rather than "capitulation"—some 5,600 troops and hundreds of soldiers' wives and children were to be marched to Boston, where they would be shipped to England and the soldiers permanently kept out of the American war. But only Burgoyne and his staff sailed to England. Congress refused to accept Gates's agreement. So the "Convention Army" eventually was marched from Boston to Charlottesville, Virginia, confined there for a while and then herded elsewhere. In a five-year odyssey of misery and despair, the Convention Army continually lost men to death and desertion. An unknown number of German prisoners freed themselves by becoming indentured to American farmers, joining the Continental Army, or simply straying away to a quiet welcome in some German-speaking village.[86]

On the night before Burgoyne and Gates signed the Convention, most Loyalists slipped into the woods and headed for Canada. Peters, suffering from a bayonet wound and a foot grazed by a musket ball, led out his son and, by his count, about 117 survivors of the Queen's Loyal Rangers.[87] Peters said Burgoyne had given him written permission to leave with other Loyalists because he was not sure whether he would be able to protect them from the Rebels when he handed the army over to Gates.[88]

Lieutenant Simmons, the New York Loyalist who had left his wife and children to march with Burgoyne, escaped with twenty-eight men, many of them teenagers. Traveling on foot and by bateaux, all of them reached Canada.[89] Loyalists were covered by the Convention, which forbade them to fight again in the war. But, reacting to the congressional rejection, Governor Guy Carleton declared that because the Convention was invalid, Loyalist veterans *could* fight again.[90]

Burgoyne's attitude toward his Loyalists mirrored the disdain that was typical of British officers. Writing to Lord Germain, the secretary of state for North America, on August 20, Burgoyne underplayed the importance of Tory soldiers, claiming to have only "about 400," of whom fewer than half were armed and could be "depended upon." The rest, he said, were "trimmers merely actuated by interest," "trimmers" being his word for the loyalty that drove them to risk their lives. An official count, given to a parliamentary investigation of Burgoyne's campaign, put the number at 680 prior to the battles near Saratoga. In fact the Queen's Loyal Rangers alone mustered 643 men, of whom 80 had been killed or captured before the Convention was signed.[91]

Neither Burgoyne nor his American foes realized that his surrender at Saratoga, hailed by Congress as a turning point in the war, produced a crisis close to home for the people of the borderland. Rebels watched Stark, Arnold, and Gates as they headed back to General Washington and the main war against General Howe. They were now on their own in their own war.

13

———

TREASON ALONG THE CHESAPEAKE

*The Country is very much draind. The Inhabitants cry out and
beset me from all quarters. but like Pharoh I harden my heart.
Two men were taken up carrying provisions into the Enemy yes-
terday morning. I gave them an hundred [lashes] each by way of
Example. . . . I am determin to forage the Country very bare.*
—*Maj. Gen. Nathanael Greene to
General Washington at Valley Forge*[1]

When Gen. Sir William Howe set forth to capture Philadelphia, he sailed from the Loyalist city of New York to a loyal
shore on Chesapeake Bay. His nineteen thousand invaders were
unopposed when they disembarked at the Head of the Elk on the
northern end of the bay. Their arrival was not a surprise. Howe's
fleet of some 250 ships had been spotted long before, and Patriots
had been able to carry off supplies that had been cached at the Head
of the Elk. But there was no organized resistance when Howe arrived or when he began his march to the Rebel capital, about fifty
miles to the northeast.[2]

Three Loyalist military units were in Howe's invasion force. And
with Howe was a prominent Loyalist: a son of William Allen, the
former chief justice of Pennsylvania and one of the wealthiest men in

the colony. William Allen, Jr., had been a member of the Pennsylvania Committee of Safety and, as a lieutenant colonel in the Continental Army, had fought in the invasion of Canada. Unable to support the idea of independence, he became a Loyalist. As an aide to Howe he would raise a corps called the Pennsylvania Loyalists and become its commanding officer with the same rank he had had in the Continental Army.[3]

Andrew Allen, a brother of William Allen, Jr., had been a delegate to the Continental Congress. Andrew had resigned because he opposed independence and would be part of the British occupation of Philadelphia, a city long controlled by rich old families like the Allens.[4] The elder Allen had gone to England as the Revolution drew near. His son John had also started out as a cautious Patriot and ended up a Loyalist.[5]

Accompanying Howe as his principal civilian aide was Joseph Galloway, once a protégé and friend of Ben Franklin. Galloway, as a delegate to the First Continental Congress, had tried unsuccessfully to get supporters for his "Plan of a Proposed Union between Great Britain and the Colonies," which would have created an American parliament and a royally appointed "president general."[6] He also called for loyalty oaths for colonial teachers, students, and lawyers, pledging "faithful obedience" to the British Parliament.[7] The failure of his plan awakened his dormant Tory leanings. He went to New York and put himself at the service of Howe.

On hand to welcome Howe to Maryland was Robert Alexander, a prominent politician who owned a large piece of land, crowned by his mansion, the Hermitage. Three days before, Alexander had been the host of a dinner for General George Washington, who knew Alexander as a Patriot and a deputy delegate to the Continental Congress. Washington, in Head of the Elk on a reconnaissance mission, expected an imminent British invasion. When he asked Alexander whether he was going to flee, Alexander said he preferred to stay.

Washington apparently did not know that Alexander had resigned as an officer in a Maryland militia after the Declaration of Independence was proclaimed. For some time, he had been conferring with

British officers aboard one of the Royal Navy warships that patrolled the bay and served as floating meeting places for the state's Tory underground.[8]

Alexander genially allowed Howe the use of the Hermitage as a temporary headquarters and turned a bit of a profit by selling the invaders livestock and grain.[9] Local farmers with Tory inclinations followed Alexander's example, supplying the invaders during the two weeks that Howe spent preparing for the march on Philadelphia.[10] Others, aided by militiamen, showed their anti-British stance by squirreling away produce and hiding their cattle, horses, and wagons.

Howe offered a general pardon to "officers and private men, now actually in arms against his Majesty," who "shall voluntarily come and surrender themselves."[11] In fact there were few armed men except for Loyalists, who had been organizing for months on Maryland's Eastern Shore. In 1776 the Rebel-controlled Maryland legislature passed a law defining Tories as traitors and making execution the punishment for treason against the state.[12] Maryland's Rebel government considered the Eastern Shore to be in insurrection. But the coastal Tories were so powerful and well organized that they were not easily tamed.

Brig. Gen. William Smallwood, the commander of the Maryland militia, called the Eastern Shore "a Place which becomes the Reception of Deserters, escaping prisoners, and most of the Disaffectted who have been expelled [from] the neighbouring states."[13] Delaware contributed many Tories to the bands formed in the Eastern Shore, where state boundaries meant little. A Tory force of two hundred was formed by men from both states. They had three cannons delivered from a Royal Navy man-of-war.[14]

Months before Howe's arrival Patriots in the area had appealed directly to Congress for military aid.[15] Congress responded with stern resolutions against Eastern Shore Tories but had few resources to offer.[16] The Maryland militia, much as its men wanted to fight Howe on Maryland's shore, lacked firearms. General Washington himself could do little more than sympathize, writing that "it gives me pleasure to hear that your people are so unanimously bent upon giving opposition to the Enemy. I wish it was in my power to furnish every man with a firelock that is willing to use one, but that is so far from being the Case that I have scarcely Sufficient for the Continental Troops."[17]

James Chalmers, a wealthy Scotland-born planter, added to the armed Tories on the Eastern Shore by raising three hundred men for a battalion he named the Maryland Loyalists. He was commissioned a lieutenant colonel. One of his officers was Capt. Philip Barton Key, whose nephew, Francis Scott Key, would write "The Star-Spangled Banner." Chalmers, as "Candidus," was famous among Loyalists for *Plain Truth*, a pamphlet he wrote as a Loyalist answer to Tom Paine's *Common Sense*.[18]

Chalmers's raising of the Maryland Loyalists augmented the recruiting drives of the three Tory units that arrived with Howe at the Head of the Elk.* Howe issued a proclamation promising land to all of "his Majesty's faithful and well-disposed subjects" who signed up for two years or the length of the war: two hundred acres for a noncommissioned officer, fifty acres for each private.

Another Howe proclamation promised protection to anyone who swore allegiance to the Crown within the next sixty days.[19] Concerned about the looting habits of his men, Howe issued warrants authorizing his provost marshal to waive courts-martial and "execute upon the Spot any Soldier or follower of the Army" caught plundering any Loyalist home or farm.[20]

Howe was still at the Hermitage when Tory spies informed him that Washington was planning to block the invaders at Brandywine Creek, about thirty miles south of Philadelphia.[21] Howe marched there, his troops harassed along the way by the few militiamen who had muskets. In Philadelphia, responding to an order from the Supreme Ex-

---

* The prestigious Queen's American Rangers, the Second Battalion of the New Jersey Volunteers, and a detachment of the Royal Guides and Pioneers. See ToriesFightingForTheKing.com for information about Loyalist military units.

ecutive Council, Patriots were arresting prominent Tories, sending some off to confinement in Virginia and offering paroles for those who promised they would not do anything injurious to "the united free States of North America."[22]

At Brandywine, Howe outmaneuvered the Continentals in a battle that raged through the day and ended with hand-to-hand combat as night was falling over a battlefield strewn with the dead and dying. The Continentals and their comrades in the Pennsylvania militia lost some one thousand men killed and wounded—about twice the casualties suffered by their foes. The Marquis de Lafayette, newly arrived to the Continental Army and about to mark his twentieth birthday, was shot in the leg but continued to fight.

Stephen Maples Jarvis, a Connecticut Tory and a Queen's American Ranger, went into battle at Brandywine. "We came in sight of the enemy at sunrise," he wrote in his journal. "The first discharge of the enemy killed the horse of Major Grymes who was leading the column, and wounded two men in the Division directly in my front, and in a few moments the Regiment became warmly engaged and several of our officers were badly wounded." (Maj. John Randolph Grymes was a descendent of one of the first families of Virginia.)[23]

The Rangers crossed the creek—"water took us up to our breasts, and was much stained with blood"—and fought into the darkness. "In this day's hard fought action," Jarvis wrote, "the Queen's Rangers' loss in killed and wounded were seventy-five out of two hundred fifty rank and file which composed our strength in the morning."[24]

A few days later Washington sent Brig. Gen. "Mad Anthony" Wayne and fifteen hundred men to the Paoli Tavern, on the Lancaster Road, where Wayne was to ambush Howe's baggage train as its wagons rumbled toward Philadelphia. Howe learned the details of Wayne's plans from local Tories[25]—intelligence that helped to produce a bloody surprise attack.

British major general Charles Grey, wanting to approach Wayne at midnight as quietly as possible, ordered his men to remove flints from their muskets so that when he ordered a bayonet charge, no Regular would accidentally fire and give an alarm. Aiding Grey in his planning was his principal aide, Capt. John André, who had a personal reason not to like Rebels. André had been captured in 1775 during the abortive invasion of Canada. Expecting to be treated like an officer merely awaiting exchange, he had been handled roughly. Tories seen befriending him were arrested, and André's guards once threatened to kill him. But through his captivity André never lost the savoir-faire and keen wit that marked him as a bright and rising officer. When he was finally exchanged, he was assigned to Grey's staff.[26]

On September 21, 1777, Grey's men, who probably had been given the password by a Continental deserter, got past Wayne's sentinels and pounced on the camp. "I stuck them myself, like so many pigs, one after the other, until the blood ran out of the touch-hole of my musket," a Hessian sergeant later wrote.[27] Seeing men hacked and pierced by cold steel, Wayne's soldiers panicked. Patriots called the battle the "Paoli Massacre" and dubbed its perpetrator "No Flint Grey." Propagandists on both sides claimed two hundred Rebels had been killed by bayonets or sabers. Eventually the likely count came down to fifty-three.[28]

Although more maneuvering and fighting would go on for a few days, nothing could now stop Howe from taking and holding Philadelphia. Hundreds of Patriots fled the city, as did members of Congress, who sought safety first in Lancaster, Pennsylvania, and then in York. Washington took his army to the hills around Whitemarsh, a small town north of Philadelphia. General Howe tried to attack him there but failed and pulled back, deciding to make British winter quarters in the comfort of Philadelphia, the largest and wealthiest city in America, and, like New York, a bastion of Loyalists.

On the night of December 11, Washington's men left the camp in the hills of Whitemarsh and began pushing wagons into the Schuylkill River and placing boards between them. Through the night the army crossed the river on the wagon bridge. Washington led his men to a hamlet called Gulph Mills, where the army camped for several days.[29]

Albigence Waldo, an army doctor from Connecticut, looked around the Gulph Mills camp and wrote in his diary: "There comes a Soldier, his bare feet are seen thro' his worn out Shoes, his legs nearly naked from the tatter'd remains of an only pair of stockings, his Breeches not sufficient to cover his nakedness, his Shirt hanging in Strings, his hair dishevell'd, his face meager. . . . exhausted by fatigue, hunger & Cold." In the face of that soldier Waldo saw the faces of all the men, whipped by a "cold & piercing" wind, who followed General Washington down a steep frozen road to a place called Valley Forge.[30]

Philadelphians hailed the arrival of Howe's army "by loudest acclamations of joy." There was no armed resistance. Tories led the invaders to Rebels still in the city, and hundreds were imprisoned. Galloway and the Allens protected upper-class Rebels they knew, leaving them to an uneasy life in a city that belonged to the British and their Loyalist friends.[31] The editor of the *Pennsylvania Evening Post* switched the newspaper from pro-Patriot to pro-Loyalist.[32]

Quakers, whose faith condemned war and earned Rebel distrust, felt more comfortable under British rule. Merchants, unenthusiastic about the Revolution, believed that the British would bring stability to the city—and payments in gold rather than near-worthless Continental currency. Young men, eager to join the winning side, signed up for the Queen's American Rangers, which needed replacements after Brandywine, and three newly formed units: the Roman Catholic Volunteers, the Maryland Loyalists, and the Pennsylvania Loyalists led by Lt. Col. William Allen, Jr.[33]

British officers easily entered high society, and upper-class families were soon inviting their amiable occupiers to parties and dances. A pretty, flirtatious teenager, sixteen-year-old Peggy Shippen, was overjoyed when a handsome British officer named John André called at her family's Society Hill mansion and sketched her.[34] "You can have no idea of the life of continued amusement," eighteen-year-old Rebecca Franks wrote a friend. ". . . . Most elegantly am I dressed for a ball this evening. . . . I spent Tuesday evening at Sir William Howe's, where we had a concert and dance." At one ball Howe shocked his hostess by appearing with his mistress, the beautiful Mrs. Joshua Loring, on his arm. (Her husband was in New York, making money as commissioner of prisoners.)[35]

Galloway became Howe's spymaster, the supervisor of police, and the superintendent of the port, controlling imports and exports and setting up regulations for pricing and selling liquor, molasses, salt, and medicines.[36] He also worked at preventing goods from reaching Washington's army. He recruited spies from the ranks of the hundreds of soldiers who were deserting Valley Forge.[37] By one estimate his intelligence network consisted of eighty agents who gathered information and twenty counterintelligence agents who ferreted out Continental Army spies. He also performed such intelligence chores as administering and recording oaths of allegiance and producing a map of all the roads in the area.[38]

About two hundred Tories served in Galloway's police force, called the "night watch." Galloway selected other Tories to control the entrances to the city, with the power to issue passports, and assigned men to disarm Rebels, act as couriers, oversee the care of prisoners, find quarters for troops (usually in abandoned Rebel homes), guide troops on missions in and around the city, and run a lottery for the relief of the poor. Galloway issued tavern licenses and supervised the takeover by Tories of stores that had been owned by Rebels. Among the new storeowners were Virginians and other Tory refugees who flocked to a Loyalist-controlled city under the protection of British troops.[39]

Galloway continually assured Howe that thousands of Loyalists were ready to spring to arms. Some did; 150 men or so joined such cavalry outfits as the Philadelphia Light Dragoons and the Chester County Light Dragoons.[40] At least 255 men enlisted in the Queen's American Rangers. A total of about fourteen hundred men joined Tory forces in the Philadelphia area. Another 400 or so enlisted in the West Jersey Volunteers, which had both infantry and cavalry units.[41]

At Valley Forge, Washington was losing men while desperately trying to find food for those who stayed. After the Royal Navy took control of the Delaware River, Howe began getting most of his supplies via ships from New York. The British also foraged, buying provisions from farmers who defied Patriot threats against people who aided the enemy. British Regulars and armed Tories attacked Rebel foraging parties. Washington, who had a highly efficient intelligence system inside Philadelphia, usually knew when Howe was sending out foragers. But, because of desertions and fears of mutiny, Washington had to limit sorties against them.[42]

The only good news at Valley Forge was the arrival in February 1778 of Friedrich Wilhelm von Steuben, a former Prussian captain on the staff of Frederick the Great. Steuben was jobless in Paris when he was discovered by Ben Franklin, a recruiting officer as well as a diplomat. Steuben immediately impressed Washington.

Steuben began by teaching the ragged, unruly men how to be soldiers who marched in formation and fired muskets on command. While his soldiers, under Steuben's tutelage, began to coalesce into a fighting force, Washington made a new move to get food for his men. In January 1778, he turned to the commander of the local militia, Brig. Gen. John Lacey, a Quaker and, at twenty-five, the youngest Patriot general.[43] Washington ordered Lacey to stop the "immense supplies" flowing to Philadelphia and to protect the meager provisions heading for Valley Forge.

The "market people," as the British suppliers were euphemistically known, included both committed Tories and apolitical war profiteers. With the addition of Tory infantry and cavalry units, Howe's foraging expeditions became bigger and bolder and persistently eluded Lacey's patrols. Many of the Tories pillaged from longtime neighbors and kidnapped others who were Rebel leaders. In one raid, twenty-six miles from Philadelphia, about forty Loyalists killed five militiamen, took thirty-two prisoners, and pillaged a wool mill, taking two thousand yards of cloth meant for Rebel uniforms.[44]

Lacey did not have enough soldiers to stop the raids or the illegal trade. His frustration showed itself one day when his men stopped a farmer who was trying to smuggle a wagonload of produce to Philadelphia. The militiamen confiscated his goods and horse, tied the farmer to a tree, and pelted him with his own eggs.[45] By the end of March, Lacey was so angry that he proposed forcibly removing everyone in a wide swath of land between the Delaware and the Schuylkill rivers. Washington, who had enough problems with civilians in the area, rejected Lacey's suggestion.[46]

Galloway's operatives, who had been carefully tracking Lacey's operations, provided Howe with information to plan a major attack on the militiamen. Howe decided to send a combined force of Regulars and Loyalists led by Maj. John G. Simcoe, regimental commander of the Queen's American Rangers.[47]

A sign on an oddly shaped piece of wood marked the Crooked Billet Tavern on the old York road about sixteen miles north of Philadelphia. Here Lacey and four hundred militiamen were camped on the morning of May 1. They awakened to discover that they were surrounded and under attack. No matter where Lacey and his men turned they were targets for Simcoe's infantrymen and cavalrymen.

Survivors fled, firing as they ran, until Simcoe broke off the attack. Carrying away everything they could find in Lacey's abandoned camp, Simcoe and his men returned to Philadelphia, where they sold their plunder and shared in the proceeds. On the battleground lay bodies of men who had been bayoneted, slashed by cutlasses, and burned. At least twenty-six militiamen were killed. Witnesses later swore in depositions that some militiamen had been killed after surrendering and that wounded men had been thrown to die into a burning stack of buckwheat straw.[48]

A week after the Crooked Billet raid, General Howe learned that his request to be relieved had been accepted and he was to be replaced by Lt. Gen. Sir Henry Clinton. British officials in London believed that the Rebels' new French allies might lead an attack on New York or that a French fleet might blockade the Delaware. So Clinton was ordered to evacuate Philadelphia and take his army to theoretically threatened New York.

To bid farewell to Howe, André—poet, artist, actor, musician, impresario—produced what he called a *mischianza*, meaning "medley" or "mixture" in Italian. The extravaganza began on the afternoon of May 18, with the boom of cannon salutes as a fleet of "Galleys, Barges, & flat boats, finely decorated" slowly moved down the Delaware River. The vessels sailed past flag-festooned ships along the wharves. Shipboard bands played "God Save the King," and cheers went up from ships and shore.

At the landing of an old riverside fort, the gala fleet disembarked its passengers—officers in their full regimentals, ladies richly gowned. They walked in a stately procession between ranks of soldiers and lines of mounted dragoons, up a gentle hill to the abandoned mansion of Thomas Wharton, a wealthy Quaker merchant who had been exiled to Virginia as a Tory.[49] The promenade ended at a large square of lawn flanked by triumphal arches and grandstands, where the ladies and officers sat. In the front row were seven beautiful young women wearing flowing, Turkish-style gowns. Each wore a jeweled turban of a different color.

As André described his creation, "A band of knights, dressed in ancient habits of white and red silk, and mounted on gray horses, richly caparisoned in trappings of the same colors," rode to the center of the lawn.[50] The chief of the knights was accompanied by two young black slaves, "with sashes and drawers of blue and white silk, wearing large silver clasps round their necks and arms." The white knights were champions for the seven young women, who included the lovely Peggy Shippen, one of André's favorite belles. He played the role of herald, proclaiming that she and the other ladies "excel in wit, beauty, and every accomplishment." If anyone doubted that, the herald declared, let him show it "by deeds of arms."

Another band of knights, dressed in black and orange, galloped up and accepted the herald's challenge. In a reenactment of a medieval joust, the chiefs of the white knights and the black knights "engaged furiously in single combat." After the jousting the revelers walked into the mansion to dance in a ballroom "filled with drooping festoons of flowers in their natural colors."

At ten o'clock all stood at the windows and gasped at fireworks bursting over the river. At midnight "large folding-doors, hitherto artfully concealed," suddenly opened and revealed a huge dining hall built by British Army engineers as a temporary addition to the mansion. Bowing before the 430 guests were "twenty-four black slaves in Oriental dresses, with silver collars and bracelets." (All the slaves in the cast were in fact slaves.)

Suddenly from the grounds came a rattle of gunfire and a flare of flame. Guests were told that it was all part of the celebration. But the roll of a drum, signaling alarm, told all the dancing officers that an attack had begun. After a few tense moments the gunfire and the alarm ceased. Cavalrymen out of Valley Forge, staging their own mischievous *mischianza*, had sneaked to nearby redoubts, poured whale oil over timber barricades, set them afire, and galloped away. Sentries fired at them, but they all escaped.[51] The party, interrupted by the short and mysterious absence of a few officers, went on until dawn.

André's flamboyant description of his production shocked many who read it in London. One newspaper called the fete "nauseous." Philadelphia Quakers were also aghast, as was a former Quaker, Edward Shippen IV, Peggy's father. He was one of many wealthy Philadelphians who, having decided to remain in the city, had tried to maintain a delicate political balance. A former royal judge and an Anglican convert, he talked like a Tory and lived like a Tory. And there was Peggy Shippen starring in the *mischianza*, which flagrantly displayed support of Howe and the British occupiers. The timing could not have been worse, for the British were about to evacuate the city, exposing people like Shippen to Rebel justice.[52]

Aware of the Continental Army's desertions and the Tory temptations to wavering Patriots, Congress decided that officers should declare their loyalty to the United States and, in a solemn oath, renounce George III, his heirs, and anyone who aided the king.[53] At Valley Forge on May 30, Maj. Gen. Benedict Arnold was one of the officers who placed his hand upon a Bible, swore the oath, wrote his name and rank on a small piece of paper with the "Oath of Allegiance" printed on it, and signed the paper.[54]

Arnold, limping from his most recent wound, could not take a field command. From spies in Philadelphia, Washington had learned that the British Army was secretly preparing to pull out of the city. He offered Arnold a new assignment, which he accepted: military governor of Philadelphia, eastern Pennsylvania, and southern New Jersey, with headquarters in Philadelphia.[55]

As the Royal Navy had often demonstrated, its ships could carry troops to their destination by sea better than their feet could on land. But General Clinton was overwhelmed by frightened Tories begging to be taken to the Tory stronghold of New York along with his army. The Tories did not want to face their Patriot neighbors after the British left. So Clinton decided he would take his troops to New York by land. The Tories—some five thousand men, women, and children—would, like their predecessors in Boston, flee by sea.[56] Men dragged carts and wagons, "laden with dry goods and household furniture, . . . through the streets . . . to the wharves for want of horses." Many of the possessions that reached the ships were thrown overboard by sailors making room for military supplies.[57]

Supplies and ammunition that the army would have loaded onto ships now had to be moved overland. When Clinton began his march to New York on June 18, 1778, he needed more than fifteen hundred horse-drawn wagons, carrying not only supplies for an army on the march but also loot taken from Patriot homes. In one of those wagons were books, musical instruments, scientific apparatus, a portrait of Ben Franklin, and other loot taken by Captain André from Franklin's home, where André had lived during the occupation of the city.[58] Soon to be promoted to major, André would serve in New York on Clinton's staff as his principal intelligence officer.

Behind the wagon train and on its flanks were Loyalist units and Tory refugees who missed the ships to New York, adding themselves and their own plunder to a line of march about twelve miles long.[59] Rebel militia cavalrymen followed the evacuees, snatching a few members of the Loyalist baggage guard and making them prisoners.

As the train passed through New Jersey, tardy Tory refugees joined the march, many of them drawn to the protection of the Queen's American Rangers.[60]

On June 19 Washington's new army, disciplined and confident, marched out of Valley Forge in pursuit of the British. Around sundown that day a small force under Major General Arnold entered Philadelphia to occupy it under military rule until the civilian government and Congress returned. Seated in a coach-and-four, Arnold rode up to the Penn Mansion, which had been Howe's headquarters. Thousands of people cheered, this time for the Patriots. Howe's men had left behind a city full of wrecked houses, gutted Presbyterian churches, and looted stores. Some grand homes had been left untouched, one of them belonging to the Shippen family.[61]

Patriots hastily formed a municipal government, which confiscated the property of Galloway, the Allens, and other leading Loyalists. Eventually Rebel officials exiled the wives and children of Tory men who had left Philadelphia; women who defied the banishment were put in the workhouse until they could give security that they would leave the state and never return.[62]

Rebel officials were empowered to collect from alleged Tories not only their weapons but even their shoes and stockings. A new ordinance authorized the seizure of the personal estates and effects of anyone who then or in the future joined Loyalist regiments or in any way aided "the King's army." College or academy faculty members, along with all schoolmasters, merchants, traders, lawyers, doctors, druggists, notaries, and clerks, could be fired and fined if they did not swear allegiance to the Patriots.[63]

On the road to New York, Pennsylvania militiamen were shadowing Clinton, sending information back to General Washington. By June 26 the departing army had reached Monmouth Court House (now Freehold, New Jersey), about thirty-five miles from ships waiting to ferry the troops and supplies from New Jersey across New York Bay to New York City. The soldiers camped around a road flanked by ravines and leading across a bridge to a stretch of swampy land. Beyond were high grounds and a better road that would speed up the march to the ships. Washington chose to strike before Howe's men reached that road.

When the two armies clashed, clouds of dust swirled through ninety-degree heat. Scattered units of British and American forces were spread across a battlefield that covered some twenty miles. Soldiers on both sides dropped from heat exhaustion. One of the Continental wings was commanded by Maj. Gen. Charles Lee, recently exchanged and back from his comfortable sojourn as a coddled captive in New York City. Lee had argued against Washington's decision to fight Clinton on the march, suggesting that he be allowed to reach New York. But Washington had made up his mind, and he ordered Lee to pounce on the British rear guard, which was separated from the main force, on June 28.[64]

Lee failed to fight. Instead he ordered a retreat. "Grating as this order was to our feelings," Pvt. Joseph Plumb Martin later wrote, "we were obliged to comply." Martin was sitting by the side of a road when an infuriated Washington rode up to Lee. Martin was close enough to Washington to hear him ask officers by whose order the troops were retreating. Washington rode off, and Martin believed he said, "Damn him!" Whatever his words, Martin thought Washington "seemed at the instant to be in a great passion." He gathered Lafayette and other officers around him and started calmly giving new orders while, Martin later wrote, British cannon shells were "rending up the earth all around him."[65]

When Washington asked the bewildered retreating soldiers if they could fight, they answered him with three cheers. "His presence stopped the retreat," Lafayette later wrote.[66] The battle ended with both Americans and British settling down for a night of rest. Before dawn Clinton ordered the rejoining of the rear unit, the wagon train, and the rest of his troops, and the reconnected army slipped away on the road to New York. Washington did not attempt to attack him again. Although the British successfully escaped, the Americans claimed a victory, marred by Lee's order to retreat. Lee's disobedience, along with his subsequent ranting to and about Washington, led to his being court-martialed. Lee's behavior inspired a duel in which he was wounded. He was dismissed from the Continental Army; and, long after his lonely death, revelations surfaced about the strategic plan he had treacherously presented to General Howe.[67]

In the spring of 1778, as better days were dawning at Valley Forge, Washington had received a routine letter from a young Pennsylvania officer reporting on legal matters concerning two civilian prisoners. After giving his equally routine reply, Washington spewed a surprising diatribe: "With respect to your future treatment of the Tories, the most effectual way of putting a stop to their traiterous practices, will be shooting some of the most notorious offenders wherever they can be found *in flagrante delicto*. This summary punishment inflicted on a few leading traitors will probably strike terror into others and deter them from exposing themselves to a similar fate."[68] Washington's outrage was understandable. Tories and Tory sympathizers had been starving his soldiers while feeding the British occupiers of Philadelphia.

A short time after suggesting that Tories be shot, Washington got news about the shadowy role that Tories and Indians were playing on the frontier. An officer, on home leave in Wilkes-Barre in northeastern Pennsylvania, learned that a combined force of armed Tories and Indians was planning an attack on a nearby settlement called Wyoming Valley, on the Susquehanna River. He rushed to York, where Congress's Board of War still had its headquarters, and asked that men be released to defend their homes. On June 19, as Washington was heading out of Valley Forge in pursuit of General Clinton, a member of the board wrote a letter to the commander in chief asking for the release of men from the threatened settlement.[69]

Already, desperate men were heading for Wyoming Valley—some

with orders, some without, some led by officers who had resigned commissions, all moving faster than the letters and documents that were churning through the Continental Army and the Board of War.[70] Washington approved the sending of the men to Wyoming Valley. But he knew that soon he would have to send far more to rescue the entire frontier.

14

———

VENGEANCE IN THE VALLEYS

*I have had a meeting with several of the principal Chiefs of the
Seneca Nation. . . . There is just now a party of Senakies come
in who have had an action with a number of Rebel forces on the
Ohio, in which the Indians . . . took two prisoners & thirteen
scalps.*

—*Letter from Col. John Butler to Sir Guy Carleton*[1]

The war against the homes and farms of Pennsylvania's Wyoming
Valley was rooted in an episode during Burgoyne's Saratoga
campaign. Sir John Johnson's frontier comrade in arms, Col. John
Butler, had led the band of Indians and Tories who ambushed Brig.
Gen. Nicholas Herkimer and his men in the ravine at Oriska. Gover-
nor Carleton, impressed by Butler's performance, authorized him to
raise a corps of eight companies of Tory Rangers "to serve with the
Indians, as occasion shall require."

At least two of the companies, Carleton said, had to have men ca-
pable of "speaking the Indian language and acquainted with their
customs of making war."[2] Butler and his recruiters would eventually
enlist more than nine hundred men of all ranks. The corps, known
as Butler's Rangers, would fight frontier Rebels in New York and

Pennsylvania, in the wild Virginia territory that would become Kentucky and West Virginia, and in the Old Northwest, which would become Ohio, Indiana, Illinois, Michigan, Wisconsin, and part of Minnesota.

Many of the earliest recruits were descendants of German immigrants who lived in the Mohawk Valley. Others included Loyalists who called themselves Refugees. They had entered Canada from New York and Pennsylvania, leaving behind homes and farms that Rebels then confiscated. While recruiting these Refugees and Loyalist Indians, Butler created a network of spies and operatives stretching as far as New York City and Philadelphia. One of his agents guided escaped British prisoners to the safety of Canada. Butler also made use of information gleaned from deserters from the Continental Army and runaway slaves drawn to the Tories by the promise of freedom.[3] Butler's own Indians spotted Oneida and Delaware—Indians friendly to the Rebels—attempting to penetrate Butler's network.[4]

Ranger headquarters was at Fort Niagara, which the British had taken from the French in 1759. The fort stood on the western bank of the Niagara River (now Niagara-on-the-Lake, Canada). Rangers' families settled there, keeping livestock and growing wheat, oats, corn, and potatoes.[5] Strategically the fort gave Butler and his Mohawk ally, Capt. Joseph Brant, an entryway to the fertile river valleys of northern New York. Butler and Brant had met with Indian leaders and, in Butler's words to Carleton, they "express the greatest desire to join me in an attack on the Frontiers of the Rebellious Colonies."[6] The Indians had dealt with British officials for generations, building a sense of trust.

A frontier raid did not always produce a written record. In obscure places people fought, people lived or died, and tales of atrocities spread and grew by word of mouth. Raids were not battles to be chronicled in the history of a Tory regiment. But the record of one early raid does survive. In May 1778, about three hundred white and Indian Tories, led by Brant, pounced on Cobleskill, thirty-five miles or so west of Schenectady. The hamlet consisted of about twenty

families living along a three-mile stretch of a small river valley. Tories marked Cobleskill as a Rebel outpost with its own militia. Unlike many other isolated communities, the town was not fortified.

People of the embattled valleys usually lived around a fort. Typically it would have bastions jutting out at corners or along a side
as strongpoints of defense. Major forts encompassed a large-enough
area to accommodate nearby families. Emergency makeshift forts
were also built around stone houses or churches. Beyond the fort
were earthworks and perhaps a stockade enclosing tents or cabins.
During times of sowing or harvesting, when farmers were most vulnerable, families made temporary homes within the stockade or in
the fort itself. Farmers went out to their fields in the morning, armed
with their muskets or escorted by militiamen, and returned to the
fort at the end of day.

At the sound of alarm—such as three shots from a cannon—people
fled to the fort. Able-bodied men mustered, prepared to sally out and
defend the fort; older men became a fort guard. Some women cared
for babies and young children. Other women and older children prepared places where they would care for the wounded, or they built
fires for heating iron pots for melting down lead, which they poured
into bullet molds.[7]

When militia scouts reported seeing Indians near Cobleskill in May
1778, there was no local fort to run to. The captain of the militia sent
men to ask for help at the nearest fort, about ten miles away. Officers
there immediately sent thirty-three Continental Army men under
Capt. William Patrick of Massachusetts. The Continentals joined fifteen militiamen who had assembled at a house chosen as headquarters. Scouts went out, encountered two Indians, and shot one dead.

Two days later scouts reported a large force of Indians and white
men approaching Cobleskill. Patrick marched his men toward the invaders and spotted about twenty Indians. Against the advice of the
local militiamen, he pursued them. The militiamen's instinct was
right: Patrick had led his men into an ambush. Musket fire burst out
of the forest.

The ambushed soldiers and their foes took cover behind trees, firing at close range. Men fell on both sides. Patrick was fatally shot, as were two men trying to carry him away. The surviving Continentals and militiamen retreated, sprinting when they reached open ground. Five men sought refuge in a house. It was vacant, as were all the other houses, because people had run into the woods at the sound of musketry. The pursuers stopped to focus on killing the men in the house and burning it down.

The rest of the soldiers were able to flee while the raiders concentrated on plundering and then torching the houses. They also burned down barns, set haystacks afire, and killed or stole livestock.[8] Militiamen recognized some of the white Loyalists, but Butler was not among them. Later a Tory accused of helping the raiders was shot dead while attempting to escape from members of a Committee of Safety. He was one of several Tories killed by frontier Rebels dispensing their own justice.[9]

Of the estimated twenty-five raiders who were killed in the firefight, most were Indians. Militiamen arrived after the battle and buried the bodies of fourteen of Patrick's soldiers. The bodies of the five men who had entered the house had been "Butchered in the most Inhuman manner," a militia officer reported. Ten houses and barns were in ashes, and "Horses, Cows, Sheep &c. lay dead all over the fields." A total of twenty-two defenders were reported killed and six wounded. At least two were captured, including a Continental officer—whose life was said to have been spared when Brant, a Mason, saw him giving the secret Masonic sign that was an appeal for help.[10]

Until July 1778 Wyoming Valley remained untouched by the raiders— but not by bloodshed. For decades the area had been the battleground of a civil war between Pennsylvanians and people from Connecticut who had settled in the valley in 1754 and later founded Wilkes-Barre. They had based their claim on Connecticut's 1663 charter, which gave the colony land as far westward as the Pacific Coast. What became known as the Yankee-Pennamite War was punctuated by small battles and some wanton killings. The feuding people of the valley set aside the land dispute at the beginning of the Revolution but were still divided as Rebels and Tories.[11]

The bountiful valley was called the breadbasket of the Revolution, making it a prime military objective. The valley also had strategic value: If Butler and his Tory-Indian force could gain control of the valley, western New York and western Pennsylvania would be open to attack from Canada. And Wyoming Valley, rich in crops but scarce of people, looked particularly vulnerable to Butler. Because of its clouded political status, neither Pennsylvania nor Connecticut was clearly responsible for protecting the valley.

Late in June, Butler assembled about four hundred Rangers and Sir John Johnson's green-uniformed King's Royal Regiment, along with about five hundred Iroquois, for his biggest and boldest raid. The invasion began when an advance party of Indians attacked men working in fields. The Indians killed and scalped three of them and took two prisoners whom they later tortured and killed. Survivors of the attacks slipped away and carried news of the invaders to two of the forts.

The next day, July 1, two men at Fort Wintermoot, on the Susquehanna opposite Wilkes-Barre, volunteered to scout the area for the invaders. Longtime residents and members of the family the fort was named after, they were trusted neighbors. They were also secret Tories, and their scouting was for Butler, not their neighbors. They led him to a good bivouac site near, but not visible from, the fort.[12]

The Wintermoot scouts returned to the fort with an officer from Butler's Rangers. The gates were opened, and the three entered. The officer, in Butler's name, demanded surrender, promising that Butler would not harm the fort's men, women, and children. One man moved to resist. His wife was at his side, grasping a pitchfork. But they stood alone. The fort was handed over to Butler, who later appeared with most of his force and made Fort Wintermoot his headquarters. A detachment of Rangers, including some who were former local residents, went to a second, smaller fort, which also surrendered.[13]

The strongest fort in the valley, Fort Forty, along the banks of the Susquehanna, became the headquarters of the Patriots' armed defenders. The fort was named after its builders, the forty settlers who had come from Connecticut in 1770 to press that colony's claim for the valley.[14] Now it sheltered hundreds of women, children, and armed men. When Butler demanded that Fort Forty surrender, the settlers gathered there refused to give up.

On July 3 they assembled a defending force: "two hundred and thirty enrolled men, and seventy old people, boys, civil magistrates, and other volunteers."[15] As they marched from the fort, three men came galloping up on exhausted horses. They were two Continental Army officers, Lt. Phineas Pierce and Capt. Robert Durkee, and Durkee's black servant, Gershom Prince. They had ridden through the night to reach Fort Forty and report that a company of soldiers was on the way. Seeing the fort's defenders heading out to battle, they realized that the rescue force would arrive too late.

Durkee and Prince, like many people of Wyoming Valley, had been born in Connecticut and later settled in the disputed Pennsylvania land. They, along with Gershom Prince, were veterans of the French and Indian War. They had served at Valley Forge and the Battles of Brandywine and Paoli.[16] The three joined the defense force, which had left the fort not quite sure where or when the battle would be. Then they saw flames rising from Fort Wintermoot and headed toward it.

Butler, expecting the fire to lure a rescue force, had drawn them into a trap. As the Fort Forty force neared Wintermoot, they saw the invaders in a clearing, arrayed in a line for battle. Butler had taken off his uniform jacket and hat and wore a black kerchief knotted around his head.[17] The defenders formed a line themselves and advanced.

Hidden Indian marksmen cut them down. Every company officer was mortally wounded or killed at the head of his men. Indian musketry, tomahawks, and war whoops set off a rout. Durkee, wounded, was dragged away by another Continental officer who had made it to Wyoming Valley in time for the battle. Seeing that the officer had a chance to flee, Durkee told him to run on alone. The officer joined a number of other officers and men who managed to escape. As Durkee lay dying, an Indian tomahawked and scalped him.[18] Nearby was the body of Gershom Prince. Afterward a powder horn was found. Inscribed on it was

GARSHOM PRINCE HIS HORN MADE AT CROWNPOINT
SEPTM. YE 3RD DAY, 1761 PRINCE+ NEGRO HIS HORNM.[*19]

Butler, in his report on the "incursion," wrote, "Our fire was so close, and well directed, that the affair was soon over, not lasting above half an hour, from the time they gave us the first fire till their flight. In this action were taken 227 Scalps and only five prisoners. The Indians were so exasperated with their loss last year near Fort Stanwix that it was with the greatest difficulty I could save the lives of those *few* . . . indeed the Indians gave no Quarter."[20] Left unsaid was the amount of bounty paid for each scalp. In enumerating the scalps taken, Butler confined his estimate to armed foes, but local people claimed that many unarmed men, women, and children had been killed.

Adam Crysler, a Loyalist friend of Brant who became a lieutenant in Butler's Rangers, went beyond Butler's casualty figure. In his journal Crysler wrote, "I went to Wyoming, New York, where we killed about 460 of the enemy."[21] By Patriot estimates about sixty invaders were killed, but Butler reported to Carleton the deaths of only one Indian and two Rangers.[22]

After the rout Fort Forty surrendered. Under the surrender terms "property taken from the people called Tories" was to be returned, and the Tories themselves were to "remain in peaceable possession of their farms, and unmolested in a free trade throughout this settlement."[23]

Although Butler had promised that non-Tory homes would not be plundered, the raiders burned down every structure in sight. In his

---

* In 1761 Crown Point was a British fort on the west side of Lake Champlain, evacuated by the French in 1759 after the British captured Ticonderoga.

report he said he and his men "destroyed eight pallisaded Forts, and burned about 1000 Dwelling Houses, all their Mills &c., we have also killed and drove off about 1000 head of horned Cattle, and sheep and swine in great numbers, took away all the cattle they could drive, and killed the remainder."[24]

Butler later insisted that he could not control pillaging. Tales of looting, torture, and atrocities spread panic throughout the region. Survivors claimed to have seen a local Tory fatally shoot his Rebel brother as he attempted to surrender.[25] He was one of the numerous men said to have been killed after asking for quarter—a long-honored plea for mercy. One of the stories told by survivors involved a regal Indian woman who ordered sixteen captured militiamen to sit in a circle, and then, passing behind them, smashed their skulls, one by one, with her war club. Men who survived the battle identified her as Queen Esther, a matriarch of Seneca and French descent well known in the valley.[26] A powerful woman among the Seneca, she was given the title "queen" by settlers. One of her sons was identified as a raider.[27]

After the attack people swarmed out of the area, clogging roads and crowding into boats to escape down the Susquehanna. "I never in my life saw such scenes of distress," a Patriot wrote. "The river and roads leading down it were covered with men, women, and children fleeing for their lives."[28] A woman and her daughter, both widowed by the battle, walked all the way to their native Connecticut. Many other civilians temporarily fled into the marshes and mountains, where an unknown number died.[29]

Washington soon learned what had happened at Wyoming Valley. He sent Col. Thomas Hartley, like Durkee a Pennsylvanian veteran of the Brandywine and Paoli battles, to the area of the raid. Hartley's orders, as he described them, were "to make war on the savages of America instead of on Britain."[30] Leading about 250 men on a two-week expedition into Indian country near Wyoming Valley, Hartley killed at least ten Indians while losing four of his men. He rescued sixteen persons captured by Indians in raids along the Susquehanna River and destroyed four Indian towns, including one said to have been ruled by Queen Esther. He also took about fifty head of cattle, twenty-eight canoes, and an assortment of items claimed to have been stolen by the Indians. To pay for the expedition he auctioned off his acquisitions, some of which had belonged to indignant people who had to bid for their own goods.[31]

In September, Captain Brant and his Loyalist Indians reacted to Hartley's war on the savages with an attack in the upper Mohawk Valley. Brant's target was Andrustown, a fertile tract of farmland called the German Flats, after the immigrants whose axes had cleared the land. Brant especially despised Andrustown because many of its residents were militiamen who had killed his Indians at Oriska. The village was all but deserted; many of its fearful residents had moved closer to Fort Herkimer, returning to their farms only when they had to.

On September 17 a few people were in Andrustown gathering hay when a large group of Indians and Tories suddenly appeared. Indians fatally shot a father, his son, and another young man, then took their scalps. More raiders followed and, after ransacking the houses for booty, burned every building. Andrustown was never rebuilt.[32] Details of the attack reached British headquarters in New York, where an intelligence officer noted, "Indians wanted to subdue German Flats . . . indeed at all times the Indians Scalped all the Rebels they met, but Joseph [Brant] restrained them."[33]

The following month an improvised force of Continentals and militiamen avenged Andrustown by attacking Brant's headquarters, Unadilla. The large village, fifty miles west of German Flats, was at the confluence of the Susquehanna and Unadilla rivers. As soon as Brant and his Mohawks learned of the coming attack they abandoned Unadilla. "It was the finest Indian town I ever saw; on both sides of the River; there was about 40 good houses, Square logs, shingles & stone chimneys, good Floors, glass windows, etc.," a Continental officer reported. The avengers burned the forty houses, torched two thousand bushels of corn, and destroyed a sawmill and a gristmill.[34]

* * *

Vengeance would soon follow vengeance in the person of Capt. Walter Butler, John Butler's son. Walter, the spy who had been captured by Benedict Arnold, had been held prisoner in Albany under a reprieved death sentence. He was first thrown into a nasty jail infamous for its wormy food. Then, after feigning illness, he was transferred to a private Albany home that happened to belong to a Tory. He easily escaped and rode off on a donated horse. Traveling by night and aided by Tories along the way, he reached Ranger headquarters at Niagara seething with a desire for revenge. While he had been a prisoner, Guy Carleton, citing Walter's "loyalty, courage and good conduct," appointed him a captain who was to "serve with the Indians."[35]

Walter's choice for a raid, blessed by his father, was Cherry Valley, fifty miles west of Albany, and, like Wyoming Valley, a thriving farming region.[36] Walter mobilized two hundred Rangers, fifty British Army volunteers, three hundred Seneca warriors, and three hundred of Brant's Volunteers, a mix of whites and Indians. Brant led them reluctantly because, as a veteran leader, he resented being under young Walter's command.[37]

By the time a typical raid began, people, warned beforehand, had fled to safety in a fort, leaving nearby houses virtually deserted. But this raid was a surprise, and the makeshift fort guarding the area was ill prepared. Many civilians and officers—including the fort commander, Col. Ichabod Alden—lived in cabins outside the fort. Alden had ignored warnings that an attack was imminent.

On November 10, 1778, the raiders arrived silently, shrouded in a thick fog that rose from the snow-covered ground. While some attacked the fort, others surrounded the cabins, killing Alden and other officers as they rushed out to defend the fort. The raid turned into a rampage. One officer had boarded with Robert Wells, a judge who had been a close friend of Walter Butler's father. After killing the officer, raiders slew Wells, his wife, his brother and sister, three sons, and his daughter. Judge Wells was said to have been killed by a Loyalist who knew him, perhaps Walter Butler himself.[38]

People in the fort could hear screams and see flames rising from cabins and barns. The next day, after the attackers had left, men were sent out from the fort to bring in the dead. "Such a shocking sight my eyes never beheld before of savage and brutal barbarity," one of the officers wrote in his diary. He saw a "husband mourning over his dead wife with four dead children lying by her side, mangled, scalpt, and some their heads, some their legs and arms cut off, some torn the flesh off their bones by their dogs—12 of one family killed and four of them burnt in his house."[39] There were thirty-two bodies of civilians, most of them women and children. Sixteen soldiers were also killed. The burning of the valley left 182 settlers homeless.[40]

The murders were blamed on Indians, but survivors said they saw Loyalists amid the carnage.[41] A man identified as a Tory sergeant named Newbury was seen murdering a young girl by driving a tomahawk into her skull.[42] In a letter to Major General Schuyler of the Continental Army, Walter Butler later denied responsibility for the murders: "I have done everything in my power to restrain the fury of the Indians from hurting women and children, or killing the prisoners who fell into our hands." He went on to make his prisoners essentially hostages to the fate of his mother and young siblings, who had been held by the Patriots ever since Walter and his father fled to Canada in 1776.

"I am sure you are conscious that Colonel Butler or myself have no desire that your women or children should be hurt," he continued in the letter. "But, be assured, that if you persevere in detaining my father's family with you, that we shall no longer take the same pains to restrain the Indians from prisoners, women and children, that we have heretofore done."[43] Schuyler had been relieved of his commission, so the letter was answered by Brig. Gen. James Clinton, who referred to what Butler had written as a threat. But Clinton began making the arrangements that would eventually send Mrs. Butler and her children to a reunion with John Butler.

*   *   *

Stirred by outcries over the Wyoming and Cherry valley raids, Congress directed Washington to send the Continental Army against the Indians and their Tory allies. Washington responded by ordering what would be history's first large-scale attack on Indians by an American armed force. He put Maj. Gen. John Sullivan in command of an expedition "against the hostile tribes of the Six Nations of Indians, with their associates and adherents."

Washington ordered "the total destruction and devastation of their settlements, and the capture of as many prisoners of every age and sex as possible. It will be essential to ruin their crops now in the ground & prevent their planting more." He went on to suggest that his attacks be "attended with as much impetuosity, shouting, and noise, as possible" and that the troops "rush on with the war-whoop and fixed bayonet. Nothing will disconcert and terrify the Indians more than this." If any Indians tried to negotiate a peace, he said, they were to be dealt with only if they showed their sincerity "by delivering up some of the principal instigators of their past hostility into our hands: Butler, Brant, the most mischievous of the Tories."[44]

Sullivan led more than twenty-three hundred troops, most of them Continentals. At an Indian village called Tioga Point (now Athens, Pennsylvania), he rendezvoused with Brigadier General Clinton, who added his fourteen hundred Continentals to the campaign of retribution and destruction. On the Fort Pitt frontier to the west, a third, smaller expedition was ordered to head up the Allegheny River valley to begin a similar slash-and-burn foray against the western villages of the Six Nations.[45]

While Clinton was in Canajoharie, New York, waiting for his troops to assemble, local militiamen captured two Butler Rangers— Sgt. William Newbury and Lt. Henry Hare—lurking near Clinton's troops. The spies had lived in the area until they went off with recruiting parties to Canada. Tried by a court-martial, they were found guilty and sentenced to death for spying. The hangings, Clinton wrote his wife, "were done . . . to the great satisfaction of all the inhabitants of that place who were friends of their country, as they were known to be very active in almost all the murders that were committed on the frontiers." Before being hanged, Newbury, the father of six, confessed to the tomahawk slaying of a young girl during the Cherry Valley raid. Hare admitted that he had killed and scalped one of the girls slain outside Fort Stanwix in 1777.[46]

John Butler had only about three hundred Rangers to oppose the mighty Sullivan-Clinton force because his Indian allies, preferring retreat to annihilation, offered little resistance to any of the three armies.[47] Brant and Walter Butler set up an ambush to slow down Sullivan. But, outmaneuvered, the ambushers fled northward to Niagara as the troops swept through the Iroquois country, burning villages and crops, sometimes avenging past atrocities with a few of their own. Some soldiers scalped Indians they had killed. An officer told of finding two dead Indians whose bodies he then had skinned "from their hips down for boot legs . . . one pair for the Major the other for myself."[48] Civilian Rebels joined the troops one day, locked an old Indian woman and a boy in a cabin, and set it afire.[49]

After one of the few armed encounters, surviving Indians went off with two captured Continentals. Comrades later found their bodies: "The Indians . . . tied them up & whipped them prodigiously, pulled out their finger and toe nails, cut out their tongues, stuck spears and darts through them & set the Leuts [lieutenant's] head on a log with the mouth open: we could not find the other head."[50]

As part of the Sullivan expedition, a detachment marched from Fort Stanwix and crossed Oneida Lake to raze the principal settlement of the Onondaga, who had not taken either the Tory or Rebel side. Leaders of the Oneida, allies of the Patriots, begged the Continentals not to attack the Onondaga, stressing their neutrality. But expedition officers dismissed the claim. The Continentals reported slaying twelve Onondaga, destroying about fifty log houses, burning "a large quantity of Corn and Beans," and killing "a number of fine horses & every other kind of Stock we found."[51] The attack turned the Onandaga into foes of the Patriots.

Sullivan went about as far north as Genesee, New York. His expedition produced devastation as methodical as Washington had ordered. In a report to Congress, Sullivan said that "every creek and River has been traced" and "not a single town left" in the Iroquois country. His men had wiped out at least forty villages and, by Sullivan's "moderate computation," had destroyed at least 160,000 bushels of corn. In one large village soldiers cut down fifteen hundred peach trees.[52]

Indian families, fleeing before the soldiers and carrying few belongings, sought refuge at the British base at Niagara, more than one hundred miles beyond Genesee. Thousands became homeless people seeking help. By the time the first of them reached Niagara, the chill of autumn was in the air, a prelude to the most severe winter in memory. The Butlers and other Loyalists built huts around the fort for the more than five thousand Indians gathered there. Food was scarce, and hunters risked freezing to death when they sought game. Hunger, cold, and disease killed an unknown number of the Indian refugees—perhaps hundreds.[53]

Originally Washington had hoped that the western force sent up the Allegheny River valley would link up with the Sullivan-Clinton expedition in a great sweep that would subdue the entire Iroquois Nation. But, fearing that he would be overextending his forces, Washington changed his mind and left the western commander to venture on his own. The Allegheny Valley expedition, which did not lose a man, reported extensive destruction, burning down thirty-five houses, including some large enough to shelter three or four families, and putting the torch to fifteen hundred acres of corn. The western Indians also fled toward Niagara. In their flight, they left behind packets of trade pelts and other valuables, which the invaders seized as booty.[54]

The western campaign turned that frontier into a cauldron of competing foes. Virginia and Pennsylvania argued over where their boundary should be drawn. Settlers, thinking the frontier had been freed of the Indian menace, began heading westward. Among them were Tories escaping persecution and seeking the protection of territory around British-held Detroit. Patriots feared that Tories would seize lead mines, vital to Rebel ammunition production, on the west-

ern Virginia frontier. A Virginia militia force swooped down on a
Loyalist settlement near the mines and reported, "Shot one, Hanged
one, and whipt several." The Virginia House of Delegates confiscated the Loyalists' land and lauded the militiamen for "supressing a
late conspiracy and insurrection on the frontiers of this State." Skirmishes between western Tories and Rebels would continue through
the war.[55]

The Sullivan-Clinton expedition inspired small hit-and-run vengeance raids that began soon after the new year.[56] Then, in May 1780,
Sir John Johnson mobilized a force of more than five hundred men,
made up of Indians and companies of his own King's Royal Regiment. Loyalist boatmen took the raiders down Lake Champlain to
a landing below Crown Point, where they went ashore to begin an
overland trek. One detachment went directly with Johnson to his
birthplace, Johnstown, New York. At Johnson Hall his men dug up
two barrels of family silver plate that he had buried before his flight to
Canada in 1776. The treasure, carefully inventoried, went into Loyalist knapsacks for the return trip.[57]

A second detachment struck settlements south of the town, burning 120 houses, barns, and mills. They killed or wounded several
men who were considered special enemies or were simply defending
their homes. Some Tories were also killed and scalped by mistake in
a frenzy of looting and burning. At sunset, on a hill in one settlement,
lost dogs from smoldering homes joined the dogs of missing masters,
and the forsaken pack howled deep into the night.[58]

Rebel militiamen and Continentals belatedly responded to the raid.
Johnson eluded them, even though he was burdened by his silver plate
and a couple of dozen prisoners. He had also rounded up 143 Loyalist men, women, and children who had been living fearfully in Rebel
territory. Pursuers finally got on Johnson's trail as his fiery nineteen-
day invasion was ending. When Johnson, his men, and their guests
boarded bateaux for their return voyage, the pursuers were right behind. But they had neither the boats nor provisions to continue their
pursuit.

The Johnson raid raised Loyalist morale in the borderland, deprived the Continental Army of food, and aided recruitment. With their families safe at Niagara, many of the Tory men Johnson had rescued signed up for the King's Royal Regiment, completing the muster of one battalion and starting a second.[59]

Sir Frederick Haldimand, Guy Carleton's successor as "Captain General and Governor in Chief in and over our Province of Quebec in America," believed that, in the wake of Burgoyne's stunning defeat, a new invasion of Canada was likely. He saw the Rebel settlements of the frontier as a potential staging area for a strike across the border.[60] To diminish that threat—and stop the flow of grain to the Continental Army—he ordered attacks on the people and crops of the Schoharie and Mohawk valleys.

Haldimand, born in Switzerland in 1718, became an officer in the Prussian army at the age of twenty-two. On the eve of the French and Indian War, as a soldier of fortune, he joined the Royal Americans, who included deserters from, or veterans of, European armies, along with Swiss and German settlers of Pennsylvania. After distinguished service in the war, he remained in the army and in America, assuming commands in posts from Massachusetts to Florida. His varied postings and familiarity with American ways made him one of Britain's most experienced North American officials.[61]

Under his direction, Sir John Johnson assembled a main force of nearly one thousand men, including about 180 British Army Regulars, twenty-five Hessians; 150 Rangers, under Col. John Butler; about two hundred men of the King's Royal Regiment; many Tories in independent companies; and about 580 Indians. Haldimand rounded up an additional 970 men for diversionary raids near Saratoga and down the Richelieu River route to the Hudson River.[62]

Scouts went out to alert Tories along the routes, assuring them that they would be escorted to Canada and resettled in safety if they believed they had endangered themselves by aiding the invaders. A scout from the King's Royal Regiment, sent specifically to seek out Loyalists who might join the invasion, was caught, tried as a spy, and hanged, as was a Continental Army deserter who was caught recruiting Tories.[63]

Men took down their muskets and went off to join the side of their choice, leaving wives and children behind. A Mohawk Valley man wrote about his father, who left his farm to join Butler's Rangers: "It was a momentous struggle, a frightful warfare. . . . The farms were left to the care of the women, who seldom ate the bread of Idleness. . . . They spun, they wove, they knit, prepared their own flax, made their own homespun gowns, the children's dresses, they churned, made cheese, and performed all the various duties of domestic and social life . . . my father's mind was at ease about the affairs of the Farm."[64]

Three forts defended the verdant Schoharie Valley, Johnson's first objective. The Lower Fort, as it was called, had as its core a stone Dutch Reformed church. Surrounding it was a stockade encompassing about an acre of land dotted with small huts. The fort's powder was stored beneath the pulpit, and in the belfry was a platform for lookouts and rifle marksmen. Outside one of the two corner blockhouses was a tavern.[65] The Upper Fort was about fifteen miles south, near the village of Schoharie; the Middle Fort was just below Middleburg.

On October 16 Johnson's expedition camped near the Upper Fort, which was built around a farmhouse and barn, its stockade surrounding about two and a half acres. The next morning a soldier outside the fort spotted the Loyalist force and ran to give warning. A signal gun boomed. Hearing it, Johnson ordered the destruction to commence. Flames and smoke began rising from barns and deserted houses. Cattle and pigs lay dying, their cries and their blood drawing dogs and vultures.[66]

The invaders broke into houses, took what they wanted, and torched them. Some people stayed to defend their homes, which in this prosperous farmland were framed and painted wooden buildings, not log cabins.[67] The raiders moved on quickly, heading for the Middle Fort through a day that was growing gray under wind-whipped sleet and snow.[68] Johnson set up two small cannons that began firing at the fort as his Rangers and Indians cautiously approached it. After a while the firing stopped, and men in the fort saw a white flag appear in the enemy ranks. The flag bearer began walking forward, flanked by an officer in the green coat of Butler's Rangers and a fifer playing "Yankee Doodle," still a mocking tune to the ears of a Rebel.

The commander of the Middle Fort, a Continental in charge of militiamen accustomed to having their own officers, ordered the gates opened to the flag of truce. Timothy Murphy, a militiaman, defied the order. He was a sharpshooter, one of Morgan's riflemen in the Battle of Saratoga. Now he shot at the flag party—to warn them off, not to hit them. He told his stunned commander that he believed that the white flag was a ruse to allow the officer to assess the fort's garrison. If the Ranger did enter, he would see how few defenders there were. The flag party turned back, then came forward two more times, and each time Murphy fired a warning shot. Murphy's defiance undermined the authority of the Continental officer, who threatened a court-martial but finally turned over command of the fort to a militia colonel.

Johnson decided to march on, bypassing the fort to continue destroying every Rebel farm in his path—while sparing Tory property.[69] After another bivouac in the Schoharie Valley, he headed for the Mohawk Valley, pursued by frustrated and outnumbered Rebel militiamen only able to "hange on their Rear."[70] As soon as Johnson left, outraged Rebels burned the untouched Tory farms, completing the absolute destruction of the valley's crops and livestock.[71] Tories fled northward, joining Johnson's followers.

At the Mohawk River the raiders split into two detachments to loot and raze along both sides of the river, camping for the night near the town of Root. On the morning of October 19 the main force crossed the Mohawk and headed for the German village of Stone Arabia, the center of an area settled by immigrants from a part of Germany known as the Palatinate. At nearby Fort Paris, Col. John Brown mustered about three hundred men, including a few Oneida Indians, and, astride his small black horse, led them toward Johnson's force.[72] Brown was killed in an ambush. His men fled, leaving behind the fallen to be scalped.[73]

Johnson's "Destructionists," as raiders were sometimes called, kept on swooping down on farms. Among his men were settlers who had lived in these houses, built these barns, tilled these fields. But now they were Tories on a mission, and to them, somehow, this rich valley had become an alien land. A farmer, hidden in the woods with his family, watched his own farm vanish in flames. He saw the Indian Tories move on, swinging firebrands over their head until they blazed, then touching them to barns full of grain. After the Indians left, the farmer found seven hogs dead in their pen, killed by a pitchfork taken from what had been a barn.[74] And so it went, farm after farm.

British soldiers in the attack force sometimes guarded prisoners to protect them from the Indians, whose behavior was unpredictable. At one farm Indians took a woman and her seven children out of their house, then loaded them and armfuls of loot into a horse-drawn wagon. Around that time Johnson was told that Continentals and militiamen were on their way from Albany and Schenectady. He released the woman and her children, except for her fourteen-year-old son, presumably kidnapped to become a future Ranger.[75]

About nine hundred Rebels, most of them militiamen, caught up with the raiders toward the end of day. In a twilight skirmish Johnson tried to set up a battle line but failed to hold off a Rebel charge. Men of Johnson's own regiment were driven back—"running promiscuously through and over one another" in the dark, a Tory said. As the Johnson Destructionists settled for the night near the battlefield, the Rebels' commander inexplicably ordered his men to camp about three miles away. He planned to strike the next morning. By then Johnson and his force were well on the way toward Canada.[76]

The diversionary raids were as successful as the main raid. Maj. Christopher Carleton, a nephew of Sir Guy Carleton, headed a force of nearly one thousand Regulars, Indians, and Tories down the Champlain Valley into the upper Hudson River valley. Carleton was knowledgeable about Indian culture, had had himself tattooed, and at times wore a ring through his nose. He had taken an Indian mistress before marrying the sister of his uncle's wife.[77] He knew the frontier wilderness trails as well as an Indian. He was an inspiring leader in a stealthy campaign that called for night marches and fireless camps.

Early in the three-week expedition, near the southern shore of Lake George, a King's Ranger spotted about fifty Rebel soldiers leaving Fort George. A mixed force of Rangers, Indians, and Regulars surrounded the men and swiftly killed twenty-seven. Eight men were captured. The rest escaped into the woods. The fort surrendered. After looting and burning it, the raiders marched on, their prisoners carrying the Indians' plunder. By the loose rules of raiding, Indians, male and female, were given looting rights, some becoming quite discerning. One settler told of an Indian who stole plates from a dark house. Once outside, he discovered they were pewter, not silver, and disdainfully threw them away.[78]

A party of militiamen was sent to the area later to bury the dead. "We found twenty-two slaughtered and mangled men," one of them remembered. "All had their skulls knocked in, their throats cut and their scalps taken. Their clothes were mostly stripped off. . . . The fighting had been mostly with clubbed muskets, and the fragments of these, split and shivered, were laying around with the bodies."[79]

A diversionary raid aimed at Schenectady, led by a former merchant in that town, shifted to the village of Ballston after militiamen mobilized on the route ahead. In Ballston, Indians broke into cabins and killed two men. Tory officers intervened, feeling that the Indians were turning too violent. In one cabin, an officer grabbed an Indian's upraised arm, protecting an unarmed man defending his family. After the usual burning and looting, the Indians, Rangers, and Tory followers headed back to Canada with a string of prisoners.[80]

Carleton's men burned down a second Rebel fort and captured 148 Rebel soldiers, a large loss in a frontier of sparse garrisons. The raiders destroyed thirty-eight houses, thirty-three barns, six saw mills, and a grist mill. They estimated that they had torched fifteen hundred tons of hay. A detachment of Indians burned down thirty-two houses and their barns.[81] Such inventories of desolation continued, year after year, raid after raid. At least twenty-nine raids struck towns in the Mohawk and Schoharie valleys in 1781 alone.[82]

Col. Marinus Willett, a Continental Army hero, writing to Washington from Fort Herkimer in the German Flats in July 1781, estimated that one-third of the people on the New York frontier had been killed or carried off, one-third had fled the frontier but remained Patriots, and "one third deserted to the enemy."[83] Postwar records show that about 380 women became widows and some two thousand children lost their fathers. About seven hundred buildings had been burned and 150,000 bushels of wheat destroyed. Some twelve thousand farms had been abandoned.[84] There is no reliable estimate of how many prisoners were taken. Some were exchanged; some did not return until a year or more after the war. Many died. Others, lured by offers of free land, became Tories and remained in Canada.[85]

"I am really at a loss to know how to feed the troops," the senior Continental Army colonel in New York wrote to Governor George Clinton, General Clinton's brother, in September 1780.[86] Two months later, reacting to the continuing loss of grain and flour from "the country which has been laid waste," Washington told Governor Clinton that "we shall be obliged to bring flour from the southward."[87]

To the west, in the wild Ohio Country, under orders from Lord Germain in London, bands of Tories and Indians raided settlements on the Pennsylvania and Virginia frontiers and in the area that would become Kentucky. Germain, saying "it is The Kings command," had instructed an obscure colonial official to "assemble as many of the Indians of his district as he conveniently can, and placing proper Persons at their Head, to . . . restrain them from committing violence on the well affected and inoffensive Inhabitants, employ them in making a Diversion and exciting an alarm upon the frontiers of Virginia and Pennsylvania."[88]

The official was Henry Hamilton, an Ireland-born veteran of the French and Indian War, who in 1775 had been made lieutenant governor and superintendent of Indian affairs at Fort Detroit (site of today's

Detroit). Hamilton was one of five lieutenant governors appointed to manage the Province of Quebec, whose boundaries had been extended by the Quebec Act of 1774 to include an immense expanse of land between the Ohio and Mississippi rivers.

Germain also authorized Hamilton to raise Tory regiments and offer each recruit the postwar promise of two hundred acres of land. With his Tories and Indians, Germain wrote, Hamilton might be able "to extend his operations so as to divide the attention of the Rebels, and oblige them to collect a considerable Force to oppose him."[89] Following Germain's orders, Hamilton sent Tory-Indian raiders into a territory that Britain had prohibited its colonists from settling. But, defying the Crown, colonists had settled there. Their disobedience had marked them as Rebels, though there were enough Tories among them for Hamilton to find recruits.

White men had been paying bounties for enemies' scalps since the French and Indian War. But Rebels on the frontier singled Hamilton out as the "Hair Buyer," a label that found its way into numerous narratives.[90] Hamilton himself routinely mentioned scalps in his reports. Early in 1778, for example, he wrote to Governor Carleton, saying that his Indians had "brought in seventy-three prisoners alive, twenty of which they presented to me, and one hundred and twenty-nine scalps." Later that year Hamilton told Carleton's successor, Frederick Haldimand, that "since last May the Indians in this district have taken thirty-four prisoners, seventeen of which they delivered up, and eighty-one scalps, several prisoners taken and adopted [by Indians] not reckoned in this number."[91] Neither letter mentions bounties for the scalps.

One of Hamilton's operatives was Simon Girty, who had been captured by French-commanded Indians as a child in 1756 and raised by Senecas. He was freed after eight years and married a young woman who had been a captive of the Delaware tribe. Girty became an officer in a Pennsylvania militia and an Indian interpreter at Fort Pitt. When the Revolution began, by one story, Girty was confined to the fort as a suspected Tory; by another, he defected because he was disgusted with American treatment of Indians. Whatever the reason, he ended up in Detroit and became not only a captain and interpreter in the British Indian Department but also the most notorious Tory on the western frontier.[92] Like Hamilton, Girty would get a label: "White Savage." His infamy was based on his witnessing but not trying to stop the heinous torture of a Patriot—"they scalped him alive and then laid hot ashes upon his head, after which they roasted him by a slow fire."[93]

Virginia governor Patrick Henry feared that Hamilton's raids would drive settlers out of the territory and give the British control over the western frontier. One of the frontier leaders was Lt. Col. George Rogers Clark, a lanky, red-haired militia officer from the first region west of the Allegheny Mountains settled by American pioneers (later Kentucky). In 1778, with Henry's support, Clark and his force of frontiersmen headed over the mountains to capture the bases from which the raiders struck. He also wanted to establish American claims to a territory nearly as large as the thirteen colonies.[94]

Clark easily took three British outposts. But the most important— Fort Sackville in what would become Vincennes, Indiana—was retaken by Hamilton and a mixed force of Indians, Tory militia, and Regulars.[95] Hamilton decided to wait until spring to attack Clark and roll back the frontier to east of the mountains. Clark surprised him by leading about 175 men to the fort, through seventeen winter days of snow and icy streams. At Vincennes his men captured five Indians carrying American scalps. To terrify Hamilton's men Clark bound the captives and tomahawked them in full view of the garrison.[96] Hamilton surrendered the fort and was taken to Williamsburg, where Governor Thomas Jefferson, Patrick Henry's successor, called him a "butcher of men, women and children," put him in irons, and treated him as a criminal, not a prisoner of war.[97] After eighteen months, through the intervention of Washington, he was finally freed in a prisoner exchange.[98]

From New York to the Ohio country, the Indian-Tory raids on frontier settlements continued, becoming a war unto itself. To the west, in

March 1782, Pennsylvania militiamen swooped down on the missionary village of Gnadenhutten. The Delaware Indians there, converted
to Christianity, were suspected of being Loyalists. The militiamen
rounded up the unarmed Indians and killed sixty-two adults and
thirty-four children by smashing their skulls with mallets. Two boys
escaped and spread word of the massacre. In an act of vengeance three
months later, Delaware braves tortured a captive militia officer who
had nothing to do with the raid and then burned him at the stake.[99]

After the Revolution ended, Fort Detroit in the west, like Fort Ni-
agara in the east, remained a British garrison and a Tory haven. But
George Rogers Clark's thrust beyond the mountains did establish an
American claim on territory that would become Kentucky and West
Virginia. And in 1803 President Jefferson looked farther westward,
asking Congress for approval of an expedition that would travel to
the Pacific. As one of it leaders, Jefferson would pick William Clark,
George Rogers Clark's younger brother.

15

___

SEEKING SOUTHERN FRIENDS

GEORGIA AND THE CAROLINAS,
NOVEMBER 1778–OCTOBER 1780

The Whigs [Rebels] seem determined to extirpate the Tories, and the Tories the Whigs. . . . If a stop cannot be soon put to these massacres, the country will be depopulated in a few months more, as neither Whig nor Tory can live.

Maj. Gen. Nathanael Greene on warfare in the South[1]

Twice the British had tried to enter the South and tap into the expected enthusiasm of local Loyalists. And twice—at Moore's Creek Bridge in North Carolina and at Sullivan's Island in South Carolina—the attempts had failed. But after the defeat of Burgoyne and the alliance with France, British strategists again looked southward. They began to believe that a campaign in the South was vital, not only to control the major southern ports and fend off the French navy but also to mobilize the Loyalists and put them into battle. Lord Germain handed down a royal imperative: "The conquest of these provinces is considered by the King as an object of great importance in the scale of the war."[2]

As the strategists saw the South, with Charleston, South Carolina, and Savannah, Georgia, under British-Loyalist control, Redcoats and Hessians could press inland toward the unruly backcountry, leaving

Loyalist regiments in their wake to occupy Rebel-cleared territory. The Loyalists, taking over the backcountry, would aid in the reestablishment of royal governments and split off the South from the northern colonies. St. Augustine, capital of the small colony of East Florida, was under royal control and could serve as a port for landing supplies and troops.

Times had changed rapidly since 1775, when the lieutenant governor of East Florida had said that the "best friends of Great Britain are in the back parts of Carolina and Georgia."[3] Many of Georgia's friends of Great Britain had been governed by Patriots since January 1776, when Governor James Wright, fleeing his Rebel kidnappers, took the usual gubernatorial escape route by boarding an offshore Royal Navy warship. He later sailed to England, leaving behind the twenty-five thousand acres and five hundred slaves he had accumulated while serving the king.

Wright had been the victim of an audacious Rebel intelligence coup and did not know it. He had written to General Gage saying that Georgia desperately needed help because Rebels were taking over the colony. But Gage had not replied, to Wright's puzzlement. What Wright did not know was that Gage did not get that letter. The letter that Gage did receive and did not answer had assured him that all was well in the colony. South Carolina Rebels had intercepted Wright's appeal letter and substituted the perfectly crafted counterfeit.[4] When Wright fled to England, he renewed his plea for help, this time urging officials to invade Georgia and adding credence to plans for a southern strategy.[5]

General Sir Henry Clinton launched the new strategy in November 1778 by sending a force of about three thousand Redcoats, Hessians, and Tories from New York to Savannah. The Tories—New York Volunteers, two battalions of De Lancey's New York Regiment, and a battalion from Skinner's New Jersey regiment—totaled 855.[6] British Regulars stationed in Florida were to enter Georgia and link up with the New York force.

Clinton landed near Savannah on December 29, rolled over out-numbered Rebel troops, and swiftly took the city. The Rebels' stunned commander, Maj. Gen. Robert Howe of the Continental Army, crossed the Savannah River into South Carolina, leaving behind some five hundred dead, wounded, or captured.[7]

The invasion rallied militant local Tories and awakened the vengeance of Thomas Brown, a powerful Tory leader. His East Florida Rangers joined the British invaders in Savannah and on their march to Augusta, about 125 miles up the Savannah River. Along the way Brown was wounded in a skirmish that he stirred up while trying to free some jailed Tories.[8]

Brown was an unlikely revolutionary. He had started life in America as a young man of privilege, seemingly destined to be like many wealthy Loyalists who supported the king but did not shoulder a musket. Brown's father, a British merchant, outfitted a ship for his son, who, at twenty-four, persuaded more than seventy people from Yorkshire and the Orkney Islands to become indentured servants and sail with him to Georgia. His enterprise won him a royal grant of a large tract of land near Augusta, which he named Brownsborough. In August 1775, while he was in his home on the South Carolina side of the Savannah River, dozens of Sons of Liberty confronted him and demanded that he support the Revolution. Brown refused, and, in the uproar that followed, shot the ringleader in the foot.

Rebels pounced on him. One struck him with a musket butt, fracturing his skull. Others partially scalped him, tarred his legs, and held them over a fire. He lost two toes to severe burns, and became known to Rebels as Burntfoot Brown. Seething for revenge, he became an obsessed Tory, determined to lead Loyalists in a personal war against the Rebels. Two weeks after he had been attacked, he was in South Carolina challenging the authority of the Patriots' Council of Safety and beginning to recruit hundreds of men for a Loyalist force.[9]

Brown and his supporters later fled to Florida, where, commissioned as a lieutenant colonel, he recruited the East Florida Rangers from settlements along the Georgia-Florida border. He convinced Patrick Tonyn, East Florida's royal governor, that the Rangers could stage over-the-border raids and act as a home guard against invaders.[10] When General Howe began his southern campaign, Brown saw it as a signal for backcountry warfare against the Rebels and a chance for what a British official called "retributive Justice."[11]

Brown had seventy-two Rangers in the mixed force heading for Augusta under the command of British Army colonel Archibald Campbell. Most of the expedition's one thousand men were British Regulars. The rest, besides Brown's men, included New York Volunteers and a recently raised unit, the Carolina Royalists, also known as the Carolina Loyalists.[12] The quick fall of Savannah was to be followed by an easy takeover of Augusta. But Campbell, realizing he was entering territory that was more hostile, moved cautiously.

Mounted East Florida Rangers, sent ahead, reported that about one thousand Rebel militiamen held Augusta. But as Campbell neared the town, most of the Rebels crossed the river into South Carolina. He entered Augusta and began taming the area by having residents swear an oath of loyalty to the king. After taking the oath, about fourteen hundred men received pardons for their previous Rebel allegiance. Georgia, the youngest colony, was to become the first to return to royal rule. Or, as Campbell put it, he was taking "a stripe and star from the rebel flag."[13]

Campbell was to manage the next phase in the southern strategy: the mass recruitment of Loyalists into military units that would help the royal government restore control of Georgia. That would start with the arrival of a British agent, James Boyd, and his Loyalist recruits.

Boyd, a South Carolina Tory, had landed with the invaders but had gone off on his own mission. Given a colonel's commission, he was sent by Campbell into South Carolina to raise a large force of Loyalists and lead them over the border to rendezvous with Campbell in Augusta.

Boyd's first stop was near Savannah, at Wrightsborough, a Quaker and Tory settlement named after the governor. There he picked up guides who took him to an isolated site in South Carolina, where he recruited about 350 Loyalist militiamen and headed back to Georgia. Joining Boyd along the way were 250 members of the Royal Volunteers of North Carolina, a Tory military unit formed specifically to aid Campbell in his occupation of Georgia.[14]

Rebels who had been trailing Boyd's band struck as the Tories were about to cross a ford of the Savannah River. In a brief firefight about twenty Patriots were killed and twenty-six captured; Boyd lost about one hundred men to Rebel gunfire and desertion but kept moving. Living off the land by plundering Rebels' farms and getting help from Tories, Boyd made his way to Augusta. He did not know that Campbell, facing a growing Rebel presence, had retreated back to Savannah.

On February 14 Boyd set up camp at Kettle Creek, some fifty miles northwest of Augusta. Again, Rebels were on his trail. About 340 South Carolina and Georgia militiamen came upon Boyd's men as they were slaughtering stolen cattle. In a surprise attack a bullet felled Boyd, who died a few hours later. Nineteen other Tories were killed; Rebel losses were seven men killed and twenty-two wounded. Most of the Tories fled. About 270 escaped to British lines. Some 150 other Tories were eventually captured either near the scene of battle or back in their own communities. Seven were hanged.[15]

Most of Campbell's fourteen hundred oaths of allegiance were not genuine. Faced with the choice of having their property confiscated or signing, people signed—and quickly found a Patriot leader to whom they denounced the pledge.[16] Many signers showed their faith by joining Rebel militias and hunting down Tories.[17] Campbell would later complain about "irregulars from the upper country [of Georgia] under the denomination of crackers, a race of men whose motions were too voluntary to be under restraint and whose scouting disposition [was] in quest of pillage." The crackers, he reported, "found many excuses for going home to their plantations."[18]

*   *   *

The victory at Kettle Rebel Creek gave heart to the Rebels and, to the British, proof that a Loyalist call to arms would not produce an army big enough to suppress the rebellion. But the Patriot victory did not stop the southern strategy. A quixotic attempt in October 1779 to retake Savannah with a joint American-French operation ended with the French losing 635 men and the Patriots 457 while the British and Loyalist defenders saved the city at a cost of fifty-five lives. "Such a sight I never saw before," a British officer wrote. "The Ditch was filled with Dead . . . and for some hundred Yards without the Lines the Plain was strewed with mangled Bodies."[19]

Next came General Clinton's siege of Charleston, which ended on May 12, 1780, when Maj. Gen. Benjamin Lincoln bowed to civilian pleas by surrendering his army of nearly six thousand Continentals and militiamen.[20] Among the victorious Clinton troops were 175 Loyalists in a special temporary corps called the American Volunteers under the command of Capt. Patrick Ferguson of the 70th Regiment of Foot. Thirty-four of the Volunteers came from the Prince of Wales's American Regiment. Only ten would still be alive or serving within fifteen months of their sailing to the South. Most of the replacements would be deserters from the Continental Army.[21]

As Clinton was taking over Charleston, he asked James Simpson, the former South Carolina attorney general, how widespread and deep was the colony's apparent euphoria over the surrender. Simpson made his own public opinion poll and reported that the population was divided into four classes: those, especially the wealthy, who were pleased to see South Carolina again under royal rule; those who had been duped by the Rebels, regretted their failings, and now supported the king; those who were repentant ex-Rebels; and those who were still Rebels and unrepentant. Simpson summed up by saying that "in drawing a comparison between the four Classes, the number and consequence of the two first by far exceed the last." Loyalists in general, he said, were "clamourous for retributive Justice."[22]

Clinton set up a model for the governing of "conquered" territory: a military government that put trusted civilians in local offices, regulated prices on goods purchased by the army, and created militias for local defense. When Governor Wright returned he found he had little power. Clinton had, for example, appointed Simpson Charleston's superintendent of police.[23]

The militias included a home guard of older men and a regular militia of younger, unmarried men who would serve away from their homes in Georgia and North Carolina. They would be given the same pay and provisions as the king's troops.[24] All able-bodied free males, generally between the ages of fifteen and sixty, had to serve in a militia "any Six Months of the ensuing twelve" or provide a substitute. Militiamen were given ammunition, material for the sewing of a loose-fitting garment called a rifle shirt, and a musket if needed. Men could serve on horseback at their own expense. They were to be restrained "from offering violence to innocent and inoffensive People" and insulting or outraging "the Aged, the Infirm, the Women and Children of every denomination."[25]

The rules for establishing the Loyalist militias were developed by the deputy adjutant general of the British Army, the brilliant, newly promoted Maj. John André, and another new major, Patrick Ferguson, inventor of the Ferguson breech-loading rifle, which could fire four shots a minute.[26] Ferguson, born in Scotland in 1744 and a soldier since the age of fifteen, fought in the Battle of Brandywine in September 1777. While he led a rifle company whose men were firing the weapon he invented, a Continental's musket ball shattered his right elbow. He would never be able to bend that arm again. He returned to duty in May 1778 and learned to shoot, fence, and wield the saber with his left arm. He raised the American Volunteers, trained in the kind of ranger warfare developed by John G. Simcoe and Banastre Tarleton.[27]

Ferguson was appointed inspector of militia, supervising the recruitment and training of the hundreds of Tories who were signing up to serve the king. He raised a regiment of 240 men at one outpost, teaching them to follow signals from his silver whistle so they could get orders even when he was not seen. "There is great difficulty in bringing the militia under any kind of regularity," he wrote. "I am exerting myself to effect it without disgusting them."[28]

Tarleton, the brash young captor of General Charles Lee in 1776, earned fame as an aggressive leader of hard-driving cavalrymen on raids in New York, New Jersey, and Pennsylvania. As the commander of the Loyalist British Legion in the Carolinas, he fought fiercely and earned a reputation for showing no mercy in battle.

Under the terms of Charleston's surrender each surrendering Rebel was made a prisoner of war. But only officers and men of the Continental Army would be confined; all civilian males and militiamen in Charleston were allowed to return to their homes after vowing not to take up arms again. Surrendering militiamen in backcountry garrisons, such as Ninety Six, were treated the same. As Clinton said in a report to Lord Germain, "there are few men in South Carolina who are not our prisoners or in arms with us."[29]

But there were many Rebels who were armed, at large, and unbowed. In an attempt to subdue those obstinate Rebels, Clinton proclaimed in June that the men he had paroled now had to swear an oath of allegiance to the king and promise to bear arms against his enemies.[30] The proclamation inflamed many Rebels, driving them into backcountry militias that harassed British and Tory outposts in hit-and-run raids. Clinton, who had hoped to quell the South's civil war, had only made it worse by essentially ordering people to take sides. He then left for New York, putting General Sir Charles Cornwallis in charge of the newly flaring war.

Patriots and Loyalists had a word for that kind of war: "intestine." Or, as North Carolina governor Abner Nash more vividly described it, "a Country exposed to the misfortune of having a War within its Bowels."[31] In the South intestine war continually raged within the conventional war of strategy and maneuver being fought by the British Army and the Continentals. Loyalist military units raised in New York fought in that conventional war, participating in every major battle from Moore's Creek in North Carolina in 1776 to Yorktown in 1781.[32]

Intestine warfare was more than battles. There was cruelty, there were murders in the night, and there were hangings without trial.

Isaac Hayne, a wealthy planter who was an officer in the South Carolina Patriot militia, was not in Charleston when the besieged city surrendered in 1780. But he went into the city, gave himself up, and agreed to become a British subject. The deputy British commandant in Charleston assured him that he would not be ordered to fight for the king. But as the Continental Army was approaching Charleston in 1781, Hayne was ordered to take up arms. He refused and made his way to the Patriot camp, where he was commissioned a cavalry colonel.

On one of his sorties he captured Andrew Williamson, a former brigadier general of the militia and a onetime Patriot leader who became a turncoat. When the British rescued Williamson, they also captured Hayne. He was brought before Col. Nesbit Balfour, Charleston's commandant. Balfour condemned Hayne as a traitor because he had signed an oath of allegiance and then fought against the king. Brought before a tribunal but not given a trial, he was hanged in Charleston on July 31, 1781, an event that attracted a crowd of thousands.[33]

Hayne had been hanged under Cornwallis's edict, bluntly described by a British officer: "We have now got a method that will put an end to the rebellion in a short time—by hanging every man that has taken protection [agreed to parole] and is found acting against us."[34] The edict also condemned Rebels who had been captured at a trading post near Augusta after a four-day battle against Thomas Brown and his band of Tories and pro-British Indians. Some retreating Rebels were captured and killed by Indians. Thirteen others were seized by Brown's Tories and hanged from the banister of a staircase in the trading post. Rebels claimed that Brown, lying wounded, watched gleefully—a claim that was never verified.[35]

* * *

Throughout the South, the war between Tory and Rebel had been pitiless from the beginning. Letters, diaries, and petitions for pensions are studded with casual references to hangings. A militiaman from Rowan County, North Carolina, in his petition for a pension, for example, says "the company took several Scots Tories and there hung one of them." He also tells of a Tory who was shot after a hasty court-martial. The life of another was spared. But he was "condemned to be spicketed, that is, he was placed with one foot upon a sharp pin drove in a block, and was turned round . . . until the pin run through his foot. Then he was turned loose."[36]

In a rare event, a Loyalist was put on trial and accused of running his sword through wounded Rebels on a South Carolina battlefield. When a judge who had recommended mercy left the courtroom, "fathers, Sons, & brothers and friends of the slain prisoners" seized the accused, took him on horseback "under the limb of a tree, to which they tyed one end of a rope, with the other round his neck, & bid him prepare to die; he urging in vain the injustice of killing a man without tryal, & they reminding him, that he should have thought of that, when he was slaughtering their kinsmen. The Horse drawn From under him, left him suspended til he expired."[37]

Tories executed countless Rebels, whether or not they strictly came under Cornwallis's order that "every Militia Man who has borne Arms with us and afterwards joined the Enemy shall be immediately hanged."[38] Maj. Gen. Nathanael Greene, commander of the Continental Army's Southern Department, found in the South a war in which "the Whigs and Tories pursue one another with the most relent[less] Fury killing and destroying each other wherever they meet. Indeed a great deal of this Country is already laid waste & in the utmost danger of becoming a Desert," in which the fighting "so corrupted the Principles of the People that they think of nothing but plundering one another" and committing "private murders."[39]

The intestine war was stoked by the kind of combat that Banastre Tarleton waged—a slashing, burning, ruthless crusade that knew no mercy. Tarleton's American Legion grew out of the green-uniformed Queen's American Rangers, founded by John Graves Simcoe, a young

English officer. Simcoe, wanting a mounted strike force, melded the Pennsylvania Light Dragoons, the Bucks County Light Dragoons, and a band of Scotch-Irish Tories, the Caledonia Volunteers, into the British Legion, which soon became known as the Tarleton Legion.

Loyalists of North and South Carolina eagerly joined the legion, swelling its ranks at times to a force of nearly two thousand men. Among the recruits were two sons of Allan and Flora MacDonald.[40] The legion galloped off on numerous raids, destroying Rebel supply caches, foraging for food and horses, and earning fame and loathing.

When Major General Lincoln surrendered Charleston, the largest Continental force in the South was a detachment of about 350 Virginians commanded by Col. Abraham Buford. He was ordered to go to an American outpost at Camden, South Carolina, carry off what he could, destroy the rest, and then take his men to North Carolina.

Tarleton and 270 men pursued Buford. On May 29, 1780, they caught up with him at the Waxhaws, as the settlement near Waxhaw Creek was called, about twelve miles north of Lancaster, South Carolina.[41] Tarleton, claiming to outnumber Buford, asked him to surrender under terms similar to what Lincoln had accepted. Buford refused, and the battle began, possibly before Tarleton's flag of truce had been withdrawn. Tarleton's horse was shot from under him as he and his dragoons rode down on Buford's men. Undaunted, Tarleton jumped on another horse.

Exactly what happened in the battle and afterward will never be known. Buford and eighty or ninety men—most of them mounted—escaped, meaning that the killed and wounded may have totaled more than three hundred. This would be an unprecedented battle casualty rate of 70 to 75 percent. The outcry against Tarleton—"Bloody Tarleton," "Butcher Tarleton," "Bloody Ban"—was inspired by what Rebels believed happened *after* the battle.

Dr. Robert Brownfield, a surgeon attached to Buford's regiment, later wrote that a request for quarter was refused and survivors said that "not a man was spared." Tarleton's men "went over the ground plunging their bayonets into every one that exhibited any signs of life." One man had his right hand hacked off, suffered twenty-two more wounds, and was left for dead; he lived to report the bayoneting. News of the massacre swept through the South, and Rebels' retaliatory savagery was accompanied by the cry "Tarleton's Quarter" or "Buford's Quarter."[42]

After two disastrous defeats of the Continental Army in South Carolina—Maj. Gen. Benjamin Lincoln's mass surrender at Charleston in May 1780 and the rout of Maj. Gen. Horatio Gates's large force at Camden in August—Cornwallis planned a march to Virginia. Georgia was under royal rule, and vast tracts of the Carolinas were controlled by a British-Loyalist regime. By September there was no large concentration of Continental Army troops anywhere in the South. John Rutledge, the Rebel governor of South Carolina, was trying to govern his state from Hillsborough, North Carolina. And some timid Rebels were suggesting that the time had come to simply let all three of those states go back to colony status under their conquerors.[43] From the British viewpoint the king's forces and friends were snug in the South, their flanks covered by the sea to the east and the mountain barrier to the west.

Beyond the mountains was the domain of the Overmountain men, colonists who had defied King George's 1763 proclamation that prohibited settlement east of the mountains. The Overmountain men, most of them Scotch-Irish, had not paid much attention to the Revolution until the British and their Tory allies began fighting southern Rebels. Now, as General Cornwallis was advancing northward, Overmountain Rebels were attacking British and Loyalist outposts.

Ferguson and his Volunteers were protecting the western flank of Cornwallis, who was mounting an invasion of North Carolina. Deciding to challenge the Overmountain men, Ferguson sent a Rebel prisoner into the mountains with a warning: If the Rebels "did not desist from their opposition to the British arms, he would march his army over the mountains, hang their leaders, and lay their country waste with fire and sword."[44]

In outraged response, at a frontier outpost on the Watauga River called Sycamore Shoals,* the mountain Rebels quickly assembled a makeshift mounted army that grew on the trail until it numbered about one thousand men from Georgia, the Carolinas, and Virginia, with officers elected on the spot. There was no military structure or wagon train of supplies, though the group was at its core an authorized Virginia militia.[45] The men carried what they needed and prodded cattle along the way for food on the hoof. Most of them did not have muskets, but long-barreled, small-caliber American rifles, for they were more hunters than soldiers. Setting off not to fight for a nation but to defend their cabins by the streams and their patches of cotton and corn, they headed across the mountains in search of Ferguson.[46]

Ferguson, playing on the fears of local Tories, distributed a proclamation to warn them of the imminent coming of the mountain men: ". . . . I say, if you wish to be pinioned, robbed, and murdered, and see your wives and daughters, in four days, abused by the dregs of mankind—in short, if you wish or deserve to live, and bear the name of men, grasp your arms in a moment and run to camp. . . . If you chuse to be degraded forever and ever† by a set of mongrels, say so at once and let your women turn their backs upon you, and look out for real men to protect them."[47]

On October 1 Ferguson, told by spies that his pursuers were on his trail, sent off a courier with an urgent message to Cornwallis: ". . . . I am on my march towards you by a road from Cherokee Ford north of King's Mountain. 3 or 400 good sodiers part dragoons would finish this business. Something must be done soon."[48] Near the border between the Carolinas, he spread his encampment of about eleven hundred men along a rocky ridge called King's Mountain, which rose about 150 feet from the land around it. Local hunters kept the ridge cleared of trees, using it to hunt deer, which liked a forest edge. Hunting was easy; the men could drop a deer at two hundred yards.[49]

Ferguson, settling in atop King's Mountain, wrote a letter to Maj. Robert Timpany of the New Jersey Volunteers, a former schoolteacher

* Today's Elizabethton, Tennessee.
† Another version: "If you choose to be pissed upon forever and ever. . . ."

from Hackensack: "Between you & I, there has been an inundation of Barbarians, rather larger than expected." If Ferguson was worried, he did not show it. He had two women with him. One left before the shooting began.[50]

On October 7, a fast-riding detachment of about nine hundred Overmountain men, joined by other Rebels east of the mountains, found Ferguson. They dismounted, silently surrounded the ridge, and started climbing. Recent rain kept down the dust and soaked the forest's carpet of leaves, muffling their tread. Their password was "Buford." Knowing that, like them, the Loyalists would not be wearing uniforms, the Rebel soldiers put bits of white paper in their hats; Ferguson's men used sprigs of pine.

In a battle that lasted more than an hour, rifles cut down 320 Tories, killing most of them. Ferguson, chirping on his silver whistle and rallying his men, rode on his white horse from one strongpoint to another. He wore a checkered hunting shirt over his uniform, and one moment, silhouetted against the sky, he became a target for numerous marksmen. At least seven bullets passed through his body.[51] Capt. Abraham DePeyster, a New Yorker in the King's American Regiment, took over and quickly raised a white flag, which was ignored by men who had heard about Tarleton's failure to give quarter.[52] Finally officers stopped the firing and accepted the surrender.

The next day local Tories came to seek their loved ones. "Their husbands, fathers, and brothers lay dead in heaps, while others lay wounded or dying," a Rebel wrote. Burying was hasty and shallow, and so later came the wolves, the dogs of dead masters, and the hungry farm pigs.[53] The Rebels, who lost twenty-nine men, led off 698 prisoners.

On the road to a prison camp many would die. At a campfire court-martial, thirty-six were condemned to death. Nine were hanged by torchlight, three at a time, from the limb of a great oak tree. As three more were ready for the noose, the mountain men's officers managed to stop the vengeance. Another prisoner was later executed for trying to escape. An unknown number, according to one of the few survivors of the Loyal American Regiment, "worn out with fatigue, and not being able to keep up" were "trodden to death in the mire."[54]

The battle at King's Mountain was a requiem for Tories everywhere in America. The reality of the Revolution was there on that Carolina ridge: The only British subject in the battle was Ferguson. Everyone else was an American, and those who chose to fight for King George III had chosen the wrong side.

General Clinton later wrote that the battle at King's Mountain was "the first link of a chain of evils" that ended in "the total loss of America."[55] Cornwallis retreated into South Carolina, and the Continental Congress called forth Nathanael Greene to lead a new southern army, which would badger Cornwallis out of the Carolinas and into Virginia, where the last link would be forged.

16

DESPAIR BEFORE THE DAWN

PHILADELPHIA 1778 AND WEST POINT, SEPTEMBER 1780

What but the fluctuation of our army enabled the enemy to detach so boldly to the southward, in '78 and '79. . . . And what will be our situation this winter? our army by the 1st of January diminished to little more than a sufficient garrison for West point, the enemy at full liberty to ravage the country wherever they please. . . . The army is . . . dwindling into nothing.
—Gen. George Washington[1]

Washington's letter of despair went on for more than twenty-five hundred words, as he looked back to the lost Battles of Paoli and Germantown, the British conquest of Philadelphia, and the "cruel and perilous situation . . . in the winter of '77, at the Valley Forge." He wrote the letter as he was emerging from his darkest day: The treason of Benedict Arnold had been revealed only a month before.

Sometime in June 1778, soon after Arnold took command of Philadelphia, Joseph Stansbury, a glass and china dealer, furnished the dining room of the mansion that Arnold had made his home and headquarters. Stansbury knew the mansion well, for he had dined there when General Howe was in residence.[2] Stansbury was a notorious Tory who had decided not to sail away to New York with the other panicky Tories, though he certainly seemed a likely candidate for self-exile.

London-born, he had come to Philadelphia as a boy and grew into a witty young man known for his verses and song. In the first days of the Revolution he had been a tepid supporter of protests against British trade regulations. But he gradually became a Loyalist. Three months after Philadelphia celebrated the Declaration of Independence, Patriots placed him under house arrest for taunting Rebels by singing "God Save the King" and urging them to join him in the chorus.[3] His writing became Loyalist. In compassionate "Verses to the Tories" he wrote:

> *Think not, tho wretched, poor, or naked,*
> *Your breast alone the Load sustains;*
> *Sympathizing Hearts partake it;*
> *Britain's Monarch shares your pain. . . .*[4]

When Howe and his Redcoats marched into Philadelphia, Stansbury had been one of the Tories who cheered. He became a favorite of Howe's, entertaining the general with Loyalist verse and songs. In one he has the gods meet on Olympus. Jove (Jupiter) says:

> *Ye know, all ye Pow rs that attend my high Throne,*
> *Your Will to my Pleasure must bow:*
> *I will, that those Gifts which you prize as your own,*
> *Shall now be bestow'd on my Howe.*[5]

Stansbury became a confidant of Joseph Galloway, Howe's spymaster and police chief. Galloway made Stansbury a commissioner of the night watch, a manager of the lottery for the poor, and an agent in Galloway's spy network. When the Patriots returned to Philadelphia, Stansbury routinely took an oath of allegiance to the new nation and was thus officially clear of taint when he became Arnold's interior decorator.[6] Stansbury had been one of the Tory entrepreneurs who had taken over Patriots' stores when they fled the city. That act alone would have qualified him as a despised Tory, shunned by any Patriot with a memory of British occupation. Arnold may not have seen Stansbury's privately circulated poems. But it seems impossible for Arnold not to have learned of Stansbury's Loyalist connections from the reports of Patriot agents who were in Philadelphia during the British occupation.

In December 1778 Galloway received information he undoubtedly knew already: Arnold was "being thought a pert Tory" by Philadelphians. A month later came another letter saying that Arnold had "behaved with lenity to the Tories"—and was to be married to Peggy Shippen.[7] Arnold, a thirty-nine-year-old widower, wed her in April 1779, three months before she turned nineteen.

As military governor Arnold won few friends among Patriots. Under loose fiscal oversight, he had been involved in shady dealings that led to a court-martial and an official reprimand from Washington. He antagonized important members of Congress. He lavishly issued passes to Tories who wanted to go to New York. This at a time when the Patriot government was charging hundreds of families with treason against Pennsylvania. In vain Arnold opposed the execution of two Quakers for treason. One was a gatekeeper appointed by Galloway, and the other was a miller who had sold provisions to the British and acted as a guide. Arnold saw their simultaneous hanging as mindless Patriot vengeance against the pacifist Society of Friends.[8]

Sometime in the early spring of 1779 Arnold's many grievances coalesced into his own treason. He sent for Stansbury. There is no record of their meeting, but there is a letter, written to Stansbury by Maj. John André, dated May 10, 1779, two weeks after General Clinton's brilliant aide had become the director of Clinton's secret service.[9] The letter reveals that Arnold had told Stansbury that he wanted to work for the British:

> Sir . . . On our part we meet ~~ArnGen~~ Monk's ouvertures with full reliance on his honourable Intentions and disclose to him with the strongest assurances of our Sincerity. . . . His own judgment will point out the

services required, but for his Satisfaction we give the following hints . . .
Contents of dispatches from foreign abettors— Original dispatches and
papers which might be seized and sent to us— Channels thro' which
such dispatches pass, hints for securing them. Number and position of
troops, whence & what reinforcements. . . .

André made the mistake of almost writing an abbreviation of Arnold's name and rank, crossed that out, and substituted "Monk," the
code name Arnold had chosen. Apparently he saw himself not as
a traitor but as a heroic reincarnation of Gen. George Monck, who
in 1660, changing sides, helped to overthrow Parliament, end the
Cromwell rebellion, and put Charles II on the throne. Thus began
the months of Arnold-André-Stansbury correspondence that led to
Arnold's treason.

Some letters were partially written in invisible ink. Most used ci-
phers (letters, symbols, or numbers substituted for real words) or
codes (for which both writer and recipient must have a key, such as a
codebook).[10] Many letters have not been found, but there are enough
to incriminate Peggy Shippen Arnold. In one André wrote to Stansbury, instructing him to tell "The Lady" to write to André through
one of her friends. "The letters may talk of the Meschianza & other
nonsense," André said. He wanted her letters to be sent through the
sieve that was the Philadelphia–New York connection, which involved a postal service operating under an Arnold-approved flag of
truce. André designed a system that left "every messenger remaining
ignorant of what they are charg'd with."[11]

Newlywed Peggy, for example, wrote an innocent-looking note to
John André to get him to contact her husband: "Mrs. Arnold . . . is
much obliged to him for his very polite and friendly offer of being
serviceable to her . . . [and] begs leave to assure Captain André that
her friendship and esteem for him is not impaired."[12] Peggy Shippen
Arnold knew what was going on, despite her husband's later claim
that she was "as innocent as an angel."[13]

When Arnold took command of West Point, he made his headquarters at Beverley, the mansion of Col. Beverley Robinson, George

Washington's erstwhile Virginia friend who had become a Tory and commander of the Loyal American Regiment. Beverley had been confiscated after Robinson's exposure as a Tory officer. Before Arnold and André began corresponding, Robinson had written to Arnold urging him to lead the Loyalist cause as an "important service to your country." Robinson, a wealthy merchant and influential Loyalist, foresaw a defeated America becoming part of a prosperous British union that would "rule the universe . . . bound, not by arms and violence but by the ties of commerce."[14]

Arnold's treason extended beyond his offer to hand over West Point for the equivalent of about one million dollars.[15] The fall of that key fortress would have been a disaster for the Patriots. But that was to have been only the beginning, for Arnold saw himself as the leader of a British-Loyalist army that would not only punish the Rebels and win the war but also raise the esteem of Loyalist fighting men. Coincidentally just such a force, independent of Clinton and the British Army, was being proposed by William Franklin, the son of Benjamin Franklin and the deposed royal governor of New Jersey, who was fast becoming one of the most dangerous Tories in America.

Franklin became aware of Arnold's offer through Stansbury and another Tory poet drawn into the conspiracy as a go-between: the Reverend Jonathan Odell, who was both a satirist and an Anglophile ("Rise Britannia, rise bright star! / Spread thy radiance wide and far!").[16] Driven out of New Jersey for his Tory beliefs, Odell became chaplain of a Loyalist regiment on Staten Island. He also became a close friend of William Franklin, after whom Odell named his only son.[17]

The drama of Arnold's betrayal and André's capture had a Tory subplot. Robinson was aboard HMS *Vulture*, which carried André up the Hudson to his rendezvous with Arnold. Joshua Hett Smith, a lawyer and the brother of the Tory attorney general of New York, handed over his house near West Point for the meeting at which Arnold gave André intelligence and documents about the fort's defenses. Smith also helped André disguise himself with a civilian coat and hat. This was André's fatal error, making him a spy rather than a British officer who, if captured, could claim the status of a prisoner of war. He compounded his error by hiding the documents in his socks, disobeying Clinton, who had ordered him not to carry incriminating papers.[18]

The *Vulture* had been fired on by Patriot cannons. Smith, rather than helping André get back to the security of the *Vulture*, led him, on horseback, to a place on the Neutral Ground near British lines; from there André could supposedly walk to safe territory.[19] Local people warned that Tory brigands, known as Cow-boys, were in the area, raiding and robbing.

André, riding on alone, neared British lines. Three young men, in odd bits of uniform and toting muskets, stopped him. As one of them later related the encounter, André said, "Gentlemen, I hope you belong to our party."[20]

"What party?" one of the men asked.

"The lower party," André said, referring to the Loyalists. He apparently thought he was among De Lancey's Cow-boys. "Thank God, I am once more among friends. I am glad to see you. I am an officer in the British service."

But there were other outlaws in the area—Rebel plunderers, sometimes called "Skinners" because they skinned victims of their valuables. Maj. Benjamin Tallmadge, Washington's chief intelligence officer, later called André's captors "Cow-boys," who were "roving and lurking above the lines, sometimes plundering on one side and sometimes on the other."[21] Whether Skinners or Cow-boys, they found the hidden papers, beginning the string of events that exposed Arnold, who would manage to reach the *Vulture* and safety ahead of his pursuers. The three captors were given medals by Congress.

Clinton, who had a paternal affection for André, tried frantically to save his thirty-year-old aide from the gallows. Washington could find no way to change André's fate without affecting the morale of his army or his own sense of military justice. He was, however, ready to exchange André for Arnold. Clinton had his own morale problem: Loyalists were cheering Arnold. Clinton feared that if he gave up Arnold, every Loyalist in America would turn against the British.

Arnold himself sent a chilling threat to Washington: If he hanged André, "I call heaven and earth to witness that your Excellency will be justly answerable for the torrent of blood that may be spilt in consequence." Looking toward his new career in the British Army, he wrote: "I shall think myself bound by every tie of duty and honor to retaliate on such unhappy persons of your army as may fall within my power." He also remarked darkly that André's execution would put forty Patriot prisoners in South Carolina in jeopardy.[22]

Washington was not moved, though witnesses said his hand shook as he signed André's death warrant. On October 2, 1780, at Tappan, New York, Washington's headquarters about twenty miles up the Hudson from New York City, Maj. John André, in his regimental uniform, was hanged.[23]

Arnold's treason and his raw threats came at a crucial moment in the Revolution. As Washington sat at his headquarters in New Jersey in October 1780 and wrote his letter of despair, he looked to the South and saw two states slipping away. He looked to Arnold and wondered what his treachery would bring. Arnold, commissioned a brigadier general in the British Army, gave new hope to Loyalists who wanted more of a combat role than the guarding of baggage trains. As commanding officer of a new Loyalist military organization, Arnold would bring a fresh ferocity to the battlefield. So would William Franklin's Associated Loyalists. As Arnold had warned, there would be a torrent of blood.

17

―――

BLOODY DAYS OF RECKONING

NEW JERSEY AND NEW YORK, JULY 1779–MAY 1782

The late conduct of the British demons in New Jersey, in the rob-
beries, burnings, ravishments and murders, with a long catalogue
of crimes as black as hell! is a call louder than lightening . . .
against the tyrant and his bloody butchers.
　　　　—*Newspaper report on terror warfare in New Jersey*[1]

Slowly the Loyalists' hope of victory ebbed, and vengeance filled
the void. William Franklin, who once had been New Jersey's
royal governor, was by 1780 the angry, frustrated personification of
all Tories who felt forgotten by Britain. But Franklin still had power,
strangely enhanced by the fact that an infamous Rebel, Ben Franklin,
was his father. William Franklin, with his uncanny ability to attach
himself to the possessors of power, had become the confidant of John
André, who, as General Clinton's secret service chief and adjutant
general, controlled the flow of intelligence from and to Clinton. And
when André was executed, Franklin coolly moved on to André's successor, Oliver De Lancey.

In De Lancey, Franklin found an ally for a plan that went beyond
the raids that once more were bloodying the Neutral Ground. Mostly
at his own expense De Lancey had raised three Loyalist regiments of
five hundred men each, creating what became known as De Lancey's

Brigade. He was made a brigadier general, adding that military clout to the matchless power of his formidable family. As Franklin saw the future, no longer would Loyalists be hit-and-run raiders on the fringes of the war: They would mobilize into strong military units and help to win the war, which was, after all, the Loyalists' war, not Britain's.

Franklin balanced his ties to De Lancey by becoming a friend of William Smith, who had married into the Livingston family, long-time rivals of the De Lancey clan. Smith, whose father had been chief justice of New York, had known Clinton for many years. The two lived only two doors apart on Broadway. Smith, although secretly critical of Clinton's strategic decisions, was an influential adviser to the general.[2]

Franklin's hatred of Rebels stemmed from his treatment in their hands. They doubly despised him, as a traitor to his father and a traitor to the colony he had governed. And he was harder to drive from office than any other Tory governor. Unlike others he would not flee or fade away.

William Franklin was born about 1731 to an unidentified "mother not in good Circumstances" and raised by Franklin and his common-law wife, Deborah Read Franklin. "Billy," as his indulgent father called him, studied in a prestigious classical academy in Phila-delphia, enlisted in the colonial army, and grew "fond of a military Life." But he drifted into law, reading under one of his father's clos-est friends in Philadelphia, Joseph Galloway. When he returned to civilian life he worked with his father on electrical experiments, in-cluding the one that involved sending aloft a kite with a metal key attached during a lightning storm.[3]

His father, as a colonial agent in London, arranged for William to continue his law studies there. While William was in London, he basked in his father's fame and charmed royal officials. That began his political career, which, with his father's unabashed influence, pro-duced his appointment as governor of New Jersey in 1762.[4] Sailing off to his royal post, he left his father behind in London to cope with the growing crisis between the Crown and the colonies.

Beginning with his support of the Stamp Act and continuing through his colony's growing mood of rebellion, William skillfully warded off critics and held on to his post. He was proudly Loyalist, declaring that "the most heinous Rage of the most intemperate Zealots" could not "induce me to swerve from the Duty I owe His Majesty."[5] Nor, obviously, could his father.

In 1775 Ben Franklin returned from his long stay in London to take up the work of the Revolution. With him was William's son, William Temple Franklin (always called Temple), born in London around 1759. His mother, like his grandmother, was unacknowledged, except for his middle name, which suggests that he had been conceived while his father was studying at the Middle Temple court of law in London. Temple was put in a foster home, his raising and education paid for by his grandfather. Temple's father ignored him and married a wealthy, well-connected British woman in 1762, as the governorship was about to come to him.[6]

Temple accompanied his grandfather when Ben Franklin left London, arriving in Philadelphia on May 5, less than a month after the Battles of Concord and Lexington. Shortly after, Ben, William, and Galloway met at Trevose Manor, Galloway's palatial seventeenth-century home about twenty miles north of Philadelphia. The meeting banished any hope of reconciliation. Only Galloway and William Franklin had any common ground. Ben had already rejected Galloway's Plan of Union, and William Franklin had already begun defying his provincial assemblymen as they marched toward revolution.[7]

On June 15, 1776, the New Jersey Provincial Congress declared Franklin "an enemy to the liberties of this country" and ordered him arrested. Franklin, using arguments both legal and vituperative, protested. But the Continental Congress confirmed the arrest order and put Franklin in the custody of Governor Jonathan Trumbull of Connecticut. By chance he was brought before Trumbull on America's first Fourth of July.[8] Franklin partisans claimed that he was placed in the notorious New-Gate Prison (named after the London original) in East Granby, Connecticut. Prisoners were kept in cells carved into the shafts of what had been a copper mine. Many Tories were jailed there, and Washington was said to have personally sent a few "flagrant and atrocious villains" to New-Gate. Franklin, however, was not one of them.[9]

He signed a parole drawn up by Trumbull, who put him in the first of a series of residences, giving him freedom to walk around town. He started openly acting as a Tory leader, issuing long-distance pardons to Loyalists in Connecticut and New Jersey to broaden his power base. The politicking violated his parole. So Trumbull put him in the Litchfield town jail. He was there when he learned that his wife, a refugee in New York, had died. He became so melancholy that he was released to a private residence. In October 1778 he was exchanged for John McKinley, the president (governor) of Delaware, who had been captured in Wilmington after a battle in September 1777.[10] Franklin headed straight for New York City, offered his services to Clinton, and founded, without Clinton's blessing, a Loyalist movement innocuously named the Refugee Club. Members of the club showed that they were not conspirators by announcing their meetings at Hicks's Tavern in newspapers. They appeared to be no more than socializing Loyalists. But Franklin had more than chatting in mind.[11]

Loyalists, particularly those whose property had been seized by Rebels, had begun calling themselves refugees to advertise their woeful status. New York City officials appointed an "Inspector of the Claims of the Refugees" and later a four-man board "to regulate the bounty of government to the refugees."[12] Both Loyalists and Patriots used "Refugees" for Tories who had fled New Jersey to join the British in New York, often returning to their native state as marauders. Some refugees had come from as far away as Georgia to find sanctuary and work for the Loyalist cause.[13]

Franklin's youthful fondness for the military life bloomed anew in New York. He had enough sense and experience to realize there was no possibility of creating an actual Loyalist army. Clinton was absolutely in charge, signing himself "General and Commander in Chief of all his Majesty's Forces within the Colonies laying on the Atlantic

Ocean, from Nova Scotia to West-Florida, inclusive, &c. &c."[14] Clinton was not against Tory military units, and he did authorize their deployment alongside his British Regulars in battle. But the kind of military organization that Franklin envisioned was a guerrilla force that would terrorize Rebels and rouse Loyalists. Clinton spurned the use of terror.

Another former royal governor, Maj. Gen. William Tryon, gave Clinton a rationalization for terrorizing the Rebels. Tryon pointed out that the Rebels' own tactics were based on terror—"the general Dread of their Tyranny" and expectation of "our forbearance." So, Tryon went on, he joined others who "apprehend no Mischief . . . if a general Terror and Despondency can be awaked among a People already divided." Tryon wrote about his views on terror after practicing his beliefs in a devastating raid on Connecticut civilians.

On July 2, 1779, Tryon assembled about twenty-six hundred soldiers—Regulars, Hessians, and a major Loyalist military unit, the King's American Regiment. The regiment, originally known as Fanning's Corps, was led by Tryon's secretary, Col. Edmund Fanning.[15] Among the black Tory soldiers in the unit was John Thompson, the secret courier who had worked for Tryon when he ran a Loyalist underground from his cabin on board the *Duchess of Gordon.*

One of Fanning's officers in the Connecticut raiding party was Capt. John McAlpine, who had been jailed by Rebels for recruiting. His recruits, in an early display of their fidelity, had broken into the jail and freed him.[16] His would be a fighting regiment, destined to serve on many battlefields. But on this day, as they boarded ships in New York City, they were part of a terror force.

The first target was New Haven, Connecticut. At dawn, atop the campus chapel, Yale president Ezra Stiles surveyed the raiders' fleet through his telescope and later saw soldiers landing and advancing from shore. Stiles immediately gathered college records and hid them away. Tryon had planned to burn down Yale and the rest of New Haven. But Fanning, a Yale graduate (class of 1757), persuaded Tryon not to torch the town.[17]

As the invaders disembarked, Yale students joined a volunteer company of about one hundred men who rushed to West Haven to delay the invaders while women and children fled from New Haven. Fifty-one-year-old Napthali Daggett, a professor of divinity and a former Yale president, got off a few shots from his ancient fowling piece. His moments on the battlefield would go down in Yale history:

A British officer yelled, "What are you doing there, you old fool, firing on His Majesty's troops?"

"Exercising the rights of war," Daggett professorially answered.

"If I let you go this time, you old rascal, will you ever fire again on the troops of His Majesty?"

"Nothing more likely," he replied. British soldiers then bayoneted and clubbed him. They probably would have killed him if one of his former students, William Chandler, a classmate of Nathan Hale, had not intervened. Chandler, a Tory officer, and his brother, Thomas, were guiding Tryon's men during their all-day looting of the city. William and Thomas were the sons of Joshua Chandler, a Yale alumnus and a wealthy New Haven Tory whom Rebels had briefly jailed for his outspoken support of the Loyalist cause.[18]

The muskets of the Yale volunteers and militiamen slowed down the raiders for a short time. At least one Tory officer was killed and seven enlisted men wounded; twenty-three Patriots were killed and fifteen wounded.[19] Tryon made no effort to stop the looting or harassing of townspeople as his force took possession of the port. The next morning they burned ships, goods, and whatever ordnance they could find and then withdrew to their fleet. The entire Chandler family went with them. A neighbor said he entered the vacant Chandler house, untouched by looters, and saw on a dining table a meal laid out and uneaten. The Connecticut government seized Chandler's estate.[20]

Next the troops sailed for Fairfield, a smaller town about twenty-five miles to the south. Alarm cannons fired on the foggy morning of July 7, after lookouts in a small hilltop fort spotted the Tryon fleet nearing Fairfield. The fleet's pilot was a Fairfield man, George Hoyt, who would accompany them ashore and guide them as they destroyed his town.

Townspeople—mostly women, children, and elderly men—drove livestock into the woods and hid their silver in wells or within the clefts of stone walls. In midafternoon the soldiers began landing on Fairfield's beach. Two columns separated and marched to the parade ground in the center of town, where Tryon himself posted a proclamation calling on all residents to swear allegiance to the king. No one seems to have paid any attention to the piece of paper. As in New Haven, militiamen fired on the invaders with muskets and cannons. But the overwhelmingly outnumbered defenders fell back before the raiders' intense musketry and cannon fire from the offshore ships. Tryon easily took control of the town, setting up his headquarters in a large house near the parade ground.[21]

Again Tryon did not prevent his troops from terrifying the civilians, especially women. Several told later of being menaced, their silver shoe buckles and silver buttons torn away, their homes pillaged, their furniture smashed. One woman said soldiers "attempted, with threats and promises, to prevail upon me to yield to their unchaste and unlawful desires." After she "obstinately denied them my body, these men then and there dragged me to bed and attempted violence, but thanks to God there appeared that instant two persons who rescued me." After hearing what happened, Tryon posted two men to guard her from further outrages.[22]

Tryon sought out the house of Thaddeus Burr, a town judge. Finding the judge absent, Tryon demanded the judge's official papers from his wife, Eunice Burr. "I told him," she later recounted, "there were none but of very old dates, which related to the old estates. The general said, those are what we want, for we intend to have the estates. Upon which he ordered an officer to take them. . . ." After Tryon left, she said, "a pack of the most barbarous ruffians" entered the house. The woman ran into the yard, where the men threw her to the ground and began searching her for a watch they thought she had, "pulling and tearing my clothes from me in a most barbarous manner." Capt. Thomas Chapman of the King's American Regiment, whom she knew socially, appeared and stopped the assault.[23]

Tryon then ordered the burning of houses, one by one, including the Burr home. As the day was reddening with sunset and flames, a thunderstorm broke and lightning laced the sky. One resident felt it was enough to make a person believe that "the final day had arrived." The sudden rain did not quench the flames, which were visible for miles around.[24]

The Reverend John Sayre, who had come to Fairfield as its Anglican minister in 1774, was such an unremitting Loyalist that Rebels had declared him to be "a person inimical and dangerous to the interests of the United States" and exiled him to Farmington, Connecticut, fifty-six miles inland. Rebels later relented and allowed him back. Now, on this day of fiery invasion, Sayre went to Tryon and pleaded with him to stop the torching of the town. Tryon promised to spare certain homes, including Sayre's.[25] Local legend held that Tory homes were marked with black chimney stripes. Other residents and servants fled their houses. At one, raiders began their arson by snatching burning coals from a large fireplace and tossing them onto the wooden floor. As soon as they left, a slave slipped in, scooped up the coals, and threw them out, saving the house. The floor's burn marks can still be seen.[26]

The next morning Tryon withdrew, his troops burdened with loot. They left behind the smoldering embers of many houses and the bodies of several slain residents, including three men bayoneted to death and another fatally shot for ignoring a soldier's order. Sayre, his wife, and eight children all joined Tryon on the path to shore and waiting boats, as did George Hoyt, the Tory guide. Furious militiamen aimed a cannon at the home of Hoyt's kin, a suspected Tory. A militia officer intervened, and the cannon was not fired.

Tryon assigned Hessians to be his rear guard, shielding his guests and the rest of the force as it withdrew. The departing Hessians put a final touch to the raid, setting fire to Sayre's church and home, the Congregational Meetinghouse, and the Congregational minister's home.[27] Tryon and his men destroyed ninety-seven houses, seventeen barns, forty-eight stores, two schools, a county building, and three churches.[28]

The raiders sailed on to Norwalk, a few miles south of Fairfield, and began disembarking troops for an attack on that busy port. Again, after driving off a small militia force, Tryon took possession of the town. His men stole whatever they considered valuable and burned everything else, including whaleboats and small vessels in the harbor. The toll was 135 houses, two churches, eighty-seven barns, twenty-two storehouses, seventeen shops, four mills, and tons of newly harvested hay and wheat.[29]

Tryon's fleet sailed next across the Sound to Huntington, Long Island, where he was refitting and resupplying his ships for continued raiding when he received orders to cancel further raids and report back to General Clinton. Tryon's infuriated superior had two reasons to reprimand Tryon for acting "contrary to his . . . Orders." First, the casualties. For no apparent military gain, the terror raids left twenty-six of Tryon's men killed, ninety wounded, and thirty-two missing. And second, the raids produced a fuming reaction among Patriots, who would turn that anger into revenge.[30]

Tryon's raid against Danbury two years before had had a military rationale because the town was the site of a Continental Army supply depot. This time Clinton had approved attacks along the Connecticut coast because he believed they would compel Washington, whose army was in New Jersey, to send troops to the rescue, giving Clinton an opportunity to strike across the Hudson at a diminished foe.[31] Washington saw through Clinton's strategy and outwitted him by ordering Brig. Gen. Anthony Wayne to lead a raid on Stony Point, the British "Little Gibraltar" on the Hudson. Washington knew from his spies that Clinton had withdrawn men from the fort to add to Tryon's force.

On his march to Stony Point, Wayne arrested civilians to keep any Tories from warning the British. In the swift midnight attack a bullet grazed Wayne's brow. His men took more than four hundred prisoners, along with several cannon and other military booty in what was the war's last major battle in the North.[32]

The attack on Stony Point emphasized Clinton's blunder in authorizing Tryon's raids, which Clinton had vaguely viewed as harassment, not terrorism. Although he had neither suggested nor forbidden arson, in the words of his critic and adviser, William Smith, "Sir Henry wished the Conflagrations, and yet not to be answerable for them."[33]

Clinton demanded a written report from Tryon, presumably so that he could prepare an explanation for anticipated criticism from London. In his report Tryon insisted that the Rebels had to be punished for their rebellion "if possible without injury to the loyalists." By terrorizing the Rebels, he said, he had instilled in them fear of reprisals. He expected "no mischief to the public from the irritation of a few in rebellion if a general terror and despondency can be awakened among a people already divided . . . and easily impressible."[34]

If Clinton feared censure from London, he was wrong. Lord Germain approved the raids, signaling what Tryon and William Franklin already knew: The conduct of the war was changing. What Tryon had labeled "desolation warfare" had been renewed, not by Clinton's soldiers but by William Franklin and his creation, the Board of Associated Loyalists. Tryon backed Franklin's Associated Loyalists, his instrument for taking the waging of the war away from Clinton and the "Parcel of Blockheads" around him.[35]

Tryon was a Briton of high military and social class. Franklin was a Loyalist and a son who was not only rebelling against the Patriots but also against his own father. Franklin felt that as a Loyalist he better understood the realities of the war and the Rebels than the British Army or its German hirelings. He had governed one of the mightiest Tory strongholds in America. By one estimate about a third of New Jersey's population—some five thousand people—were Tories. In Bergen County, just across the river from New York City, Franklin believed he could create an army.[36] Eventually more than two thousand men would serve in the New Jersey Volunteers.[37]

While Tryon was preparing for his Connecticut raids, Franklin presented to him—as commander of the provincial forces in America—a plan that would make the Board of Associated Loyalists an independent, quasi-military force whose operations would "distress the Enemy in any Quarter not expressly forbid by the Commander in Chief."

The distress would include pillage, for Franklin adopted a practice of Tryon's desolation warfare: The raiders had the right to keep all "the plunder they take, which is to be only from rebels."[38] Tryon endorsed the plan, which called for battalions of about 250 men and a fleet of ships and whaleboats. Nine prominent New York and New Jersey Loyalists began recruiting for Franklin.[39]

Public notice of Franklin's organization came on December 30, 1780, in *Rivington's New-York Gazetteer*, which published the "Declaration of the Board of Directors of Associated Loyalists." The group, it noted, had been established "for embodying and employing such of his Majesty's faithful subjects in North America, as may be willing to associate under their direction, for the purpose of annoying the sea-coasts of the revolted Provinces and distressing their trade, either in co-operation with his Majesty's land and sea forces, or by making diversions in their favor, when they are carrying on operations in other parts."[40]

The announcement made clear to any discerning reader that the Associated Loyalists were to wage an independent war. Franklin's Associators, as they were called, got British Army benefits but not British Army control. The Associators would be commanded by their own officers, recommended by the board and commissioned by Clinton. He would furnish them with arms, ammunition, and rations, along with vessels that they would crew. If they were sick or wounded, they would be treated in royal army hospitals. Each Associator would "receive a gratuitous grant of Two Hundred Acres of Land in North America." In a reluctant concession to Franklin, Clinton also gave him the right to handle his own prisoners, rather than place them in the regular British Army prisoner-of-war system.[41]

The ten-man board of directors, headed by Franklin and approved by Clinton, included Josiah Martin, who had succeeded Tryon as governor of North Carolina. Back in 1775 Martin had deliriously foreseen stamping out rebellion in his colony with a Loyalist force numbering in the thousands, especially "throughout all the very populous western counties of this Province."[42] Also on the board was Brigadier Timothy Ruggles, who had pioneered the idea of organizing armed Loyalists. Other members were Daniel Cox, who had been a royal councillor in New Jersey, and George Duncan Ludlow, victim of a Rebel home plundering and superintendent of police on Tory Long Island.[43]

George Leonard, who had been a volunteer Tory combatant in the Battle of Lexington, was the maritime member of Franklin's board. Leonard had gone to England and won from the king himself approval of a new Loyalist organization, the Loyal Associated Refugees, which provided the ships, boats, crews, and forces for seagoing raids.[44] Franklin's fleet intensified the whaleboat war being waged in the Sound, where Rebels in Connecticut and Refugees in Long Island raided each other's shores, plundering and kidnapping.

In a short time more than four hundred Loyalists became Associators. Franklin began feeling like an important placeman again. But he also felt "shackled," because, despite Germain's endorsement of the Board of Associated Loyalists, Clinton insisted on approving every mission that Franklin and his board planned. In reality, however, there seems to have been little British control over the amphibious Associators' raids and privateering along the New Jersey coast.

New Jersey had become an odd kind of battleground, on which vengeful partisans warred while their respective armies remained essentially above the fray. Since the Battle of Monmouth, the war had been deadlocked in the North, each army so entrenched that it dared not attack the other. But once again the armies were foraging in the Neutral Ground.

Occasionally there were large-scale foraging raids. In March 1780, for instance, about three hundred British Regulars and Hessians attacked Hackensack, "a large and beautiful settlement consisting of about two hundred houses," as Johann Conrad Dohla, a Hessian soldier, wrote in his diary.

All the houses "were immediately broken into and everything ruined. . . . All the male[s] were taken prisoners, and the town hall and some other splendid buildings were set on fire. We took considerable booty, money, silver pocket watches, silver plate and spoons, as well as furniture, good clothing, fine English linen, good silk stockings, gloves, and neckcloths, as well as other expensive silks, satins, and other materials."[45] Dohla was disappointed because he and his men had to abandon so much loot after being harassed by Rebel militiamen. Still, they did carry off about fifty or sixty men for future prisoner ransoming and exchanging.[46]

"There was no trusting of the inhabitants, for many of them were friendly to the British, and we did not know who were or who were not," wrote Private Joseph Plumb Martin, that intrepid chronicler of the Revolution. Assigned to a Continental Army guard force along the New Jersey shore, Martin told of a night when armed Refugees came ashore, killed a Continental guard, and burst into a house where guardsmen were quartered. "After they had done all the mischief they could in the house," Martin wrote, "they proceeded to the barn and drove off five or six head of Mr. Halstead's young cattle, took them down upon the point and killed them, and went off in their boats, that had come across from the island for that purpose, to their den among the British."[47]

Vicious warfare was continual between armed Loyalists and the Rebel militias—along with an assortment of freebooters, Cow-boys, Skinners, highwaymen, and robbers who knew no other cause than the filling of their own purses. A peculiar New Jersey breed was the Pine-Banditti, who operated out of the dense forests of the Pine Barrens in the southern coastal plain. Typically justice was dispensed without judges. Three Pine-Banditti, for example, were simply put to death after being caught by Rebel militiamen in Monmouth County.[48] Joseph Mulliner, the most notorious of the Pine-Banditti, raided from a base near Little Egg Harbor. Mulliner, a newspaper reported, "made practice of burning houses" and robbing "all who fell in his way, so that when he came to trial it appeared that the whole country, both whigs and tories, were his enemies." Mulliner was hanged for high treason against New Jersey on August 8, 1781.[49]

By one reliable count, from the time of the Battle of Monmouth to the end of the war, there were 266 raids and other clashes in New Jersey and 108 attacks or other incidents between June 1778 and 1783 along the state's coast or on its rivers. Most actions were Rebel privateer attacks on British shipping.[50]

The Associators staged many hit-and-run operations, but none of them was militarily significant because Clinton, a professional dealing with an amateur, manipulated the British Army bureaucracy to thwart Franklin. Then, in December 1779, General Clinton sailed south with about seven thousand British, Hessian, and Loyalist troops to launch the southern strategy, once again anticipating a rising of Loyalists. Before leaving, Clinton put a Hessian officer, Lt. Gen. Wilhelm von Knyphausen, in command. If Knyphausen agreed, Franklin, Smith, and Tryon would now have their chance to show that Loyalists could rise in New Jersey, too.

Washington and the Continental Army were in winter quarters in Morristown, New Jersey, about thirty-five miles west of New York. Hunger—and mutiny—stalked the encampment. Starving, freezing men were trying to live through a winter so cold that, for the first time in memory, both the Hudson River and Long Island Sound were frozen solid.[51] "At one time it snowed the greater part of four days successively, and there fell nearly as many feet deep of snow," wrote Private Martin, who had lived through a milder winter at Valley Forge. ".... We were absolutely, literally starved. ... I saw several of the men roast their old shoes and eat them."[52]

An attack on Washington's dwindling mutinous army looked easy on the map: Cross from Staten Island to Elizabethtown Point, which was within British lines. Then across about ten miles of fairly friendly territory to Springfield, which was at the threshold of the Hobart Gap, a pass through the Watchung Mountains. Behind those natural ramparts was Washington's encampment. An enemy force massing at Springfield would surely draw him out to a decisive battle.

William Smith, Franklin's ally and Clinton's secret enemy, became the intelligence chief for a possible thrust toward Springfield. Smith interviewed deserters from the Continental Army and Loyalist spies. Another source was Christopher Sower, a Pennsylvanian who was a member of the Board of Associated Loyalists. He said that he had been told that Loyalists from his state and from Maryland would flock to join a British Army marching to attack Washington.[53] All this positive information was presented to Maj. George Beckwith, a British Army intelligence officer. He translated for Knyphausen, who could not speak English.

Some of the intelligence pouring in to Beckwith was false. George Washington, who often acted as spymaster of the Continental Army, ordered the commander at West Point to "magnify the present force on the North [Hudson] river, but keep it within the bounds of what may be thought reasonable or probable." Whatever other misinformation Washington may have spread is not known. But documents show he was getting good intelligence from his own agents. As early as March 11 he had received an intelligence report that "the enemy have it in contemplation to pay us a visit (and in a very short time)." He was also aware that "the enemy have taken up a large number of vessels (it seems for an expedition against this quarter). All the houses on the western and northern sides of Staten Island are taken for barracking troops."

In anticipation of an attack he ordered the inspection of "certain Signals established for alarming the Militia in case of a serious movement." These were tall log towers filled with brush and set afire to bespeak alarm. Seeing the flames—or hearing alarm cannons boom on Mount Hobart and elsewhere—militiamen would muster at pre-arranged sites.[54]

Knyphausen kept authorizing sporadic raids while building up forces for a major offensive, a change in tempo that Washington's agents had noted. Finally, on the night of June 6, 1780, with Clinton still in the South, Knyphausen sent six thousand men, including Loyalist units, into New Jersey. The force outnumbered Washington's army nearly two to one.[55] Washington kept most of his Continental men in Morristown, leaving defense of their homeland to the New Jersey Continental Brigade and local militiamen. Responding to the mustering alarms, they tracked the invaders, sniping at them as they slowly headed west, finally stopping at a village called Connecticut Farms (now Union).

The pastor of the Presbyterian church in Connecticut Farms was the Reverend James Caldwell, who had moved there from Elizabethtown (now Elizabeth) after Tory raiders torched his church and tried to kidnap him. When Knyphausen's Tories arrived in Connecticut Farms, the minister was in Morristown serving as a chaplain. His wife, Hannah, was home with their nine children. Militiamen and civilians were firing from their houses at the invaders. The invaders began setting fire to the houses and to Caldwell's church. Someone—perhaps a militiaman, perhaps a Redcoat, perhaps a Tory—fired through the window of Caldwell's house, killing Hannah Caldwell instantly.

The next morning, Knyphausen, shadowed by militamen, withdrew to Elizabethtown Point. He had just learned that General Clinton, having taken Charleston, South Carolina, was sailing home with most of his men. For the next thirteen days the invaders camped in Elizabethtown, fought off skirmishing militiamen, and awaited Clinton. Newspapers spread the word about the torching of Connecticut Farms and the killing of Hannah Caldwell, denounced as murder by "one of the barbarians."[56] The possibility of an accidental shooting was quickly eclipsed by propaganda and outrage over the Tory murderers.

On June 23 Knyphausen set out for Springfield again, this time under orders from Clinton, who would take his own southern army up the Hudson, hoping to split Washington's forces and destroy the Continental Army. Rebel fury, rather than Tory fervor, rose along the path that Knyphausen was now taking for the second time. New Jersey Volunteers and the Queen's American Rangers, marching in the advance guard, got no new Tory recruits. This time, however, British artillery backed up the infantry, and fierce fighting forced Washington to send more Continentals into battle at Springfield, about a dozen miles from Morristown.

Stymied, Knyphausen pulled back, setting fire to houses as he withdrew. Enraged militiamen, some of whom saw their own homes burning, pursued the retreating rear guard, which included Queen's Rangers and a German unit of Jägers, who suffered heavy casualties. In the two battles, some twenty militiamen and Continentals were killed. About three hundred of Knyphausen's men were killed, wounded, or missing. The commander of the Jägers, writing home to Germany, summed up the seemingly pointless battles: "I regret from the depths of my heart that the great loss of the Jägers took place to no greater purpose."[57]

But the battles had achieved unexpected results. The Loyalists lost their hold on much of New Jersey, as Patriots in Springfield immediately demonstrated: They ordered their Tory neighbors to get out of town with only whatever they could carry. The Springfield exiles became refugees and headed for Staten Island, well behind the invaders, who were going to the same place.[58] The fiasco should have wiped out William Franklin and his Board of Associated Loyalists, but his influence among Loyalists extended beyond New York and New Jersey.

Sometime in the spring of 1781 the Associated Loyalists sent an agent to Maryland to set in motion a plot aimed at raising a large force of Loyalists who would spearhead a British strike into Chesapeake Bay, peeling off Maryland and Virginia from the North. Local Patriots learned of the plot and obtained documents bearing the names of the Maryland Tories involved. In July 1781 seven of the accused plotters were put on trial before a three-man tribunal, which found them guilty of high treason against Maryland.

After telling the men to make their peace with God, the presiding judge handed down the sentence: "You shall be carried to the gaol of Frederick town and be hanged therein; you shall be cut down to the earth alive and your entrails shall be taken out and burnt while you are yet alive, your heads shall be cut off, your body shall be divided into four parts and your heads and quarters shall be placed where his excellency the Governor shall appoint. So the Lord have mercy upon your poor souls." Four of the sentences were commuted, but three were hanged. Contemporary accounts do not confirm the drawing and quartering, except to say the three men "suffered the full vigor of the law."[59] In Delaware jurors condemned another plotter, "seduced by the instigation of the Devil as a false rebel and traitor," and ordered him executed in a similarly gruesome manner.[60]

It would take an international incident to finally bring down Franklin's Associators. The incident traced to a day in September 1780 when two notorious New Jersey terrorists fought a gun battle. Joshua Huddy, a captain in the Monmouth County militia, was in a house in the village of Colts Neck, about a dozen miles west of Long Branch. Surrounding the house was Colonel Tye, a runaway slave, and his gang of armed Tories. Huddy called himself a refugee hunter, using New Jersey Patriots' favorite term for targeted Tories. And Tye, a pillaging Tory guerrilla, was the kind of refugee that Huddy hunted.

In November 1775, after Lord Dunmore offered freedom to any slave who joined the British, Tye, twenty-one years old and six feet tall, fled to freedom from his master in Shrewsbury, New Jersey.[61] He somehow made his way to Virginia and became a member of Dunmore's Ethiopian Regiment. When the regiment disbanded, Tye returned to New Jersey, where he fought in the Battle of Monmouth. He then became a renegade refugee, leader of a quasi-military gang of Continental Army deserters and fugitive ex-slaves who looted Rebel homes and farms, kidnapped Rebel and militia leaders, and stole livestock and provisions for the British Army. British officers gave Tye the honorary rank of colonel and dubbed his gang the Black Brigade.[62]

One of the brigade's white commanders was Maj. Thomas Ward of the Loyal Refugee Volunteers, whose slaves included at least one of his black soldiers. Their headquarters was a timber-built fortress at Bull's Ferry, New Jersey, across the Hudson from New York City. On their numerous raids they stole cattle, gathered intelligence, and helped fugitive slaves and refugees reach British lines. On wooded Rebel land they harvested firewood, vitally needed for the cook fires and fireplaces in the vast British establishment in New York. Ward was a law unto himself. Three of his black soldiers were hanged after a British Army court-martial found them guilty of murdering a man; Ward was accused of ordering the killing but was never charged.[63]

Washington, irritated by the Bull's Ferry raids, sent Brigadier General Wayne and about one thousand troops to wipe out the Loyalist base. About seventy Loyalists—a number of them probably Franklin's Associators—were in the crude fortress, built against a sheer cliff. Wayne cannonaded the redoubt and tried to storm it. Finally, after fifteen of his men were killed and forty-nine wounded, he withdrew. Twenty-one Loyalists were killed or wounded.[64] The rest lived on to continue pillaging and terrorizing the area.

Huddy, meanwhile, accused of murdering several refugees, became a target for Tory vengeance. One of his victims was a Tory hanged from a tree limb after being charged with spying for the British. When asked about the hanging, Huddy said that all he had done was "slush" (grease) the rope and pull it.

In the summer of 1780 Huddy took to the sea. The Continental Congress gave him a privateer's commission to "set forth . . . in a warlike manner" against the British in "the Armed boat called *Black Snake*."[65] He had been a privateer for about a month, when, one day, an hour before dawn, Colonel Tye and his men surrounded Huddy's house. One of them broke a window, hoping to get in, seize Huddy, and get out quickly. The breaking glass awakened Huddy, who grabbed a musket and began firing.

The raiders pulled back. A musket was fired from another window, then another. Tye was convinced that some of Huddy's men were with him. Inside the house Huddy was acting like several shooters, moving from window to window, firing muskets. They were loaded and handed to him by the only other occupant, a young woman variously described as his housekeeper and mistress.

One shot hit Tye in the wrist. He bandaged the wound and continued the attack, ordering some men to creep up to the house and set it afire. Huddy shouted that he would surrender if they extinguished the fire. They did, and he walked out. Tye led him away to a waiting whaleboat, which would take him through "the lines," a wavering boundary along the New Jersey coast that marked the border between refugees and Rebels. But militiamen, rushing toward the sound of musketry, saw the motley refugee gang pulling away and fired at the boat. In the confusion Huddy jumped overboard and escaped.[66]

Gangrene developed in Tye's wound and he died. Stephen Blucke, a free black from Barbados, succeeded to Tye's title and command, and the refugee raiding went on. To fight the refugee gangs, a group of citizens formed the Monmouth County Committee of Retaliation, headed by David Forman, a brigadier general in the state militia and a former judge. The committee, unattached to militias or the Continental Army, resembled Franklin's unrestrained Board of Associated Loyalists.

Acting as a shadow government, the Committee of Retaliation plundered and murdered refugees, often settling private grudges in the name of the Patriots. Forman owned a piece of property in Freehold that was known as the Hanging Place, where at least a dozen Loyalists were hanged before Forman saw reason to leave the state. The state legislature condemned the Retaliators as "utterly subversive of the Law" but could not stop them.[67] Refugee and Retaliator raids continued until the end the war.

On March 20, 1782, a mixed force of Franklin's Associators and Pennsylvania Tories set off in whaleboats from New York City, escorted by the *Arrogant*, an armed brigantine, and sailed for the village of Toms River, New Jersey, about sixty miles south. They landed at the mouth of the river on Saturday, March 23, and were met by a group of local armed Tories. Just before dawn they all set off for their objective: a blockhouse at the edge of the village.

The blockhouse was a stockade made of logs about seven feet tall, pointed on top, forming a square with no openings. The only way to enter or leave was by a scaling ladder. Every few feet there was an opening in the logs just large enough to sight and fire a musket. The blockhouse included a barracks and a partially underground room that was the powder magazine. On each of the corners was a small, pivoting cannon.

The blockhouse guarded a saltworks, which produced salt from ocean water under a contract with the Continental Army. A vital military commodity, the salt was used in curing meat for shipment to soldiers. The workers, exempted from militia service, were armed and expected to aid in the defense of the works, a prime objective of Associated Loyalist raids. The saltworks also provided nearby residents with a chance to become war profiteers, buying a bushel of salt at fifteen dollars and then selling it to the Continental Army at Morristown for thirty-five dollars.[68]

The commander of the Toms River blockhouse was Joshua Huddy in a new incarnation as a captain of a state regiment attached to the Continental Army. When the raiders approached and demanded surrender, the defenders responded with muskets and cannons. In a short firefight several raiders were killed or wounded. But Huddy could see that he and his men were outnumbered about four to one. After seven defenders were killed or mortally wounded, Huddy surrendered the blockhouse. He was taken prisoner, along with sixteen others, four of them wounded. The Tories set fire to the blockhouse, killed a local militiaman, and then torched the saltworks and the entire village of Toms River.

The prisoners were put aboard the *Arrogant* and taken off to the Sugar House Prison in New York City, where British prisoners of war were held. But Clinton had reluctantly given Franklin control over his prisoners. Capt. Richard Lippincott of the New Jersey Volunteers, on assignment from the Board of Associated Loyalists, transferred Huddy to a British warship in New York Harbor. Lippincott was carrying out secret verbal instructions from Franklin.

A few days later Lippincott and a party of other Tories were rowed out to the warship. Lippincott went aboard, took custody of Huddy, put him in the boat, and beached at a desolate stretch of shore near Sandy Hook. Lippincott walked Huddy to a makeshift gallows, put a noose around his neck, pointed to a barrel under the gallows, gave him a piece of foolscap, and told him he could write his will.

Using the barrelhead as a desk, Huddy made his will. A note on the back of the foolscap says, "The will of Captain Joshua Huddy, made and executed the same day the Refugees murdered him, April 12th, 1782." He then climbed onto the barrel and a black Tory—probably one of Blucke's men—kicked it over. When Huddy was dead, someone attached to his body a statement that began, "We, the refugees" and ended, "Up goes Huddy for Philip White."[69]

Philip White had been a refugee raider captured by Rebels in a skirmish at Long Branch, New Jersey. According to the Rebel account, White was killed while attempting to escape. One of White's guards was the son of a man White had killed, and so questions arose about the circumstances of White's death. But Huddy was not involved. He was in prison in New York when White was killed. Clearly, up went Huddy because Franklin wanted a sacrificial atonement for White's death.[70]

Even the Presbyterian minister who preached at Huddy's funeral joined other Rebels in a demand for retribution. The roar of outrage quickly reached George Washington, who called the hanging an "instance of Barbarity." Backed by Congress, Washington wrote to Clinton, warning that he would execute a British prisoner if Clinton did not turn over Lippincott. Clinton responded by ordering that Lippincott be court-martialed.

Washington ominously directed that a British officer of similar rank to Huddy be selected by lot from prisoners of war and sent to the Continental Army encampment in Chatham, New Jersey. Thirteen officers confined in Pennsylvania were selected. Each drew from a hat a piece of paper. Twelve papers were blank. The paper with "unfortunate" written on it was drawn by Capt. Charles Asgill of the 1st Regiment of Foot, the twenty-year-old son of Sir Charles Asgill, a former lord mayor of London.

It was also an unfortunate selection for Washington. Not only was Asgill from an illustrious family and famed regiment but his prisoner status was tied to an article of the surrender agreement at Yorktown, which stipulated that surrendered prisoners would be protected.[71] Washington had accepted the surrender of General Cornwallis at

Yorktown on October 19, 1781, ending the British southern campaign and effectively ending the war.

By May 1782, when the Huddy crisis was flaring, informal talks between the British and an American Commission led by Ben Franklin were under way in Paris. If Washington ordered Asgill's execution, William Franklin's diehard Loyalists could wreck the peace negotiations.

While the Lippincott court-martial was going on, Gen. Sir Guy Carleton replaced Clinton as commander of British forces in America, and Clinton sailed for home. Carleton, condemning "unauthorized acts of violence," disbanded the Board of Associated Loyalists. But he could do nothing about the court-martial, whose members included such leading Loyalist officers as Brig. Gen. Cortlandt Skinner, founder of the New Jersey Volunteers, and Col. Beverley Robinson, commander of the Loyal American Regiment. Not waiting for the outcome of the court-martial, Franklin left for England, never to return.

Carleton had been in command little more than a month when the verdict came in. The court found that Lippincott was convinced that "it was his Duty to obey the Orders of the Board of Directors of Associated Loyalists." He said he had not committed murder and thus was not guilty.[72] The verdict stunned Washington. He knew that he had to make good his threat of retaliation, a decision that "has distressed me exceedingly," he said in a letter.[73] Then, unexpectedly, came a reprieve for both Asgill and Washington.

Asgill's mother, Lady Asgill, had written a pleading letter to the Comte de Vergennes, the French foreign minister, asking him to intercede. Vergennes sent his own plea to Washington, along with the mother's letter. Washington, touched by both maternal love and French diplomacy, submitted the appeal to Congress, which told Washington to free Asgill. Because of the long time it took for letters to travel, Asgill was not released until November.[74] By then a preliminary peace treaty had been negotiated.

The last major event of the war in the North came in September 1781 when Benedict Arnold burned down New London, about twelve miles from his Connecticut birthplace. There was an apocalyptic air to this final act in Arnold's long war—five years in the Continental Army, one year in the British Army—as he led turncoats to their deaths on his native soil. The core of his invasion force was his American Legion, which included more than two hundred deserters from the Continental Army.[75] Arnold's force of about seventeen hundred men also included New Jersey Volunteers and some refugees attached to the Volunteers.

Local Tories—called "friends to Government" in Arnold's report to Clinton—told him that the two forts guarding New London Harbor were undermanned. But the Patriots fought ferociously, even resorting to spears when Arnold's force stormed Fort Griswold. Lt. Col. Abraham Van Buskirk, a founder of the Volunteers back in New Jersey, entered the fort. In his report, Arnold said that most of the Patriot officers, among them the fort commander, Lt. Col. William Ledyard, "were found dead in Fort Griswold, and 60 wounded, most of them mortally." Arnold was glossing over an atrocity.[76]

A bayoneted survivor, lying in a puddle of blood at Fort Griswold, regained consciousness and later told what had happened: "The first person I saw afterwards was my brave commander, a corpse by my side, having been run through the body with his own sword." Other witnesses said that Ledyard was bayoneted to death as, surrendering, he handed his sword to Van Buskirk. Other soldiers, identified as Tories, bayoneted the surrendering soldiers and the wounded. "After the massacre," the bayoneted survivor remembered, "they plundered us of every thing we had, and left us literally naked." About 115 Patriots were killed. One out of every four men in Arnold's force was killed or wounded, one of the highest casualty rates of any battle in the war.[77]

As Arnold was destroying New London, a French fleet, commanded by Admiral François de Grasse, was defeating a British squadron off Yorktown, Virginia. The victory would give the French control of the Chesapeake and trap General Cornwallis's army on the Yorktown Peninsula. A combined army, led by Washington and French general Jean-Baptiste de Rochambeau, was on the way to besiege and defeat Cornwallis.

Cornwallis's surrender on October 19 at Yorktown meant the Revolution was almost certainly over. But the civil war went on. In Pennsylvania, where two Quakers had been hanged earlier in the Revolution, 490 people were accused of high treason and put on a "Black List."[78] New York's Committee (later, Commission) for Detecting and Defeating Conspiracies had tried more than one thousand people for Loyalist activities, and trial was tantamount to conviction, a heavy fine, and confiscation of property.[79] Delaware, perceiving a rebellion within the state itself, charged forty-eight Tories with treason against the state; the ringleader, convicted of murder, was finally hanged in 1788.[80]

Along the New York frontier, after General Cornwallis's surrender, Maj. Walter Butler led a group of Rangers and Indians into the Mohawk Valley on a routine foray for prisoners and plunder. He was killed during a skirmish with Patriot pursuers. A Continental officer later wrote that "the inhabitants expressed more joy at the death of Butler than the capture of Cornwallis."[81]

AND THEY BEGAN THE WORLD ANEW

YORKTOWN, OCTOBER 1781–NEW YORK CITY, APRIL 1783

> *. . . If cats should be chased into holes by the mouse,*
> *If mammas sold their babies to gypsies for half a crown,*
> *If summer were spring and the other way round,*
> *Then all the world would be upside down.*
>   *—Nursery rhyme, "The World Turned Upside Down"*[1]

During the negotiations for the surrender of Cornwallis at York-town, a British officer brought up the uneasy topic of Loyalists, for he knew that many armed Tory units were among the troops who had laid down their arms. He also knew that those troops were not in the British Army and not affected by the terms of surrender. Finally negotiators hammered out an article of capitulation that attempted to give them legal immunity: "Natives or Inhabitants of different Parts of this Country . . . are not to be punished on Account of having joined the British Army." Washington, well aware of the anti-Tory laws passed by every state, took a look at the article and insisted that those laws be somehow acknowledged. So an awkward sentence was added: "This Article cannot be assented to being altogether of Civil Resort."[2] Those words would be the grounds for hunting down the fifteen hundred Tories suspected of having served at Yorktown.

Still, a way was found to get some of them out safely. In another article of capitulation, Washington allowed the Royal Navy sloop *Bonetta* to be "permitted to sail without Examination . . . ," ostensibly so that Cornwallis could send a report to New York.[3] Lt. Col. John G. Simcoe was allowed to board because he was in poor health. Some of his Rangers also slipped aboard, as did other Tories who rowed out as the *Bonetta* set sail. Among them were ex-Rebels who had served in state militias, viewed by Patriots as deserters who could be executed for taking up the king's arms. The fate of other Tories at Yorktown is not known. Some men were caught and executed for desertion. Some Loyalist muster rolls show the surrender of 241 officers and men of the British Legion, including a lieutenant colonel.[4]

That would seem to be Lt. Col. Banastre Tarleton, who had commanded the British force at Gloucester Point, across the river from Yorktown. A captain, a lieutenant, and two enlisted men of the New Jersey Volunteers were also captured at Gloucester. They presumably became prisoners, like the men of the British Legion, indicating that at least some Loyalist soldiers were treated as prisoners of war. Tarleton, who feared he would be attacked at the time of the surrender, did reach New York and eventually England, as did Simcoe. Many Rangers and veterans of Tarleton's Legion, presumably using Tory safe houses along the way, did get to New York and would sail away to Nova Scotia during the British evacuation of the city.

In the days before and after the surrender, many of the runaway slaves who had fled to Cornwallis were dying of smallpox. As the British troops had marched north from the Carolinas, slaves had joined the invaders in a bid for freedom. Many carried smallpox and died along the road. "Within these days past," a Connecticut soldier wrote, "I have marched by 18 or 20 Negroes that lay dead by the way-side, putrifying with the small pox. . . . infecting the air around with intolerable stench & great danger."[5] Cornwallis, who had put the former slaves to work on his redoubts, could not feed them as the siege tightened. So he sent them out to fend for themselves. Private Joseph Plumb Martin, ever a witness, wrote, "During the siege, we saw in the woods herds of Negroes which lord Cornwallis . . . had turned adrift, with not other recompense for their confidence in his humanity than the small pox for their bounty and starvation and death for their wages. They might be seen scattered about in every direction, dead and dying, with pieces of ears of burnt Indian corn in the hands and mouths, even of those that were dead."[6]

Soldiers then, as well as Benjamin Franklin later, speculated that the British were deliberately using dead and dying blacks to infect the Continental Army with smallpox. Buttressing the speculation in her 2001 work, *Pox Americana*, the historian Elizabeth A. Fenn quotes from a letter written to Cornwallis by Maj. Gen. Alexander Leslie: "Above 700 Negroes are come down the River in the Small Pox," he wrote. "I shall distribute them about the Rebell Plantations."[7] Under Washington's edicts the soldiers of the Continental Army were inoculated, but few militiamen were. It will probably never be known whether Leslie's suggestion was carried out and, if so, whether any Rebels were struck down.

By the time Sir Guy Carleton succeeded Clinton in May 1782, there was no doubt that the war was lost. Carleton promptly ordered the evacuation of Savannah and Charleston—"a deplorable necessity in consequence of an unsuccessful war."[8] The evacuation of Savannah, which came in July, was supposed to be primarily a military operation, carried out under terms laid down by Major General Wayne, who had arrived in Georgia in January with orders to drive the British out of the state.

As in Boston in 1776 and Philadelphia in 1778, Tories wanted to leave with the British soldiers. And Wayne, who had fought Tories on his road to Savannah, had to deal with Tory militiamen, who were not leaving. Georgia, the only colony to be conquered by the British, had reestablished royal militias; they had to be disbanded.

Wayne had already broken the royal militias by encouraging desertions to Rebel militias. He made sure that the deserters were well treated when they changed sides, inspiring further defections. Working closely with Wayne, Patriot governor John Martin, in a proclamation of February 1782, promised two hundred acres of land, a cow, and two breeding swine to every Tory who defected. Scores of Tory militiamen immediately deserted. Copies of the proclamation, translated into German, were slipped into Hessian encampments, and twenty-six deserted immediately. British officers in the South lost whatever confidence they might have had in Georgian and Hessian soldiers.[9]

Shortly before the evacuation of Savannah, Wayne promised that he would exert every influence "in my power with the Civil Authority, that all past offences (except murder) shall be buried in Oblivion"— providing that the militiamen enlisted in the Georgia Continental Infantry for two years or for the duration of the war in Georgia.[10] More than two hundred militiamen joined, switching sides with ease.[11] Eventually many of the civilians in Savannah's swelling refugee population chose to go to New York with the British Army. Wealthy families could book passage to England. Others, particularly families with large numbers of slaves, chose to go on army transports to Jamaica or the British colony of East Florida.

Some two thousand white Tories and five thousand slaves left Savannah. Most of them headed for St. Augustine in Florida; others sailed to Charleston.[12] In Florida the newcomers could expect grants of land, "upon which," in the words of the Loyalist historian Thomas Jones, "they sat down and began the world anew."[13] Nearly seven thousand Tories from Georgia and the Carolinas were already in East Florida, which since 1775 had been giving free land to Loyalists. Many of them had been driven there by confiscation-and-banishment laws.[14]

In December 1782 came the evacuation of Charleston, which completed the British withdrawal of all troops from the southern colonies. Thousands of civilians chose to leave with the troops. By one accounting, 9,127 Charleston Loyalists became exiles. A total of 1,278 whites went to Jamaica, accompanied by 2,613 slaves. East Florida was chosen by 1,615 white evacuees, who took with them 3,826 slaves. England was the destination of 274 whites; they sailed to England with 56 slaves, who probably would be called house servants in England. Twenty men took 350 slaves to the sugar plantations of St. Lucia in the West Indies. About 190 people, with 50 slaves, went to New York, and 417 men, women, and children, with 53 slaves, picked a new destination being suggested by Carleton: Nova Scotia.[15]

The evacuations were preludes to the spreading dread among Loyalists that they and Britain, having lost the war, now were about to lose the peace. For Britain, life would go on much as before, but America's Loyalists now were aliens and, to the most vindictive winners, enemies. "The Rebels breathe the most rancorous and malignant Spirit everywhere," a New York Tory wrote. "Committees and Associations are formed in every Colony and Resolves passed that no Refugees shall return nor have their Estates restored. The Congress and Assemblies look on tamely and want either the Will or the Power to check those Proceedings. In short, the Mob now reigns as fully and uncontrolled as in the Beginning of our Troubles and America is as hostile to Great Britain at this Hour as she was at any Period during the War."[16]

On March 25, 1783, *Rivington's New-York Gazetteer* reported that a ship had arrived at Philadelphia, "in thirty-five days from Cadiz, with dispatches to the Continental Congress, informing them that . . . the preliminaries to a general peace . . . was signed at Paris, in consequence of which hostilities by sea and land were to cease." When the news became official a few days later, instead of cheers there were "groans and hisses . . . attended by bitter reproaches and curses upon their kind for having deserted them."[17]

Finally came the details: The new residents of Florida learned that Spain, which had entered the war as an ally of France, was rewarded by getting back its former possession, Florida. British subjects living there had until March 1785 to clear out. The Indians along the frontier, who had expected rewards of lands for helping the British, learned that the treaty gave the United States virtually all the land east of the Mississippi River, except for British possessions in Canada and Spanish territory in Florida. And Loyalists learned that the treaty's words dealing with their future were meaningless. Congress, the treaty said, would "earnestly recommend" to the states that Loyalists' "estates, rights, and properties" be restored. But the states had already confiscated the property of tens of thousands of Tories and, because there was no national government, the states had no need to accept any recommendations from Congress.

New York City and Long Island, which had been Tory citadels, now became exits to exile. The wealthiest Tories had gone to England long before, and most reports of life there were dire. Transplanted Americans were treated as Americans, not former or new Britons. Hannah Winslow, who had left Boston with the evacuation ships in 1776 and had expected to return when the war ended, had a typical lament: She did not like where she was living but knew she could not return because "my unhappy fate is fix'd."[18] One outstanding Loyalist who did not leave was James Rivington, the Tory publisher. Beginning around the time that France became a Rebel ally, Rivington became an ally, too—a secret one, spying for Washington, who personally paid him in gold at the end of the war. He lived in New York until his death in 1802.[19]

Some wealthy Loyalists chose exile in England, though they knew Loyalists were not welcome there. Still, merchants, expecting an awakening of transatlantic trade, chartered ships, loaded them with the contents of their warehouses, and sailed off with their families. The merchants had no worries about breaking any trade laws. Carleton had ordered port officials to grant the evacuees "particular permissions" to leave New York and make whatever profits they could.[20]

Carleton had assured the Loyalists that his troops would not leave New York City until the departure of all of the civilians who wanted to go. And Carleton knew where they should go: Canada. The land that awaited them in Nova Scotia had once been the land of the Acadians, a French people expelled in the 1750s in a cruel deportation. By pouring Loyalists into sparsely populated Nova Scotia, Britain would assure the creation of an English-speaking region that would offset the French-speaking, Roman Catholic Quebec region.

The Reverend John Sayre had run off from his church and home in Fairfield during Tryon's raids on the Connecticut coast. He and his wife and eight children, "destitute of food, house, and raiments," joined the refugees in New York, and he became a propagandist for Carleton's Canadian plan.[21] In April 1783, he went to the large Loyalist refugee community at Eaton's Neck, Long Island. During a worship service in a schoolhouse, he preached about Nova Scotia. Each Loyalist family, he told the congregation, would get two hundred acres of land. They would also receive warm clothing, a year's worth of provisions, and a house—which would be in the form of planks, nails, and even window glass, all waiting for them in storage in Nova Scotia. Along with those supplies would come an iron plow and other farming equipment. A world anew.

Within days dozens of people in Lloyd's Neck decided to go, joining the official evacuation, which would begin in April and for which Carleton had found 183 ships to sail in fleets from New York. On Long Island 471 heads of families were divided into sixteen companies and assigned to Royal Navy transport ships. Each company had a captain and two lieutenants. They were to maintain order on the ship and, when the ship arrived in Nova Scotia, apportion tracts in a land they had never seen. That would basically be the system used for the entire evacuation, which would last until November.[22]

Soon after Sayre preached, HMS *Union* sailed into Huntington Bay and, on April 11, began taking on 209 passengers—108 of them children. About half of the passengers were from Connecticut. The rest were from Rhode Island, New York, Massachusetts, New Jersey, and Pennsylvania. Only twelve of the sixty-one white males had served in Loyalist regiments. One of them was Tom Hyde, a twenty-seven-year-old ex-slave who had run away from his master in Fairfield, Connecticut, and joined British forces.[23] Hyde was traveling as the servant of Fyler Dibblee, a Connecticut lawyer who had fled to Long Island only to be kidnapped in a whaleboat raid. He had been held for several months and then released in a prisoner exchange, one of the rituals of the whaleboat war.

On April 16 the *Union* sailed down Long Island Sound to New York City and ten days later joined a fleet of twenty ships headed for Nova Scotia. On May 11, the *Union* entered the port of Paartown (now Saint John, New Brunswick) on the north shore of the Bay of Fundy, at the mouth of the Saint John River. Three scouts left the ship in search of suitable home sites, followed a week later by the rest of the passengers. Most of them settled on a long peninsula at a village called Kingston. "A happier people never lived upon the globe," one of the passengers later wrote. ". . . Here with the protection of kind providence we were perfectly happy, contented and comfortable."[24]

Back in New York City, Rear Admiral Robert Digby, who had recently arrived to command the Royal Navy in America, became Carleton's close ally. Digby supervised the sailing of the armada carrying Tories out of America. An admiral who had often commanded warships in battle, he now found himself shepherding voyages of transports like the *Union* because, as he said in one of his orders to a ship captain, "I think too much cannot be done for their [Loyalists'] assistance."[25] He reminded another captain about a phrase in article 7 of the peace treaty: ". . . his Britannic Majesty shall with all convenient speed, and without causing any destruction, or carrying away any Negroes, or other property of the American inhabitants, withdraw all his Armies, Garrisons, and Fleets from the said United States." The phrase Digby cited was about "not carrying away Negroes."

Thousands of slaves and ex-slaves were pouring into New York. Many of them found shelter in Canvas Town, an area burned out in the 1775 fire. All were claiming the right to sail to Nova Scotia because they were Loyalists who had served in British forces and had been freed. But white masters had no interest in article 7; they wanted their slaves back.

Slave catchers went to New York to claim their property, even seizing ex-slaves on the streets. Boston King, born a slave on a plantation near Charleston, had joined the British, carrying dispatches and working aboard a Royal Navy warship. As the war was ending, King made his way to New York City, where he married Violet, a fellow runaway who had fled a master in North Carolina. Along with all the other ex-slaves in New York, they were terrified when "we saw our old masters coming from Virginia, North Carolina, and other parts, and seizing upon their slaves in the streets of New York, or even dragging them out of their beds."[26] And then the word began to spread: Ships were sailing off to Nova Scotia with ex-slaves aboard.[27] Carleton had his own interpretation of article 7: Negroes who had served British forces were not property. They were the king's subjects.

Carleton ordered Brig. Gen. Samuel Birch, commandant of New York, to arrange for the certification of any black who claimed British service and thus the right to join the Nova Scotia fleets. Birch had a printer (probably Rivington) produce passes that could be issued to freed blacks. A typical pass said:

> *This is to certify to whomsoever it may concern, that the Bearer hereof,*
> *_____ , a Negro, resorted to the British Lines, in consequence of*
> *The Proclamations of Sir William Howe, and Sir Henry Clinton, late*
> *Commanders in Chief in America; and that the said Negro has hereby*
> *His Excellency Sir Guy Carleton's Permission to go to Nova-Scotia, Or*
> *wherever else _____ may think proper.*

> *By Order of Brigadier General Birch*[28]

George Washington, a fourth-generation slave owner, was shocked to learn how Carleton was interpreting article 7. The topic came up at the first and only meeting of the two men, on May 6, 1783, at a house at Washington's headquarters in Tappan, New York. Washington said he had three points to raise: the timetable for the withdrawal of British forces from New York, the release of prisoners, and the point that he wished to raise first: "The Preservation of Property from being carried off, and especially the Negroes." Carleton said that some Negroes had already left—660 black men, women, and children—on April 27. Washington "appeared to be startled."[29]

Washington told Carleton he was violating the treaty. Carleton countered that the former owners of the former slaves could be compensated, and he described what became known as *The Book of Negroes*, then consisting of a handwritten list of black passengers leaving New York. Each one was identified by name, age, and physical description.

Some names included a former owner's name and residence. At least one once belonged to Thomas Jefferson.[30] Three appear to have run away from Washington's Mount Vernon.[31] The list was compiled by General Birch's men, who also gave out the passes. American "commissioners" were supposed to be present for the embarking of all ships. So presumably they, unlike Washington, were aware of the passes and the list.[32] Copies of the list that Carleton referred to were bound, creating *The Book of Negroes*. One copy of the book is in the British National Archives, another in the Nova Scotia Archives.*

In a formal letter to Carleton, written on the day of the meeting, Washington said it was not up to him to decide whether the treaty had been violated. But he added that he was ready "to enter into any Agreements, or take any Measures, which may be deemed expedient to prevent the future Carrying away any Negroes or other property of the American inhabitants."[33]

No measures were taken, and the sailing of ex-slaves continued. Eventually, nearly three thousand ex-slaves would sail to Nova Scotia and be granted land, usually small, hardscrabble tracts, in contrast to the better, and usually bigger, tracts granted to white Loyalists.

Many of the black Loyalists settled on the outskirts of a landing that became Shelburne. By the end of 1783, Shelburne's population was 7,922, of whom 1,521 were freed slaves. They named their community Birchtown, after the general who signed the ship passes. One of its leaders was Colonel Stephen Blucke, the man who succeeded Colonel Tye in the New Jersey raiders.[34]

The last recorded black evacuees were former members of those New Jersey raiders known as the Black Brigade, along with their kin: forty-

---

* There is no copy of *The Book of Negroes* in the National Archives in Washington. Copies of the lists exist in a file containing miscellaneous records of the Continental Congress.

seven men, thirty-seven women, and sixteen children. They are listed in *The Book of Negroes* as "Inspected on the 30th of November . . . on Board the fleet laying near Statten Island."[35]

In 1792, unhappy about their treatment in Nova Scotia, 1,196 black Loyalists decided, with British help, to seek a new life in Sierra Leone, a West Africa territory controlled by Britain. Sixty-seven people died during the voyage. The rest of the African returnees, including Boston King and his wife Violet, landed at Freetown and began a new country.

Within a year after the war ended, about one hundred thousand Americans left their homes. Most of them went to Canada. The rest chose England, Scotland, or British possessions in the West Indies. Within a generation the new Canadians had spread across the vast British dominion, taking with them the virtues and the visions that they and their ancestors had had as American colonists. Granted large tracts of land, they transformed a wilderness into a vibrant nation. Many became prosperous farmers or started mercantile dynasties. "Seldom had a people done so well by losing a war," a Canadian historian wrote.

Today, four to six million Canadians—about one-fifth of the population—claim a Tory ancestor. Many Canadians believe that their nation's traditional devotion to law and civility, the very essence of being a Canadian, traces back to being loyal, as in Loyalist.

Below the border live the people who started another country, built by Rebels. Within a generation, those Rebels would begin to forgive—and forget—the Tories. They would call the Revolution a war between Americans and the British, losing from their collective memory the fact that much of the fighting had been between Americans and Americans.

Acknowledgments

My wife, Scottie, has been part of every book I have written. On this book she did more than support me. She was at my side numerous times in archives from England and Northern Ireland to New England and Florida. When I started writing and was going through my research folders, several times I pulled out a piece of paper bearing an image of her fingers, holding down a document on a copy machine in some archive. She and I worked together, copying thousands of documents and traveling thousands of miles. Again and again she steered me to information and ideas I would never have found if my hands had been the only ones on those copied documents. And, when the first draft of my manuscript was done, she read it and raised questions. I responded with varying degrees of grace—and now the thanks that I heartily express here.

Scottie was not the only family member who worked with me. Our son Roger abandoned his own writing to come to the aid of Dad, digesting enormous amounts of information about Loyalist military units and helping in many ways. Our daughter, Connie, provided German translations. Our son Chris helped me at the archives in his state, North Carolina, and our granddaughter, Victoria, did the same in the archives of her state, South Carolina. Our grandsons Aaron and Jonny used their computer skills to work on the bibliography. Our son-in-law, Jim Witte, contributed incisive analysis as a twenty-first-century sociologist looking at the lives of eighteenth-century Loyalists and Patriots. And my grandnephew Jeff Riling made use of his collegiate databases to gather material.

Old friends helped in a myriad of ways: Thanks to Norman Polmar, Paul Dickson, Penny Daly, Chuck Hyman, Bob Stock, Bob Shogan, Bob Poole, Jim Srodes, Bill Dudley, Ray Longden, Stephanie Cooke, Judy Folkenberg—and Lori Annahein, who kept my computer alive. Rob Cowley, to whom this book is dedicated, came up with the idea of Tories after I had wandered the eighteenth century examining other possible topics. And when I settled on Tories, my agent, Carl Brandt, provided not only encouragement but also ideas from his own knowledge of the subject.

Todd W. Braisted, founder of the On-Line Institute for Advanced Loyalist Studies, answered numerous inquiries, especially those concerning Loyalists who served in military units. Paul J. Bunnell, UE, editor and founder of the *Loyalist Quarterly Newsletter*, also educated me about contemporary Loyalists—and why they put "UE" after their names. I was luckily directed to http://personal.nbnet.nb.ca/halew/, the marketplace of R. Wallace Hale, who has gathered thousands of pages of Loyalist material on a series of CDs. Susan Swiggum of the Ship List website (http://www.theshiplist.com/) allowed me to tap into her extensive knowledge of Loyalist ships' names and passenger lists.

Stephen Eric Davidson especially helped me understand the pride of present-day Loyalists. His work on the Loyalist past is a model for genealogists, for he adds human details and family stories to the "begat, begat, begat" of traditional genealogies. My introduction to that invaluable kind of genealogy came from a pro, Sharon Sergeant, adjunct professor at Boston University and an indefatigable tracker of people of the past.

By gleaning information from genealogies, I defied the belief of an historian who, writing in the early twentieth century, said that he eschewed family recollections as sources because they rest on "the lowest rung on the ladder of evidential credibility." I found that it is on that rung that the understanding of Tories and Rebels begins.

Many people with Loyalist ancestors helped me, including Bill Jarvis, who led me to my first Tory warrior, Stephen Maple Jarvis. Russell Moe's ancestor, Walter Barrell, produced the definitive list of Boston's Loyalist evacuees. I learned about the Acadians of Canada from Jerry Bastarache, an Acadian descendant whose ancestors, Pierre and Michel Bastarache, were deported in 1755 from land that Tories would be given three decades later. Others who shared their genealogical research with me included Don Chrysler, Pat Kelderman, Bob Moore, Eric Nellis, and Elsie Pyonen.

Along the way Scottie and I found dozens of archivists, researchers, and volunteers who sent us to gems in their collections. Our greatest treasure hunt was at the Harriet Irving Library at the University of New Brunswick in Fredericton, Canada—the fountainhead of Loyalist information, where we were helped by Kathryn Gilder and Janelle Sweatnam. The Loyalist Research Network, administered by Dr. Bonnie Huskins of the Department of History, operates out of the university. Other Canadian archivists and aides who helped us were Kate Richardson, Ava Griffin Sturgeon, and Eric Allaby on Grand Manan Island; Sheryl Stanton at the Admiral Digby Museum, at Digby, Nova Scotia; and Finn Bower at the Shelburne County Museum in Shelburne, Nova Scotia. We also visited archives in Saint John, Halifax and learned about the archeological research at Birchtown, Nova Scotia, a black Loyalist settlement. I received answers to my many questions from Robert L. Dallison, a retired lieutenant colonel in the Canadian Army, author of *Hope Restored*, and former director of Kings Landing Historical Settlement.

We received help and hospitality from Paddy Fitzgerald and Christine Johnston at the Centre for Migration Studies in Omagh, Northern Ireland. Michael Barton and Kim MacDonald initiated us at Britain's National Archives at Kew. Tom Mann, Dave Kelly, Abby Yochelson, and James Hutson provided insights at the Library of Congress. Lewis Bushnell, associate director of the Cambridge Historical Society, showed us Tory Row, and Lance Kozikowski showed us New-Gate Prison in East Granby, Connecticut. We also wish to thank Katherine A. Ludwig of the David Library of the American

Revolution; Brent Brackett, curator of Tannenbaum Historic Country Park in Greensboro, North Carolina; Christa Dierksheide at Thomas Jefferson's Monticello; and J. L. Bell, whose Web site (http://boston1775.blogspot.com/) is full of fascinating and meticulously researched stories about revolutionary Boston. Rachel Dorfman and her brother Isaac also helped me keep track of material.

Finally I profoundly thank my original editor, Elisabeth Kallick Dyssegaard, for her faith and patience.

Notes

PREFACE: "LITTLE LESS THAN SAVAGE FURY"

1. Charles B. Todd, *The History of Redding* (New York: Grafton Press, 1906), pp. 35–36. Todd publishes a gruesome account of the hanging but concludes that it was based on faulty memory.
2. Harold C. Syrett, ed., *The Papers of Alexander Hamilton* (New York: Columbia University Press, 1961), p. 529.
3. Jared Sparks, *The Writings of George Washington*, vol. 5 (Boston: Little, Brown, 1855), p. 222. Letter to Col. Walter Stewart, January 22, 1778.
4. William Spohn Baker, *Itinerary of General Washington June 15, 1775 to December 23, 1783* (Philadelphia: J. B. Lippincott, 1892; reprint, Whitefish, MT: Kessinger Publishing, 2007), p. 101.
5. W. O. Raymond, *The United Empire Loyalists* (Saint Stephen, NB: Saint Croix Printing and Publishing Co., 1893), p. 5.
6. Stephen Jarvis, "An American's Experience in the British Army," *Connecticut Magazine* 11 (Summer-Autumn 1907), pp. 191–215, 477–490. Note that the magazine treats Jarvis's service in Tory regiments as being in "the British Army."
7. A true copy from the Minutes of the Committee of Inspection for the Town of Stamford, attested by John Haight, Jr., committee clerk. http://www.stanklos .net/index.php?act=para&psname=CORRESPONDENCE%2C%20 PROCEEDINGS%2C%20ETC&pid=4392; accessed 1/12/2009.
8. Jarvis Munson, Petition for Indemnity as a Loyalist, presented in Saint John, New Brunswick, March 27, 1786. (Courtesy of Bill Jarvis.)
9. Robert McCluer Calhoon, *The Loyalists in Revolutionary America 1760–1781* (New York: Harcourt Brace Jovanovich, 1973), p. 282.

10. Wallace Brown, "Loyalist Military Settlement in New Brunswick," in *The Loyal Americans: The Military Role of the Loyalist Provincial Corps and Their Settlement in British North America*, gen. ed. Robert S. Allen (Ottawa: National Museum of Man, 1983), pp. 81–90.

11. David H. Villers, "Loyalism in Connecticut, 1763–1783" (Ph.D. diss., University of Connecticut 1976), p. 280. Villers bases his estimate of one thousand on the reports of Loyalist recruiters and on testimony in Connecticut courts.

12. Jarvis-Powell Family Papers 1767–1919, Loyalist Collection, University of New Brunswick, Canada, MIC-Loyalist FC LFR.J3P5P3.

13. He is identified in Royal Ralph Hinman, *A Historical Collection from Official Records, Files, &c., of the Part Sustained by Connecticut, During the War of the Revolution* (Hartford, CT: E. Gleason, 1842), p. 137. He is not named in James Shepard, "The Tories of Connecticut," *Connecticut Quarterly* 1, no. 3 (July, August, and September 1895).

14. Jarvis Munson petition.

15. Jarvis, "Experience." (Unless otherwise stated, information about Stephen Jarvis's life comes from this source.)

16. Shepard, "The Tories of Connecticut."

17. Mary Beth Norton, *The British-Americans* (Boston: Little, Brown, 1972), p. 10.

18. These are some of the occupations listed in documents showing the confiscation of properties of Pennsylvania Loyalists accused of treason. Samuel Hazard, ed., "Minutes of the Supreme Executive Council of Pennsylvania," *Colonial Records of Pennsylvania* (Philadelphia: Theo Fenn & Co., 1852), vol. 11, p. 353.

19. W. Stewart Wallace, *The United Empire Loyalists* (1914; reprint, Whitefish, MT: Kessinger Publishing, 2004), p. 26; Walter Bates, *Kingston and the Loyalists of the "Spring Fleet" of 1783* (1899; reprint, Fredericton, NB: Nonentity Press, 1980), p. 28.

20. Gordon S. Wood, *The Radicalism of the American Revolution* (New York: Vintage Books, 1991), p. 176.

21. The estimate of eighty thousand appears consistently in writings in Canada, the repository of innumerable Loyalist documents. For example, United Empire Loyalists' Association of Canada: http://www.learnquebec.ca/export/sites/learn/en/content/curriculum/social_sciences/documents/loyalistoverview.pdf; accessed 4/13/2009. Also, Canadian National Army Museum, "The War for America," http://www.national-army-museum.ac.uk/exhibitions/warForAmerica/page5.shtml; accessed 4/13/2009. Some Loyalist historians favor the "not less than 100,000 souls" estimate because it comes from an influential eyewitness. The source is Thomas Jones, *The History of New York During the American Revolution* (New York: New-York Historical Society, 1879), vol. I, p. 260.

22. Author's count of the battles in the War Chronology of the On-Line Institute for Advanced Loyalist Studies, http://www.royalprovincial.com/history/chronology/chrono.shtml; accessed 3/5/2009.

23. James Srodes, *Franklin* (Washington, DC: Regnery Publishing Co., 2002), p. 276.

24. Lorenzo Sabine, *Biographical Sketches of Loyalists of the American Revolution: With an Historical Essay* (Boston: Little, Brown & Co., 1864), p. 65.

25. Paul H. Smith, "The American Loyalists: Notes on Their Organization and Numerical Strength," *William and Mary Quarterly* 25 (1968): 259–277.

26. 2,205,000: U.S. Bureau of the Census (*1998 World Almanac and Book of Facts*, p. 378); 2,780,4000: Digital History online (http://digitalhistory2.uh.edu/eta.cfm?eralID=3&smtID=4). Accessed 3/23/2010.

27. Charles Francis Adams, *The Works of John Adams, Second President of the United States,* vol. 10 (Boston: Little, Brown, 1856), pp. 193–197.

28. Ibid., Letter to Morse, December 22, 1815, p. 193.

29. Worthington Chauncey Ford, ed., *The Writings of George Washington*, vol. 7 (New York: G. P. Putnam's Sons, 1890), p. 345. "The half tories," Washington wrote, ". . . might prove very useful instruments."

30. Henry Steele Commager and Richard B. Morris, eds., *The Spirit of 'Seventy-Six,* vol. 1 (Indianapolis: Bobbs-Merrill, 1958), p. 325.

31. Adams, *The Works of John Adams*, p. 197.

CHAPTER 1: TWO FLAGS OVER PLYMOUTH

1. Franklin was questioned in February 1766. Adolphus Egerton Ryerson, *The Loyalists of America and Their Times*, vol. 1 (Toronto: William Briggs, 1880), p. 310.

2. Whether "Plymouth Rock" was *actually* trod by Pilgrims is not known. According to the Pilgrim Hall Museum in Plymouth, "There are no contemporary references to the Pilgrims' landing on a rock at Plymouth."

3. The December 22 date was wrong. The landing occurred on December 11, according to the British calendar, which did not recognize the new calendar decreed in 1582 by Pope Gregory XIII, who added ten days to the Julian calendar. Catholic countries accepted the change. But anti-Catholic Britain resisted the change until 1751. By then the change had increased to eleven days. The Old Colony members forgot that the difference was only ten days in the previous century, adding eleven days to December 11, 1620, celebrating on the twenty-second instead of the twenty-first.

4. Ryerson, *The Loyalists of America and Their Times*, p. 22. The work is dedicated, with permission, to Queen Victoria.

5. W. O. Raymond, *The United Empire Loyalist* (Saint Stephen, NB: Saint Croix Printing and Publishing Co., 1893), p. 3; John Fiske, *The Beginnings of New England* (Boston: Houghton, Mifflin, 1889), pp. 192–193.

6. L. Carle Duval, "Edward Winslow, Portrait of a Loyalist" (graduate thesis in history, University of New Brunswick, Canada, 1960), p. 9; Atlantic Canada Virtual Archives, *The Winslows: Edward Winslow*. http://atlanticportal.hil .unb.ca/acva/en/winslow/family/biography.php; accessed 4/20/2009.

7. Information on the club and its members: William T. Davis, *History of the Town of Plymouth* (Philadelphia: J. W. Lewis & Co., 1885), and *Ancient Landmarks of Plymouth* (Boston: A. Williams and Company, 1883); "Plymouth in the Revolution" (Pilgrim Hall Museum); James H. Stark, *The Loyalists of Massachusetts and the Other Side of the American Revolution* (Salem, MA: Salem Press, 1910). These sources draw their information primarily from *Proceedings Massachusetts Historical Society*, 2nd series, III (1886–1887), in which "The Records of the Old Colony Club" are reprinted, pp. 381–444. See also http:// www.pilgrimhall.org/Rev3.htm; http://www.oldcolonyclub.org/ClubHistory/ occhist1.htm; accessed 12/28/2008.

8. Patrick Henry, condemning the Stamp Act, ended his speech by saying, "Tarquin and Caesar each had his Brutus, Charles the First his Cromwell, and George the Third"—here the Speaker of the House shouted, "Treason!" while Henry continued—"may profit by their example. If this be treason, make the most of it." Henry later apologized to the House and declared his loyalty to the king.

9. S. E. Morison, *Sources and Documents Illustrating the American Revolution, 1764–1788 and the Formation of the Federal Constitution* (Oxford, UK: Clarendon Press, 1923), p. 33.

10. Page Smith, *A New Age Now Begins*, vol. 1 (New York: McGraw-Hill 1976), pp. 281–285.

11. Colin Nicholson, *The "Infamas Govener"* (Boston: Northeastern University Press, 2001), p. 177.

12. Moses Coit Tyler, *The Literary History of the American Revolution 1763–1783*, vol. 1 (New York: G. P. Putnam's Sons, 1897), p. 240.

13. John Clark Ridpath, *James Otis, the Pre-Revolutionist* (Chicago: University Association, 1898; reprint, Whitefish, MT: Kessinger Publishing, 2004), p. 107.

14. Circular produced by Hancock, September 23, 1768.

15. Paul Revere, "Sons of Liberty Bowl," Museum of Fine Arts, Boston.

16. "The Signer," *Time*, July 4, 1976.

17. Nicholson, *The "Infamas Govener,"* p. 52.

18. Ibid., p. 167.

19. *Encyclopedia of American Foreign Relations*, s.v. "Gage," http://www.american foreignrelations.com; accessed 4/20/2009. (The modern U.S. counterinsur-

.gency doctrine calls for twenty security troops and police per thousand inhabitants.)

20. From the caption to a copperplate print by Revere, advertised in the *Boston Gazette and Country Journal*, April 16, 1770.

21. *Encyclopedia of American Foreign Relations*, http://www.americanforeignrelations.com; accessed 4/20/2009.

22. Smith, *A New Age Now Begins*, p. 305.

23. Alexander Winston, "Firebrand of the Revolution," *American Heritage* 18, no. 3 (April 1967).

24. Carol Troyen, "A Choice Gallery of Harvard Tories," *Harvard*, March-April 1997; Margaretta M. Lovell, *Art in a Season of Revolution* (Philadelphia: University of Pennsylvania Press, 2007), p. 58.

25. Michael G. Kammen, *Colonial New York: A History* (New York: Oxford University Press, 1996) p. 270.

26. John Adams Diary, August 14, 1769: "Adams Family Papers: An Electronic Archive." Massachusetts Historical Society. http://www.masshist.org/digitaladams; accessed 12/28/08.

27. Nicholson, *The "Infamas Govener,"* p. 203.

28. John Gorham Palfrey and Francis Winthrop Palfrey, *History of New England* (Boston: Little, Brown, 1890), pp. 404–406; Nicholson, *The "Infamas Govener,"* pp. 201–204.

29. D. Hamilton Hurd, *History of Plymouth County* ... (Philadelphia: J. W. Lewis, 1884), p. 141.

30. Mary Archibald, *Gideon White, Loyalist* (Halifax: Petheric Press, 1975); p. 2; "Biography of Edward Winslow Junior," Winslow Papers, University of New Brunswick. The papers, consisting of more than 3,600 items and 11,000 pages, cover a period from 1695 to 1866. The biography quotes from *Biographical Sketches of Graduates of Harvard University*, by John Langdon Sibley.

31. Richard B. Morris, *Encyclopedia of American History* (New York: Harper Collins, 1996).

32. *Boston Chronicle,* January 15, 1770.

33. Christopher's last name is variously spelled, and his age is given as fourteen— by a pro-Loyalist newspaper—and twelve in other contemporary sources. J. L. Bell, a Massachusetts writer on the Revolution, tracked down church baptismal records showing that Christopher was baptized in Braintree on March 18, 1759. Typically, in that age of high infant mortality rates, babies were baptized within days of birth. That would mean he was almost certainly ten years old when he was murdered. See http://boston1775.blogspot.com/2006/05/christopher-seider-shooting-victim.html; accessed 12/28/08. Bell also reported on the Thursday crowd-forming phenomenon, the Seider autopsy, and the examination of the

other boy, age unknown. See http://boston1775.blogspot.com/search/label/
Samuel%20Gore.

34. James K. Hosmer, *The Life of Thomas Hutchinson, Royal Governor of the Province of Massachusetts Bay* (Boston: Houghton Mifflin, 1896), p. 93.

35. Stark, *The Loyalists of Massachusetts*, pp. 422–423.

36. *Boston Evening Post*, February 26, 1770; Robert C. Kuncio, "Some Unpublished Poems of Phillis Wheatley," *New England Quarterly* 43, no. 2 (June 1970), pp. 287–297.

37. John Singleton Copley's half-brother Henry Pelham, a Loyalist, accused Revere of plagiarizing Pelham's own depiction of the event, "as truly as if you had plundered me on the highway." *Letters & Papers of John Singleton Copley and Henry Pelham, 1739–1776* (Boston: Massachusetts Historical Society, 1914); Henry Pelham to Paul Revere, March 29, 1770, p. 83.

38. Howard Zinn and Anthony Arnove, *Voices of a People's History of the United States* (New York: Seven Stories Press, 2004), p. 82. "Teague" was a contemptuous British term for an Irish person.

39. William T. Davis, *History of the Town of Plymouth* (Philadelphia: J. W. Lewis & Co., 1885), p. 86.

40. Tyler, *The Literary History of the American Revolution*, vol. 2, p. 194.

41. Benjamin Woods Labaree, *The Boston Tea Party* (New York: Oxford University Press, 1966), p. 118 (cited as Hutchinson to Tryon, November 21, 1773, Hutchinson Letterbooks 27, pp. 572–574, Massachusetts Archives, Massachusetts Historical Society transcripts).

42. The Tea Act of 1773, http://www.americanrevolution.com/TeaAct.htm; accessed 4/20/2009.

43. Francis S. Drake, *Tea Leaves, Being a Collection of Letters and Documents Relating to the Shipment of Tea to the American Colonies in the Year 1773 . . .* (Boston: A. O. Crane, 1884; digitally reproduced by the Gutenberg Project, 1/15/2008).

44. Stark, *The Loyalists of Massachusetts,* p. 165.

45. Sir Theodore Martin, *A Life of Lord Lyndhurst* (London: John Murray, 1883), p. 6.

46. Stark, *The Loyalists of Massachusetts,* pp. 216–217.

47. Labaree, *The Boston Tea Party,* pp. 80–128.

48. *Massachusetts Gazette and Boston Weekly News-Letter*, December 23, 1773.

49. Winslow Papers, "Biography of Edward Winslow Junior," http://www.lib.unb.ca/winslow/sibley.html#7#7; accessed 4/20/2009.

50. Abigail Adams to Mercy Otis Warren, December 5, 1773, Adams Family Correspondence, 1:88, Massachusetts Historical Society, http://www.masshist.org/adams/quotes.cfm; accessed 4/20/2009.

51. Drake Labaree, in his definitive account, paraphrases the Adams quotation but accepts the "teapot" quotation from an eyewitness account.

52. Labaree, *The Boston Tea Party,* p. 141; Alfred F. Young, *The Shoemaker and the Tea Party* (Boston: Beacon Press, 1999), p. 43.

53. John Adams Diary, December 16, 1772–December 18, 1773. http:www.masshist .org/digitaladams/diary; accessed 3/18/2010.

54. Thomas A. Bailey and David M. Kennedy, eds., *The American Spirit,* vol. 1 (Lexington, MA: D. C. Heath and Co., 1994), pp. 130–131.

55. John R. Alden, *A History of the American Revolution* (New York: Alfred A. Knopf, 1969), pp. 139–140.

56. Peggy M. Baker, "'Let the Celebration begin!' Patriots, Pilgrims & the Old Colony Club," Pilgrim Hall Museum, Plymouth, http://www.pilgrimhall.org/ RevOCC.htm; accessed 4/20/2009.

57. Davis, *History of the Town of Plymouth,* pp. 83–87; Baker. In 1880 Plymouth's Pilgrim Society moved the top piece of the rock to the bottom of the rock, which was under a canopy. The date 1620 was cut into the rock.

CHAPTER 2: ARMING THE TORIES

1. A ballad of 1774. Esmond Wright, ed., *The Fire of Liberty* (London: Folio Society, 1983), p. 154.

2. "Edward Winslow," *Dictionary of Canadian Biography Online,* http://www.bio graphi.ca/009004-119.01-e.php?BioId=36839; accessed 4/20/2009.

3. L. Carle Duval, "Edward Winslow, Portrait of a Loyalist" (M.A. thesis, University of New Brunswick, 1960), p. 15; Sabine, *The American Loyalists,* vol. 2, p. 445.

4. Duval, "Edward Winslow, Portrait of a Loyalist," pp. 14–15.

5. John Langdon Sibley, *Biographical Sketches of Graduates of Harvard University in Cambridge, Massachusetts* (Cambridge: Charles William Sever, University Bookstore), 1873.

6. Henry Steele Commager, *Documents of American History* (New York: F. S. Crofts & Co., 1940), pp. 71–76.

7. George Bancroft, *Bancroft's History of the United States,* vol. 7 (London: Charles Bowen, 1834), chap. 10, p. 37.

8. *Biography of Edward Windslow Junior,* p. 136; Winslow Papers, University of New Brunswick.

9. Jarvis Munson petition.

10. Peter Oliver's *Origin and Progress of the American Revolution: A Tory View.* Douglass Adair and John A. Schutz, eds., p. 114.

11. James Grant Wilson, John Fiske, and Stanley L. Klos, eds., *Appleton's Cyclopedia of American Biography* (New York: D. Appleton and Company, 1887–1889), available at virtualology.com. Scholars in the twentieth century discovered errors in some of *Appleton's* biographies; Bowdoin's is accurate.

12. Allen Johnson, ed., *Readings in American Constitutional History, 1776–1876* (Boston: Houghton Mifflin, 1912), p. 34.

13. Ryerson, *The Loyalists of America and Their Times*, vol. 1, p. 408.

14. David Hackett Fischer, *Paul Revere's Ride* (New York: Oxford University Press, 2004), p. 42; Ryerson, *The Loyalists of America,* p. 408.

15. James Thacher, *A Military Journal of the American Revolution: From the Commencement to the Disbanding of the American Army* (Boston: Cotton & Barnard, 1827), p. 15. Thacher, a witness to the incident, served as a surgeon through the war.

16. Oliver, *Peter Oliver's Orgins and Progress of the American Revolution,* pp. 115–116.

17. *Boston Gazette,* July 4, 1774.

18. Paine's son, William, would become apothecary to the British forces in America. http://www.worcesterart.org/Collection/Early_American/Artists/badger/lois_o/painting-discussion.html; accessed 4/28/2009.

19. William Lincoln, *The History of Worcester, Massachusetts, from its Earliest Settlement to September, 1836* (Worcester: Charles Hersey, 1862), p. 86.

20. David McCullough. *John Adams* (New York: Simon & Schuster, 2001), p. 42.

21. Lincoln, *The History of Worcester,* pp. 86, 87; Sabine, *The America Loyalists,* p. 376.

22. Calhoon, *The Loyalists in Revolutionary America,* p. 227.

23. Ibid., p. 87.

24. Stark, *The Loyalists of Massachusetts,* p. 123.

25. Frank Moore, *Diary of the American Revolution,* vol. 1 (London: Sampson Low, Son & Co., 1860), p. 27.

26. Van Tyne, *The Loyalists in the American Revolution,* pp. 33–34, 40.

27. Stark, p. 131.

28. Marshfield Massachusetts History. http://www.marshfield.net/History/marl.htm; accessed 4/29/2009.

29. Justin Winsor, *History of the Town of Duxbury, Massachusetts, with Genealogical Registers* (Boston, 1849), pp. 123–146. Winsor was superintendent of the Boston Public Library from 1868 to 1877 and went on to become librarian of Harvard University.

30. John K. Mahon, *History of the Militia and the National Guard* (New York: Macmillan, 1983), p. 36.

31. Winsor, *History of the Town of Duxbury,* pp. 123–146.

32. Adams, *The Works of John Adams,* pp. 194–195.

33. Leonard wrote anonymously as "Massachusettensis" in letters addressed to "The inhabitants of the province of the Massachusetts-Bay," published in the *Boston Gazette* from December 1774 to April 1775. John Adams believed Jonathan Sewell, Massachusetts attorney general and longtime friend, was Mas-

sachusettensis. Adams reacted using the pseudonym "Novanglus." The letters were later published in pamphlet form.

34. "A Biography of Daniel Leonard," *An Outline of American History,* published by the U.S. Information Agency, http://www.let.rug.nl/usa/B/dleonard/leonard.htm; accessed 9/23/2008.

35. Stark, *Loyalists of Massachusetts,* p. 326.

36. Sabine, *The American Loyalists,* p. 708; Calhoon, *The Loyalists in Revolutionary America,* pp. 277–278.

37. Sabine, *The American Loyalists,* p. 708.

38. Ernest F. Henderson, "Laws of Richard I Concerning Crusaders Who Were to Go by Sea," *Select Historical Documents of the Middle Ages* (London : George Bell and Sons, 1896); see www.yale.edu/lawweb/avalon/medieval/richard.htm; accessed 4/29/2009.

39. Quoted in the *Pennsylvania Gazette,* June 29, 1774; the citation is from Benjamin H. Irvin, "Tar and Feathers in Revolutionary America," http://revolution.h-net.msu.edu/essays/irvin.feathers.html; accessed 4/24/2009.

40. Edward L. Pierce, "The diary of John Rowe, a Boston mercant, 1764–1779," Proceedings of the Massachusetts Historical Society. Second Series, vol. 4, 1885–1886, p. 82.

41. Thacher, *A Military Journal of the American Revolution*, p. 16.

42. McCullough, *John Adams,* p. 77; Adams, *Works of John Adams,* vol. 4, p. 8.

43. Cambridge Historical Commission, *Cambridge During the Revolution* (Cambridge, MA: Cambridge Discovery, 1975).

44. Lucius Robinson Paige, *History of Cambridge, Massachusetts, 1630–1877* (Boston: H. O. Houghton & Co., 1877), p. 169, quoting from a letter of Frederika Baroness Riedesel.

45. Fischer, *Paul Revere's Ride,* pp. 44–46; "Boston1775": http://boston1775.blogspot.com/2007/09/gen-gage-secures-provincial-powder.html; accessed 4/22/2009.

46. Rowe, *The Diary of John Rowe,* p. 284. Accompanying the diary is a paper presented by Edward L. Pierce, who identifies Rowe as a wealthy merchant, "prudent enough to keep up pleasant personal relations with both sides."

47. Berkin and Sewall, *Odyssey of an American Loyalist*, p. 106.

48. Sabine, *American Loyalists*, p. 495.

49. Rowe, *The Diary of John Rowe*, p. 88.

50. Winslow Papers, letter to Gage from Edward Winslow, Jr., and others, March 7, 1775.

51. William Clarke to Joseph Patten, August 6, 1774, quoted in Catherine S. Crary, ed., *The Price of Loyalty* (New York: McGraw-Hill, 1973), p. 26.

52. Letter from an anonymous Loyalist in Marshfield to "a gentleman in Boston," January 24, 1775, Winsor, *History of the Town of Duxbury,* p. 128.

53. Richards, *History of Marshfield,* vol. 1, p. 128.

54. Ibid., pp. 128, 117.

55. Mary Standish Paine, "Patriots and Tory Marshfield," (Marshfield, MA: Marshfield Historical Society, 1924). Daniel Webster bought the estate from a Thomas descendant in 1832 and lived there the rest of his life.

56. Winslow Papers, New Brunswick University, microfilm 1, 47.

57. Ibid.; Richards, *History of Marshfield,* vol. 1, pp. 118, 121.

58. Richards, *History of Marshfield,* vol. 1, p. 121.

59. Duval, "Edward Winslow, Portrait of a Loyalist," p. 16.

60. "Plymouth in the Revolution," Plymouth Hall Museum. http://www.pilgrim hall.org/Rev4.htm; accessed 4/28/2009.

61. Wilbur H. Siebert, "Loyalist Troops of New England," *New England Quarterly* 4, no. 1 (January 1931), pp. 108–147; see especially p. 110.

62. Sabine, *The American Loyalists,* p. 322.

63. "Evidence in the Claim of Thomas Gilbert," Transcript of the Manuscript Books and Papers of the Commission of Enquiry into the Losses and Services of the American Loyalists, National Archives (formerly Public Records Office of England), vol. 28, p. 213.

64. Ibid.

65. Siebert, "Loyalist Troops of New England," p. 108.

66. Jones, *Loyalists of Massachusetts,* pp. 224–225.

67. "Hardwick, Massachusetts," pp. 358–359 in *Nason and Varney's Massachusetts Gazetteer* (1890) reproduced at http://capecodhistory.us/Mass1890/Hardwick 1890.htm; accessed 6/11/2009.

68. *Autobiography of John Adams,* part 1, through 1776, sheet 1 of 53. "Adams Family Papers: An Electronic Archive," Massachusetts Historical Society. http://www.masshist.org/digitaladams/; accessed 4/24/2009.

69. Adams, *The Works of John Adams*, vol. 2, p. 67.

70. Sabine, *The American Loyalists,* p. 584, quoting a note to the "Printers of the Boston Newspapers," December 22, 1774.

71. Ibid., p. 585.

72. Lucius R. Paige, *History of Hardwick Massachusetts with a Genealogical Register* (Boston: Houghton Mifflin, 1883), pp. 73, 83.

73. Ibid., pp. 74–76, quoting *Boston Evening Post,* December 26, 1774.

74. Winsor, *History of the Town of Duxbury*, p. 128.

75. New Hampshire Historical Society exhibit, based on Charles Lathrop Parsons, *The Capture of Fort William and Mary, December 14 and 15, 1774* (Durham: Published by the society, 1903; reprint, paper delivered at the seventy-seventh annual meeting of the New Hampshire Historical Society, Durham, March 1974).

76. Brian C. Cuthbertson, *The Loyalist Governor* (Halifax, NS: Petheric Press, 1983), pp. 21–22.

77. "Revolutionary War in Georgia," *New Georgia Encyclopedia*. http://www
   .georgiaencyclopedia.org/nge/Article.jsp?id=h-2709; accessed 4/26/2009.

78. Evangeline Walker Andrews, ed., *The Journal of a Lady of Quality* (New Haven:
   Yale University Press, 1921), p. 149. The journal was written by a woman from
   Scotland who was visiting in North Carolina in 1775.

79. http://www.history.org/Almanack/people/bios/biohen.cfm; accessed 4/27/2009.

80. http://www.uppercanadahistory.ca/uel/uel2.html; accessed 4/27/2009.

CHAPTER 3: FLEE OR FIGHT

1. Letter from Copley to Pelham, August 6, 1775, *Letters & Papers of John Singleton Copley and Henry Pelham, 1739–1776* (Boston: Massachusetts Historical
   Society, 1914), p. 348. Copley, who had left Boston in 1774, was writing from
   Italy in response to a letter from his half-brother Henry Pelham, who was still
   in Boston.

2. John Trumbull, *M'Fingal, A Modern Epic Poem.* (Boston: John G. Scobie,
   1826), canto 1, p. 31. The first canto of *M'Fingal* appeared in 1776; the first
   publication of the entire poem was in 1782.

3. George Atkinson Ward, ed., *Journal and Letters of Samuel Curwen, an American in England, from 1775 to 1783* (Boston: Little, Brown, 1864), p. 26.

4. Ibid., p. 34.

5. Adams, *Letters & Papers of John Singleton Copley and Henry Pelham,* pp. 318,
   322; *A plan of Boston in New England with its environs. . . . With the military
   works constructed in those places in the years 1775 and 1777* (London: Henry Pelham, 1777). Included on a corner of the map is a rendering of the pass issued
   to Pelham so he could visit British fortifications. Pelham sent his drawn map
   to London, where Francis Jukes engraved it using aquatint, a new medium.

6. Norton, *The British-Americans*, p. 36.

7. Siebert, "Loyalist Troops of New England," p. 111.

8. Edward Livingston Taylor, "Refugees to and from Canada and the Refugee
   Tract," *Ohio Archaeology and Historical Society Publications* 12 (1903), p. 222;
   Van Tyne, *The Loyalists in the American Revolution,* pp. 43–44.

9. Stark, *The Loyalists of Massachusetts*, pp. 343–346.

10. Samuel Francis Batchelder, *The Life and Surprising Adventures of John Nutting,
    Cambridge Loyalist* (Cambridge, MA: Cambridge Historical Society, 1912),
    p. 55. Biographical details come from the same source, pp. 55–63.

11. Ibid., pp. 65, 66.

12. "The Diary of Lieutenant John Barker," *Journal of the Society for Army Histori-
    cal Research* [London] 7 (1928), p. 84.

13. *Boston Gazette*, December 12, 1774.

14. John R. Galvin, *The Minute Men* (Dulles, VA: Brassey's, 1996), p. 73.

15. John Bakeless, *Turncoats, Traitors and Heroes* (Philadelphia: J. B. Lippincott Co., 1959), p. 28.

16. Frank J. Rafalko, *A Counterintelligence Reader*, chap. 1, National Counterintelligence Center, http://fas.org/irp/ops/ci/docs/ci1/ch1a.htm; accessed 4/25/2009.

17. Eric W. Barnes, "All the King's Horses . . . and All the King's Men," *American Heritage Magazine* 11, no. 6 (October 1960).

18. Charles M. Endicott, *Account of Leslie's Retreat at the North Bridge, on Sunday, Feb'ry 26, 1775* (Salem: William Ives and George W. Pease, printers. 1856), p. 16. The actual date, by the modern calendar, is February 27.

19. Fischer, *Paul Revere's Ride,* p. 58.

20. Ibid. Fischer credits the "lest" quote to a letter, dated March 1 and signed by Revere and several other Boston Sons of Liberty, to the Sons of Liberty in New York. The letter promises to send weekly reports of "the Earliest and most authentic intelligence" from Boston.

21. Endicott, *Account of Leslie's Retreat at the North Bridge,* p. 20.

22. Galvin, *The Minute Men,* p. 79.

23. The events in Salem are based on Endicott, *Account of Leslie's Retreat at the North Bridge;* Galvin, *The Minute Men;* Fischer, *Paul Revere's Ride,* and Ralph D. Paine, *The Ships and Sailors of Old Salem* (Chicago: A. C. McClurg, 1912), pp. 154–158. Endicott and Paine use *recollections* of witnesses. Also, "Regiment of Troops Under the Command of Colonel Leslie," *American Archives* 4, vol. 1, p. 1268. http://dig.lib.niu.edu/amarch/; accessed 4/24/2009. Documents held by the Essex Institute Historical Collection, whose Revolutionary War material encompasses Essex County, where Salem is located.

24. Sabine, *The American Loyalists,* vol. 2, pp. 265–266.

25. Paine, *The Ships and Sailors of Old Salem,* p. 155.

26. The words come from a reconstruction of the Leslie-Barnard encounter in "The Memorial services at the centennial anniversary of Leslie's expedition to Salem, Sunday, February 26, 1775, on Friday, February 26, 1875," digitized at http://www.archive.org/stream/memorialservices00sale/memorialservices00 sale_djvu.txt; accessed 4/25/2009.

27. Gage to Dartmouth, August 27, 1774, quoted by Fischer, *Paul Revere's Ride,* pp. 80–81.

28. "Instructions of General Gage to Captain Brown and Ensign D'berniere," American Archives 4, vol. 1, p. 1263. http://lincoln.lib.niu.edu/cgi-bin/amarch/ getdoc.pl?/var/lib/philologic/databases/amarch/.1579; accessed 3/30/2010.

29. Description of the mission: "Narrative of Ensign D'berniere," *American Archives,* ibid.

30. The tavern still stands on the Old Post Road in Weston. Information from the Golden Ball Tavern Museum, housed in the tavern. http://www.goldenball tavern.org; accessed 4/25/2009.

31. *The Narrative of General Gage's Spies, March, 1775,* with Notes by Jerome Carter Hosmer (Boston: Reprinted from the Bostonian Society's Publications, 1912), pp. 30–31.

32. Rafalko, *A Counterintelligence Reader.*

33. Fischer, *Paul Revere's Ride,* p. 159.

34. *Collections of the Worcester Society of Antiquity* (Worcester, MA: 1892). Vol. 13, p. 204.

35. D. Michael Ryan. "Times and Tribulations for Concord's Tories," *Concord Magazine*, August–September 1999.

36. Ibid., Fischer, *Paul Revere's Ride*, pp. 205–206; George Tolman, "John Jack, the Slave, and Daniel Bliss, the Tory," *Collections of the Worcester Society of Antiquity* (1892), Vol. 13, pp. 251–256.

37. Tolman, p. 254.

38. D. Michael Ryan, *Concord and the Dawn of Revolution* (Salem, MA: History Press, 2007), p. 33.

39. Alden, *A History of the American Revolution*, p. 170.

40. The Earl of Dartmouth to Gov. Thomas Gage January 27, 1775, *Secret Documents on the American Revolution*, Peter King, Department of History, Carleton University, Ottawa, Canada. http://http-server.carleton.ca/~pking/; accessed 5/4/2009.

41. Ibid.

42. Ibid.

43. *The Manuscripts of the Earl of Dartmouth*, vol. 2, *American Papers*, pp. 256, 275 London: Historical Manuscripts Commission, Fourteenth Report, Appendix, Part X. Her Majesty's Stationery Office, 1895.

44. Fischer, *Paul Revere's Ride*, p. 80.

45. Arthur B. Tourtellot, *Lexington and Concord: The Beginning of the War of the American Revolution* (New York: W. W. Norton & Company, 1959), p. 87.

46. "Journal of Dr. Belknap," *Proceedings of the Massachusetts Historical Society (Selected from the Records, 1858–1860)* (Boston: 1860), p. 85. (Dr. Jeremy Belknap was a Patriot and clergyman who served as a chaplain in the Revolution.)

47. Fischer, *Paul Revere's Ride,* p. 97.

48. *King John* 3:1. Fischer builds a strong case (pp. 96–97; 386–387). He points to "her husband's decision to send her away after the battles, and the failure of their marriage" (p. 387).

CHAPTER 4: "TO SUBDUE THE BAD"

1. John Thomas Scharf, *History of Westchester County* (Philadelphia: L. E. Preston & Co., 1886), p. 257. Circular letter to all New York counties issued by the Committee of Observation, New York Provincial Legislature, April 28, 1775.

2. Fischer, in his exhaustive analysis of the force sent to Concord (*Paul Revere's Ride,* pp. 313–315), says the total number of troops was somewhere between eight hundred and nine hundred. He also mentions an unknown number of "Loyalist volunteers."

3. Ibid., p. 240.

4. Ibid., pp. 114, 315; Tourtellot, *Lexington and Concord,* p. 114.

5. "George Leonard," *Dictionary of Canadian Biography.* [http://www.biographi .ca/EN/ShowBio.asp?BioId=37088; accessed 7/5/2009.

6. Sabine, *The American Loyalists,* p. 153.

7. "A Fragment of the Diary of Lieutenant Enos Stevens, Tory, 1777–1778," *New England Quarterly* 2, no. 2 (June 1938), p. 377. Stevens, of Charlestown, New Hampshire, traveled to New York in 1777 to volunteer for service for the king. He wrote that he saw Beaman on Staten Island in May 1777.

8. Henry & Sarah Ballinger Chiles Family, http://www.henrychiles.com/i97.html; accessed 11/24/2009. Joseph Beaman's name appears on Lancaster militia rolls.

9. Daniel Murray appears as the captain, and his brothers as members, on the muster roll of a company of Governor Wentworth's Volunteers taken at Flushing on Long Island on October 16, 1777. http://www.royalprovincial.com/ Military/Musters/govwent/gwvmurrl.htm.

10. "George Leonard," *Dictionary of Canadian Biography.* http://www.biographi .ca/EN/ShowBio.asp?BioId=37088; accessed 7/5/2009.

11. Ann Gorman Condon, "The Mind in Exile: Loyalty in the Winslow Papers"; lecture, November 18, 1998 at Archives and Special Collections, Harriet Irving Library, University of New Brunswick, Canada. http://www.lib.unb.ca/win slow/mind.html; accessed 10/14/2008.

12. From a report written by Lieutenant William Sutherland of the 38th Regiment to General Gage, April 26, 1775. http://www.revolutionarywararchives .org/lexington.html; accessed 7/5/2009. Also, Fischer, *Paul Revere's Ride,* p. 127, where Murray is identified as the guide.

13. Galvin, *The Minute Men,* p. 100. Orders from General Thomas Gage to Lieut. Colonel Smith, 10th Regiment of Foot, Boston, April 18, 1775.

14. Letter from Rachel Revere to Paul Revere, undated; the Clements Library, University of Michigan. http://www2.si.umich.edu/spies/letter-1775apr.html; accessed 3/25/2010.

15. Fischer, *Paul Revere's Ride,* pp. 174–183, recounts the riders' hasty visit to Adams and Hancock. They kept on the move, avoiding Boston and finally reappearing in Philadelphia to attend the Continental Congress in May. Dorothy and Aunt Lydia soon left for safety in Fairfield, CT. On August 28 Hancock and Dorothy Quincy were married in Fairfield's First Congregational Church.

16. Tourtellot, *Lexington and Concord* (p. 20), puts the number at thirty-eight;

Fischer, *Paul Revere's Ride* (p. 189), says others were falling into the line, putting the total at "no more than sixty or seventy militia . . . perhaps less."

17. Fischer, *Paul Revere's Ride,* pp. 127, 200.

18. "Depositions Concerning Lexington and Concord, April 1775," [Affidavit] No. 4, Lexington, April 25, 1775, *Journals of the Continental Congress,* Library of Congress, http://memory.loc.gov/learn/features/timeline/amrev/shots/concern .html; accessed 4/29/2009.

19. Major Pitcairn to General Gage, April 26, 1775, excerpted in Lillian B. Miller, *"The Dye Is Now Cast"* (Washington, DC: Smithsonian Institution Press, 1975), p. 74.

20. Fischer, *Paul Revere's Ride,* p. 191; Affidavit of Elias Phinney, *History of the Battle at Lexington* (Boston: Phelps & Farnham, 1825), p. 38.

21. Elizabeth Ellery Dana, ed., *The British in Boston: Being the Diary of Lieutenant John Barker of the King's Own Regiment from November 15, 1774 to May 31, 1776* (Cambridge: Harvard University Press, 1924), April 19, 1775, p. 32.

22. The account of the Concord battle is based on Fischer, *Paul Revere's Ride,* pp. 203–232; Tourtellot, *Lexington and Concord,* pp. 159–174; Don Higginbotham, *The War of American Independence* (Boston: Northeastern University Press, 1983), pp. 60–64; and Craig L. Symonds and William J. Clipson, *A Battlefield Atlas of the American Revolution* (Mount Pleasant, SC: Nautical & Aviation Publishing, 1986), pp. 14–15.

23. Barrett's inventory was on a document found among his papers, according to Richard Frothingham, Jr., *History of the Siege of Boston, and of the Battles of Lexington, Concord, and Bunker Hill* (Boston: Little, Brown, 1872), p. 102.

24. Massachusetts Historical Society, *Proceedings* 56, p. 89, quoted by Miller, *The Dye Is Now Cast,* p. 77. Emerson's grandson, Ralph Waldo Emerson, wrote the "Concord Hymn"'s enduring words, "Here once the embattled farmers stood; And fired the shot heard round the world."

25. "For many years," Fischer writes (*Paul Revere's Ride,* p. 406), "the town of Concord threw a shroud of silence 'round this event." He mentions several suspects.

26. British Statement on the battle, Whitehall, June 10, 1775, as published in Albert Bushnell Hart, ed., *American History Told by Contemporaries* (New York: Macmillan & Co., 1901), vol. 2, p. 549.

27. Fischer, *Paul Revere's Ride,* pp. 320–324.

28. Tourtellot, *Lexington and Concord,* p. 174.

29. Charles Knowles Bolton, ed. *Letters of Hugh, Earl Percy, from Boston and New York, 1774–1776* (Boston: Charles E. Goodspeed, 1902), p. 31.

30. Fischer, *Paul Revere's Ride,* p. 240. Winslow himself is the source of the horse incident: Winslow Papers, http://atlanticportal.hil.unb.ca/acva/en/winslow/family/biography.php; accessed 4/29/2009.

31. General Order, quoted in Sabine, *The American Loyalists,* p. 482.

32. James H. Stark, *The Loyalists of Massachusetts and the Other Side of the American Revolution* (Boston: J. H. Stark, 1910), p. 228.

33. Allen French, *General Gage's Informers: New Material Upon Lexington and Concord, Benjamin Thompson as Loyalist and the Treachery of Benjamin Church, Jr.* (Ann Arbor: University of Michigan Press, 1932), pp. 57–58.

34. Frank Warren Coburn, *The Battle of April 19, 1775* (Lexington, MA: Published by author, 1912), p. 116.

35. Frederick Mackenzie's journal, published as *A British Fusilier in Revolutionary Boston*, edited by Allen French (Cambridge, MA: Harvard University Press, 1926), cited in Mark Urban, *Fusiliers* (New York: Walker & Co., 2007).

36. Fischer, *Paul Revere's Ride,* p. 241.

37. Charles Stedman, *history of the origin, progress, and termination of the American War*, vol. 1, p. 118, as quoted by Lossing, *Signers of the Declaration of Independence,* p. 529. Stedman was an officer with Percy.

38. Urban, *Fusiliers,* p. 27.

39. Mackenzie journal, *A British Fusilier in Revolutionary Boston*, April 18–21, 1775, as quoted in John H. Rhodehamel, ed., *The American Revolution: Writings from the War of Independence* (New York: Penguin Putnam, Library of America, 2001), p. 9.

40. Coburn, *The Battle of April 19, 1775*, p. 156.

41. Percy to Lieutenant General Edward Harvey, April 20, 1775, Bolton, *Letters of Hugh, Earl Percy*, p. 52.

42. Oliver Ayer Roberts, *History of the Military Company of the Massachusetts* (Boston: Alfred Mudge & Son, Printers, 1898), vol. 3, p. 79.

43. Evidence in the claim of Samuel Gilbert, October 13, 1786, transcript of the Manuscript Books and Papers of the Commission of Enquiry into the Losses and Services of the American Loyalists, vol. 17, pp. 37–38.

44. Freetown Historic Districts Database, http://www.assonetriver.com/preservation/dist_period.asp?P=FED; accessed 4/29/2009.

45. Sabine, *The American Loyalists,* pp. 322–323; Freetown Historic Districts Database.

46. Tourtellot, *Lexington and Concord,* p. 236, quoting the Warren letter in *Historical Collection of the Essex Institute,* vol. 36, p. 19.

47. Paine, *The Old Merchant Marine* (New Haven: Yale University Press, 1921), pp. 52–53.

48. Srodes, *Franklin,* pp. 257, 259.

49. Paine, *The Ships and Sailors of Old Salem* (Chicago: McClurg, 1912), p. 165.

50. Artemas Ward, "Memoir of Major General A. Ward," *New England Historical & Genealogical Register* 5, no. 3 (July 1851), p. 272.

51. Charles Martyn, *The Life of Artemas Ward* (New York: Artemas Ward [the general's great-grandson], 1921), p. 40.

52. Connecticut Society of the Sons of Liberty. http://www.connecticutsar.org/patriots/israel_bissell.htm; accessed 4/29/2009.

53. The exact figure is not known. Sources give estimates ranging from 15,000 to 20,000. In *The Continental Army* (Washington, DC: Center of Military History, 1983), p. 18, Robert K. Wright, Jr., writes, "The Massachusetts Provincial Congress had set the minimum force needed to meet the British threat at some 30,000 men. By July a substantial portion of that total had assembled around Boston."

54. Thomas Weston, *History of the Town of Middleboro, Massachusetts* (Boston: Houghton Mifflin, 1906), p. 239. In the hospital Sturtevant became infected with smallpox and died on August 18.

55. Martyn, *The Life of Artemas Ward,* p. 168.

56. http://www.ushistory.org/libertyBell/timeline.html; accessed 4/29/2009.

57. Willard Sterne Randall, *Benedict Arnold: Patriot and Traitor* (New York: Barnes & Noble, 2003), pp. 78, 83. Reprint, William Morrow, 1990.

58. Miller. *"The Dye Is Now Cast"* (Washington, DC: Smithsonian Institution Press, 1975, pp. 106–107.

59. "The Diary of Lieutenant John Barker," *Journal of the Society for Army Historical Research* 7 (1928), p. 103.

60. Frothingham, *History of the Siege of Boston,* p. 114.

61. Broadside, Gilder Lehrman Institute of American History, document number GLC04781. http://www.gilderlehrman.org/search/display_results.php?id=GLC04781; accessed 4/29/2009.

62. Frothingham, *History of the Siege of Boston,* p. 115.

63. Ibid., p. 116.

64. A. J. Langguth, *Patriots* (New York: Simon & Schuster, 1988), pp. 217–271; Thomas J. Fleming, *Now We Are Enemies* (New York: St. Martin's Press, 1960), pp. 9–11.

65. Royal Navy History. http://www.royal-navy.mod.uk/server?show=nav.1301& outputFormat=print; accessed 4/29/2009. Patrick O'Brian used the real *Lively* as a command for his fictional hero Jack Aubrey.

66. Paul K. Walker, *Engineers of Independence: A Documentary History of the Army Engineers in the American Revolution, 1775-1783* (Washington, DC: U.S. Government Printing Office, 1981), pp. 51–57.

67. James Thacher, *Military Journal of the American Revolution: From the Commencement to the Disbanding of the American Army* (Boston: Cottons & Barnard, 1827), p. 71.

68. Allen French, *The Siege of Boston* (New York: Macmillan Co., 1911), pp. 260–261.

69. Van Tyne, *The Loyalists in the American Revolution,* p. 51, publishes the anec-
     dote but does not identify Willard. Frothingham does identify him, *History
     of the Siege of Boston,* p. 126. Slightly different words are attributed to Willard
     in Alexander Graydon and John Stockton Littell, eds., *Memoirs of His Own
     Time: With Reminiscences of the Men and Events of the Revolution* (Philadel-
     phia: Lindsay & Blakiston, 1846), p. 424. The biographical information about
     Willard comes from Henry S. Nouse, "The Loyalists of Lancaster, Massachu-
     setts," *Bay State Monthly* 1, no. 6 (June 1884).

70. Letter from Charles Stuart to Lord Bute, April 28, 1776, as quoted in Com-
     mager and Morris, eds., *The Spirit of 'Seventy-Six,* vol. 1, pp. 181–182.

71. Frothingham, *History of the Siege of Boston,* p. 151.

72. This account is based on Alden, *A History of the American Revolution,* pp. 178–
     182; Fleming, *Now We Are Enemies,* pp. 17, 80–92, 132–137, 234–240, 313–314;
     John Fortescue, *The War of Independence* (Mechanicsburg, PA: Stackpole Books,
     2001 [originally published in 1911 as part of the second, revised edition of vol. 3
     of *A History of the British Army*]), pp. 8–14; Langguth, *Patriots,* pp. 271–277;
     Symonds and Clipson, *A Battlefield Atlas,* p. 19; and Burgoyne's description.
     http://americanrevwar.homestead.com/files/bunker.htm; accessed 4/29/2009.
     Accounts vary on casualty numbers; the numbers here are based on a consensus
     of sources.

73. George F. Scheer and Hugh F. Rankin, *Rebels and Redcoats: The American
     Revolution Through the Eyes of Those Who Fought and Lived It* (New York: Da
     Capo Press, 1957), p. 62.

74. Timothy Ruggles, founder of the Loyalist American Association, refers to the
     firewood duties in a letter, written on January 13, 1776. http://www.royalprovin
     cial.com/military/rhist/loyaa/laalet2.htm; accessed 3/8/2009.

75. A short biography (Archibald, *Gideon White Loyalist,* p. 3) says only that
     "when the Battle of Bunker Hill exploded, Gid was in Boston and fought as a
     volunteer with the British Army."

76. "John Coffin," *Dictionary of Canadian Biography Online*. http://www.biographi
     .ca/009004-119.01-e.php?&id_nbr=3321&&PHPSESSID=uug6tgdqile5b9qs5
     edh9li2c3; accessed 5/1/2009.

77. Piers Mackesy, *The War for America: 1775–83* (Lincoln: University of Nebraska
     Press, 1993), p. 88.

CHAPTER 5: THE WAR FOR BOSTON

1.   James Thacher, *A Military Journal During the American Revolutionary War*
     (Boston: Cottons & Barnard, 1827), p. 23. Dr. Thacher joined the Continental
     Army in 1775 and served during the entire war.

2. Albert Bushnell Hart, *American History Told by Contemporaries*, vol. 2, p. 550.
3. Copley, *Letters & Papers*. Massachusetts Historical Society, 1914, p. 318.
4. "Military pass," Library of Congress, Printed Ephemera Collection, portfolio 38, folder 16.
5. Frothingham, *History of the Siege of Boston,* p. 237.
6. Ibid.
7. "Maps of North America, 1750–1789," Library of Congress. The map and Pelham's pass are shown at http://www.ushistoricalarchive.com/cds/boston.html; accessed 3/25/2010.
8. Copley, *Letters & Papers*, pp. 320, 322.
9. "An association, proposed to the loyal citizens," Library of Congress, Printed Ephemera Collection, portfolio 38, folder 35. http://memory.loc.gov/cgi-bin/query/h?ammem/rbpebib:@field%28NUMBER+@band%28rbpe+03803500%29%29; accessed 3/25/2010.
10. Frothingham, *History of the Siege of Boston,* p. 29.
11. Wilbur H. Siebert, "Loyalist Troops of New England," *The New England Quarterly*. Vol. 4, No. 1 (January 1931), pp. 113–114.
12. Frothingham, *History of the Siege of Boston,* p. 235.
13. Ibid., p. 208.
14. Justin Winsor, ed. *The Memorial History of Boston* (Boston: James R. Osgood and Co., 1882), p. 77.
15. Stark, *Loyalists of Massachusetts*, p. 94.
16. Batchelder, *The Life and Surprising Adventures of John Nutting,* p. 68.
17. "Letter from Falmouth to Watertown, May 11, 1775," *American Archives,* series 4, vol. 2, pp. 552–553, as cited in Richard Irving Hunt, "The Loyalists of Maine" (Ph.D. thesis, University of Maine, 1980), p. 116. (Hereafter, Hunt.)
18. Hunt, pp. 114–121.
19. John Howard Ahlin, *James Lyon, Patriot, Preacher, Psalmodist* (Machias, ME: Centre Street Congregational Church, 2005), p. 10.
20. Journal of the Second Provincial Congress of the Massachusetts Bay Colony, February 1, 1775, p. 86 (as cited by Cahill in citation below).
21. This detail occurs in the report to the Reverend Lyon, as chairman of the Committee of Safety and Correspondence, sent to the Massachusetts Provincial Congress, June 14, 1775.
22. Thomas P. Cahill, *A Short Sketch of the Life and Achievements of Captain Jeremiah O'Brien of Machias, Maine* (Worcester, MA: Harrigan Press, 1936), pp. 1–8; Jack Coggins, *Ships and Seamen of the American Revolution* (New York: Courier Dover Publications, 2002), pp. 13–16.
23. Sabine, *Biographical Sketches*, vol. 1, p. 595.

24. Ibid., vol. 2, pp. 382–83.

25. Frothingham, *History of the Siege of Boston,* p. 208.

26. "Propaganda hand bills," Library of Congress, Printed Ephemera Collection, portfolio 38, folder 30.

27. History Place documents: http://www.historyplace.com/unitedstates/revolu tion/proclaims.htm; accessed 3/25/2010. Alden, *A History of the American Revolution,* p. 215.

28. George Washington to John Augustine Washington, July 18, 1755, George Washington Papers at the Library of Congress, 1741–1799, http://lcweb2.loc .gov/cgi-bin/ampage?collId=mgw2&fileName=gwpage001.db&recNum=89; accessed 12/24/2008.

29. Edmund S. Bouchier, ed., *Reminiscences of an American Loyalist.* (New York: Houghton Mifflin, 1925), pp. 109, 113.

30. Information on the volunteers comes from the On-Line Institute for Advanced Loyalist Studies (http://www.royalprovincial.com/), citing documents from the British National Archives, including Headquarters Papers of the British Army in America and the Orderly Book of Sir William Howe.

31. Liberty Tree: *Constitutional Gazette.* September 9, 1775, in Frank Moore, ed., *Diary of the American Revolution; From Newspapers and Original Documents* (New York: C. Scribner, 1860), p. 131. Court-martial: http://www.royal provincial.com/military/courts/cmwilliams3.htm; accessed 4/29/2009. The court-martial convicted him of manslaughter. In a senseless fight he had hit a fellow officer on the head with a chair.

32. On-Line Institute for Advanced Loyalist Studies: http://www.royalprovincial. com/Military/musters/loyamregt/mrlarmain.htm: accessed 4/29/2009.

33. Sabine, *Biographical Sketches*, vol. 2, p. 252.

34. *The Encyclopedia of Canada*, vol. 2, p. 98.

35. Sabine, *The American Loyalists,* p. 167.

36. Siebert, "Loyalist Troops of New England," p. 134.

37. Sabine, *The American Loyalists,* p. 545.

38. http://www.royalprovincial.com/Military/Musters/loyamassoc/laaofficers .htm; accessed 3/24/2010.

39. "Transactions 1917–1919," Publication of the Colonial Society of Massachusetts (Boston: Colonial Society of Massachusetts, 1920), vol. 20, p. 14; Francis S. Drake, *The Town of Roxbury* (Boston: Municipal Printing Office, 1908), p. 412.

40. Lucius Robinson Paige, *History of Hardwick, Massachusetts* (Boston: Houghton Mifflin, 1883), p. 92.

41. Kemble Orderly Book, Head Quarters, Boston, December 7th, 1775, pp. 207–271. New-York Historical Society Collections 1883, as published by the On-Line Institute for Advanced Loyalist Studies: http://www.royalprovincial. com/military/rhist/loyirsh/livform.htm; accessed 4/29/2009.

42. James Thomas Flexner, *Washington the Indispensable Man* (Boston: Little, Brown, 1969), p. 18.

43. Papers of George Washington, http://gwpapers.virginia.edu/documents/revolution/itinerary/all.html; Harvard Library biography of Samuel Langdon, http://oasis.lib.harvard.edu/oasis/deliver/~hua04005; accessed 3/25/2010. Vassall had died in 1769. In 1775 his widow fled to Boston and then to her estates in Antigua. The mansion was the home of Henry Wadsworth Longfellow from 1837 until his death in 1882.

44. Frothingham, *History of the Siege of Boston,* p. 213.

45. Stockbridge Indian Tribe History, http://www.accessgenealogy.com/native/tribes/stockbridge/stockbridgehist.htm; accessed 4/29/2009.

46. Letter sent to eastern tribes by the Massachusetts Provincial Congress, May 17, 1775, excerpted in *Glimpses of the Past,* chap. 45, "The Passamaquoddies and the Revolutionary War." *Glimpses* is a series of articles that appeared in early 1890s in the *Saint Croix Courier,* published in St. Stephen, New Brunswick, Canada. See http://members.shaw.ca/caren.secord/locations/NewBrunswick/Glimpses/Intro.html

47. "Thomas Gage," *Dictionary of Canadian Biography Online,* http://www.biographi.ca/009004–119.01-e.php?&id_nbr=1895&&PHPSESSID=bqbfnkp59hqcj4r7c5srdbkhj1; accessed 5/22/2010.

48. Frothingham, *History of the Siege of Boston,* pp. 227–228. The rifleman's uniform was described in the double-named *St. James Chronicle or the British Evening Post,* September 9, 1775.

49. Ibid.

50. General Washington to General Gage, Headquarters, Cambridge, August 11, 1775; General Gage to General Washington, Boston, August 13, 1775. *American Archives,* Correspondence, Proceedings, Etc. http://colet.uchicago.edu/cgi-bin/amarch/documentidx.pl?subpart_id=S4-V3-P01-sp01&sortorder=doc_id; accessed 3/25/2010.

51. Frothingham, *History of the Siege of Boston,* p. 235.

52. David C. Hsiung, "Food, Fuel, and the New England Environment in the War for Independence, 1775-1776," essay at University of Georgia Workshop in Early American History and Culture, December 1, 2006. http://www.uga.edu/colonialseminar/Hsiung%20Essay.pdf; accessed 3/25/2010. (See also article with same title in *New England Quarterly,* December 2007, Vol. 80, No. 4, pp. 614-654.)

53. Ibid. Also, "HMS *Rose* and the American War of Independence," Maritime Historical Studies Centre, University of Hull, http://www.hull.ac.uk/mhsc/FarHorizons/Documents/HMSRose.pdf; accessed 5/1/2009.

54. William Bell Clark, ed. *Naval Documents of the American Revolution* (Washington, DC: Naval Historical Center, 1966), vol. 2, pp. 324–326.

55. Ahlin, *James Lyon, Patriot, Preacher, Psalmodist,* p. 11.

56. "Dispatch from Capt. Henry Mowatt [*sic*] to Vice Admiral Graves about the destruction of Falmouth," Maine Memory Network, Maine Historical Society, http://www.mainememory.net/media/pdf/6775.pdf; accessed 3/25/2010.

57. Nicholas Barker, "An Historical Geography of Middle Street, Portland, Maine, 1727 to 1977, http://www.geocities.com/middle_street_portland_maine/1807 .htm; accessed 3/9/2009.

58. Hsiung, "Food, Fuel, and the New England Environment."

59. Benjamin H. Hall, *History of Eastern Vermont* (Albany, NY: J. Munsell, 1865), vol. 2, pp. 602–603; text of commission, p. 607.

60. John J. Duffy and Eugene A. Coyle, "Crean Brush vs. Ethan Allen: A Winner's Tale," *Vermont History,* vol. 70, nos. 3 and 4 (Summer/Fall 2002), p. 103.

61. Hall, *History of Eastern Vermont,* pp. 610–612.

62. Frothingham, *History of the Siege of Boston,* pp. 328, 282; On-Line Institute for Advanced Loyalist Studies: Ruggles to Captain Francis Greene, January 13, 1776. Great Britain Public Records Office (Now National Archives), Audit Office, Class 13, vol. 45, folio 484.

63. Frothingham, *History of the Siege of Boston,* p. 282.

64. Jacqueline Barbara Carr, *After the Siege,* (Boston: Northeastern University Press, 2005), p. 28; Elizabeth A. Fenn. "The Great Smallpox Epidemic of 1775–82," *History Today*, vol. 53, issue 8 (August 2003), p. 1.

65. Howe order, October 28, 1775, as published at http://www.royalprovincial .com/military/rhist/loyaa/laaproc.htm; accessed 3/25/2010.

66. Winsor, *Memorial History of Boston*, vol. 3, pp. 163. 77; "Orders to be observed during the Time of Fires . . . ," signed by Howe, Library of Congress, Printed Ephemera Collection, portfolio 38, folder 26a.

CHAPTER 6: INTO THE FOURTEENTH COLONY

1. William Kingsford, *The History of Canada* (Toronto: Roswell & Hutchinson, 1893), vol. 6, book 19, p. 5. Montgomery's letter was dated December 6, 1775.

2. William Wood, *The Father of British Canada: A Chronicle of Carleton* (Toronto: Chronicles of Canada, 1916), p. 17. Also, Albert Henry Smith, ed. *The Writings of Benjamin Franklin* (New York: Macmillan Co. , 1907), vol. 10, p. 296.

3. William Renwick Riddell, *Benjamin Franklin and Canada* (Toronto: Published by the author, 1923); p. 52, where Riddell quotes French dictionaries. His work is available at http://www.archive.org/stream/benjaminfranklin00ridduoft/ benjaminfranklin00ridduoft_djvu.txt; accessed 5/5/2009.

4. "Indian Department List of Men," undated but including an internal date (June 4, 1775), lists thirty-nine officers and men, the cadre of identified agents. (Public

Records Office [British National Archives], War Office, class 28, vol. 10, folios 399–400.) See the On-Line Institute for Advanced Loyalist Studies, http://www .royalprovincial.com/military/rhist/dian/dianretn.htm; accessed 5/9/2009.

5. Instructions to General Schuyler, June 27, 1775. *American Archives*, S4-V2-p1855. http://lincoln.lib.niu.edu/cgi-bin/philogic/showrest_?conc.6.1.28739.300.399 .amarch; accessed 3/25/2010.

6. "George Washington to the Inhabitants of Canada, September 6, 1775." John C. Fitzpatrick, ed., *The Writings of George Washington from the Original Manuscript Sources, 1745–1799*, vol. 3, pp. 478–480.

7. Thomas C. Haliburton, *History of Nova Scotia* (Halifax, NS: Joseph Howe, 1829; facsimile edition, Belleville, ON: Mika Publishing, 1973), p. 483.

8. *Journals of the Continental Congress, 1774–1789*, vol. 1, pp. 105–113, available at http://press-pubs.uchicago.edu/founders/documents/v1ch14s12.html; accessed 5/12/2009.

9. D. Peter MacLeod, "Revolution Rejected: Canada and the American Revolution," commentary for an exhibit of that name in the Canadian War Museum in Ottawa, http://www.warmuseum.ca/cwm/games/expo/background_e.html; accessed 5/4/2009.

10. "Thomas Walker," *Dictionary of Canadian Biography Online.* http://www .biographi.ca/009004-119.01-e.php?&id_nbr=2204&&PHPSESSID=ychzfqk vzape; accessed 5/7/2009.

11. James Han Nay, *History of New Brunswick* (St. John, NB: John A. Bowes, 1909), pp. 101–102.

12. Edward Livingston Taylor, "Refugees to and from Canada and the Refugee Tract," *Ohio Archaeological and Historical Quarterly* 12, vol. xii (1903), p. 224. http://publications.ohiohistory.org/ohstemplate.cfm?action=detail&Page= 0012224.html&StartPage=24&EndPage=328&volume=12&newtitle=Volume %2012%20Page%2024; accessed 5/12/2009.

13. Clare Brandt, *An American Aristocracy, the Livingstons* (Garden City, NY: Doubleday & Co., 1986), p. 112.

14. Hal T. Shelton, *General Richard Montgomery and the American Revolution: From Redcoat to Rebel* (New York: New York University Press, 1996), p. 36.

15. Brandt, *An American Aristocracy,* pp. 111, 117. Livingston did not contribute a word to the Declaration. Nor did he sign it. He was not in Philadelphia on August 2, 1776, when delegates signed the formal embossed version of the Declaration. Instead his cousin, Philip Livingston, signed and became known in the family as "Philip the Signer."

16. Brandt, *An American Aristocracy,* pp. 111–112; Michael P. Gabriel, *Major General Richard Montgomery* (Madison, NJ: Fairleigh Dickinson University Press, 2002), p. 50.

17. Eugene R. Fingerhut and Joseph S. Tiedemann, *The Other New York* (New York: SUNY Press, 2005), p. 130.

18. George Washington Papers at the Library of Congress, 1741–1799, Series 4. General Correspondence, 1697–1799. Philip J. Schuyler to Canadian Citizens, September 5, 1775.

19. Tice's tavern is mentioned in a pension application for Joseph Lobdell, a Continental Army veteran. Lobdell said he had been captured by Indians who sold him to Tice. See http://morrisonspensions.org/lobdelljoseph.html; accessed 5/6/2009.

20. Page Smith, *A New Age Now Begins* (New York: McGraw-Hill, 1976), vol. 1, p. 601.

21. "Ethan Allen," *Encyclopedia Americana* (Danbury, CT: Grolier, 1996); John R. Alden, *A History of the American Revolution,* p. 201.

22. Harry Stanton Tillotson, *The Beloved Spy* (Caldwell, ID: The Caxton Printers, Ltd., 1948), p. 30.

23. Willard Sterne Randall, *Benedict Arnold* (New York: Barnes & Noble, 2003), pp. 134–135. Reprint of William Morrow, 1990.

24. Ibid., p. 151.

25. Ibid., p. 152.

26. Anthony Nardini, "The American Defeat at Quebec," Publications of Villanova, http://www.publications.villanova.edu/Concept/2004/The%20American%20 Defeat%20at%20Quebec.pdf; accessed 5/8/2009.

27. Randall, *Benedict Arnold,* p. 187.

28. "Guy, Carleton, 1st Baron Dorchester," *Dictionary of Canadian Biography Online.* http://www.biographi.ca/009004-119.01-e.php?&id_nbr=2310&interval= 25&&PHPSESSID=ahmvo4mq2g15edj3vs2gden730#il; accessed 5/9/2009.

29. Edwin Martin Stone, ed., *The Invasion of Canada in 1775, including the Journal of Captain Simeon Thayer* (Providence, RI: Knowles, Anthony & Co., 1867), p. 18.

30. "Thomas Walker," *Dictionary of Canadian Biography Online,* http://www .biographi.ca/009004-119.01-e.php?&id_nbr=2204&interval=25&&PHPS ESSID=uug6tgdqile5b9qs5edh9li2c3.

31. Randall, *Benedict Arnold,* p. 146; William Renwick Riddell, *Benjamin Franklin and Canada* (Toronto: Published by the author, 1923), p. 31, http://www .archive.org/stream/benjaminfranklin00ridduoft/benjaminfranklin00riddu oft_djvu.txt; accessed 5/6/2009.

32. "David Franks," Jewish Virtual Library, http://www.jewishvirtuallibrary.org/ jsource/biography/Franks.html; accessed 5/6/2009.

33. P. H. Bryce. "The Quinte Loyalists of 1784," *Papers and Records* of the Ontario Historical Society (Toronto: Ontario Historical Society, 1931), vol. 27, pp. 5–14, http://my.tbaytel.net/bmartin/quinte.htm; accessed 5/8/2009.

34. "Military Units—Loyalist Units," United Empire Loyalists' Association of Canada, http://www.uelac.org/Military/Young-Emigrants.php; accessed 5/5/2009.

35. Donald I. Stoetzel, *Encyclopedia of the French & Indian War in North America, 1754–1763* (Westminster, MD: Heritage Books, 2008), p. 439. Also, "Sir Frederick Haldimand," *Dictionary of Canadian Biography Online,* http://www .biographi.ca/009004-119.01-e.php?&id_nbr=2445; accessed 5/9/2009. The Royal Americans' regiment was officially the 62nd Regiment of Foot, later renumbered as the 60th (King's Royal Rifle Corps).

36. Robert L. Dallison, *Hope Restored* (Fredericton, NB: Goose Lane Editions, 2003), p. 62.

37. Ibid.

38. Report from General Wade to King George I, December 10, 1724, http:// www.highlanderweb.co.uk/general.htm; accessed 6/10/2009. Wade had traveled through the Highlands to gather intelligence about the loyalty of the clans. http://www.highlanderweb.co.uk/general.htm; accessed 6/10/2009.

39. Evangeline Walker Andrews and Charles McLean Andrews, eds., *Journal of a Lady of Quality* (New Haven: Yale University Press, 1922), appendices, p. 258.

40. "Allan Maclean," *Dictionary of Canadian Biography Online,* http://www. biographi.ca/009004-119.01-e.php?&id_nbr=2041&&PHPSESSID=ahmvo 4mq2g15edj3vs2gden730; accessed 5/9/2009.

41. Mary Beacock Fryer, *Allan Maclean, Jacobite General* (Toronto: Dundurn Press, 1987), p. 130.

42. Randall, *Benedict Arnold,* p. 208.

43. William Wood, *Father of British Canada* (Toronto: Chronicles of Canada, 1916), p. 72.

44. Samuel B. Griffith, *In Defense of the Public Liberty* (Garden City, NY: Doubleday & Co., 1976), pp. 248–249.

45. Shelton, *General Richard Montgomery,* p. 138.

46. William Kingsford, *The History of Canada* (Toronto: Roswell & Hutchinson, 1893), vol. 6, p. 5.

47. Randall, *Benedict Arnold,* p. 223.

48. Craig L Symonds and William J. Clipson, *A Battlefield Atlas,* p. 23.

49. Alden, *A History of the American Revolution,* p. 207; William Wood, *Father of British Canada,* p. 112.

50. William Wood, p. 120.

51. Lee Enderlin, "The Invasion of Canada during the American Revolution," *Military History,* August 1999, vol. 16, no. 3. Some Canadian estimates of invader casualties range as high as three hundred.

52. Shelton, *General Richard Montgomery,* pp. 3, 4.

53. Randall, *Benedict Arnold,* pp. 230, 235; Albert Henry Smith, *The Writings of Benjamin Franklin,* vol. x, p. 295.

54. Symonds and Clipson, *A Battlefiield Atlas,* p. 25.

55. David Wilson, *The Life of Jane McCrea* (New York: Baker, Godwin & Co., Printers, 1853), p. 76. Skene supporters suggested that Skene had kept her unburied because he wanted to take her back to her ancestral burying ground somewhere in Scotland or Ireland. Clan Skene genealogy shows Andrew's birth as 1753 and gives him the military rank of major. http://www.clanskene.org/newsletters/April%202007.pdf; accessed 5/10/2009.

56. Stephen Howarth, *To Shining Sea: A History of the United States Navy 1775–1991* (New York: Random House, 1991), pp. 26–28.

57. Symonds and Clipson, p. 25.

58. Randall, *Benedict Arnold,* pp. 157–189, 205, 221, 237.

59. Symonds and Clipson, p. 21.

60. Lorenzo Sabine, *Biographical Sketches of Loyalists of the American Revolution: With an Historical Essay*, rev. ed. (Boston, Little, Brown & Co., 1864), vol. 2, pp. 439–441. Sabine notes Winslow's confusing military rank at the time: As commander of this particular force he was a lieutenant colonel; he was also a half-pay (comparable to reservist) captain in the British Army, and a major general in the royal militia.

61. Grand-Pré National Historic Site. http://www.grand-pre.com/Histoireen.html; accessed 5/12/2009.

62. Stark, *The Loyalists of Massachusetts,* p. 65. Also, "Jonathan Eddy," *Dictionary of Canadian Biography Online*, http://www.biographi.ca/009004-119.01-e.php?&id_nbr=2391&interval=25&; accessed 5/12/2009.

63. J. W. Porter. "Memoir of Colonel Jonathan Eddy, of Eddington, Maine," *Bangor History Magazine*, September 1888. Also, Ford, *The Writings of George Washington*, pp. 497–498.

64. *Journal of the Continental Congress,* February 16, 1776, p. 155, http://memory.loc.gov/ll/lljc/004/0100/01550155.gif; accessed 5/12/2009.

65. Porter, "Memoir of Colonel Jonathan Eddy," pp. 44–47, which includes Eddy's own account, sent to the Massachusetts legislature.

66. René Chartrand, *American Loyalist Troops 1775–84* (Westminster, MD: Osprey Publishing, 2008), p. 23.

67. "Jonathan Eddy," *Dictionary of Canadian Biography Online.*

68. George Washington Drisko, *Narrative of the Town of Machias* (Machias, ME: Press of the Republican, 1904), p. 57.

69. Ford, *The Writings of George Washington,* p. 381.

CHAPTER 7: THE FAREWELL FLEET

1. Margaret Wheeler Willard, ed., *Letters on the American Revolution, 1774–1776* (Boston: Houghton Mifflin, 1925), p. 119. The letter was written on May 26, 1775.

2. Stark, *The Loyalists of Massachusetts,* p. 410.

3. Frothingham, *History of the Siege of Boston,* p. 303.

4. Ibid., p. 304.

5. Clarence F. Winsor, *The Memorial History of Boston* (Boston: James R. Osgood, 1881), vol. 3, proclamation reproduced, p. 97.

6. Attributed to letter of witness, William Gordon, D.D., *The History of the Rise, Progress, and Establishment, of the Independence of the United States of America . . .* (London: Printed for author, 1788), vol. 2, p. 198.

7. Hall. *History of Eastern Vermont,* vol. 2, p. 616.

8. Winsor, *Memorial History of Boston*, p. 163.

9. Newell Diary, http://boston1775.blogspot.com/search/label/Timothy%20Newell.

10. Inventory made by order of Thomas Mifflin, quartermaster-general of the Continental Army, March 18 and 19, 1776, published in *History of the Siege of Boston,* p. 406.

11. Winsor, *Memorial History of Boston*, p. 164.

12. Excerpt from a Lord Dartmouth letter to Howe, dated August 2, 1775, Frothingham, *History of the Siege of Boston,* p. 302.

13. Letter from Washington to his brother John Augustine, March 31, 1776, excerpted in Sabine, *American Loyalists*, p. 13. (For "last trump," see 1 Cor. 15:51, 52.)

14. Frothingham, *History of the Siege of Boston,* p. 302; Winsor, *Memorial History of Boston*, p. 164.

15. David DeVoss, "Divided Loyalties," *Smithsonian*, January 2004; "Historical Narratives of Early Canada," http://www.uppercanadahistory.ca/uel/uel4.html; accessed 3/26/2010.

16. Norton, *The British-Americans*, quoting from Caner Letterbook.

17. Massachusetts Historical Society, King's Chapel Records, 1686–1942, available at http://www.masshist.org/findingaids/doc.cfm?fa=fa0249&hi=on&within=1&query=caner&submit=Search#firstmatch; accessed 3/26/2010.

18. Siebert, "Loyalist Troops of New England," p. 117.

19. Sabine, *Biographical Sketches*, vol. 1, p. 211. Walter's son, Theodore, became a merchant in what was British Guiana (now Guyana) and spent his final years in Saugerties, New York. As custodian of the family papers, he donated them to Columbia University. Another Barrell, Samuel, a relative, is acknowledged in the Historical Society publication of the names. (Private communication from Russel Moe, a descendant.)

20. Sabine, *Biographical Sketches*, vol. 1, p. 544; Loyalist Collection, MIC-Loyalist FC LFR H8H4A3, University of New Brunswick, Canada, http://www.lib.unb.ca/collections/loyalist/seeOne.php?id=647&string=; accessed 5/14/2009.

21. For information on Hannah, thanks to J. L. Bell, author of the blog Boston 1775, devoted to the "history, analysis, and unabashed gossip about the start of

the American Revolution in Massachusetts." See http://boston1775.blogspot
.com/search/label/Hannah%20Loring%20Winslow; accessed 5/14/2009.

22. Winslow Papers, http://www.lib.unb.ca/winslow/sibley.html#41; accessed
5/14/2008.

23. Sabine, *American Loyalists*, pp. 470, 556.

24. Wentworth biography, Nova Scotia history, http://www.blupete.com/Hist/
BiosNS/1764-00/Wentworth.htm#rfn8; "Sir John Wentworth," *Dictionary of
Canadian Biography*.

25. "Governor Wentworth's Volunteers Report of Formation," October 16, 1777,
On-Line Institute for Advanced Loyalist Studies. http://www.royalprovincial
.com/Military/rhist/govwent/gvwform.htm; accessed 5/16/2009.

26. Norton, *The British-Americans*, pp. 100, 101, 220. The John Ervings settled in
Wales, as did a number of refugees.

27. Stark, *The Loyalists of Massachusetts*, p. 458.

28. Sabine, *Biographical Sketches*, vol 2, p. iii. The preface of James J. Talman,
ed., *Loyalist Narratives from Upper Canada* (Toronto: Champlain Society in
Canada, 1946), mentions illiteracy and the lack of time and paper. A modern
collection of writings came in 1946, when the Champlain Society of Toronto
published the narratives of twenty-five Loyalists in an edition of 550 copies.

29. Edmund Duval Poole, *Annals of Yarmouth and Barrington (Nova Scotia) in
the Revolutionary War* (Yarmouth, NS.: Reprinted from *The Yarmouth Herald*,
1899, p. 6.

30. Fitzpatrick, *The Writings of George Washington from the Original Manuscript
Sources*, vol. 4, p. 456. Letter to Joseph Reed, April 1, 1776.

31. Lawrence B. Evans, ed., *Writings of George Washington* (New York: G. P. Putnam's Sons, 1908), p. 55. Letter from Cambridge, March 31, 1776.

32. *Boston Gazette* advertisement, February 20, 1767, reproduced in Henry M. Brooks,
The Olden Time Series (Boston: Ticknor and Company, 1886), vol. 4, p. 40.

33. *Proceedings of the Massachusetts Historical Society* (Cambridge: John Wilton &
Son, 1879), p. 68.

34. Jolley Allen, *An Account of Part of the Sufferings and Losses of Jolley Allen, a
Native of London,* Jolley Allen Minute Book, Massachusetts Historical Society
(Republished: Boston: Franklin Press/Rand, Aver & Co., 1883). See also Rob-
ert J. Cormier, "The Ordeal of Jolley Allen: A Tory Merchant of Boston," *New
England Journal of History* 61 (Spring 2005), pp. 1–26.

35. Frothingham, *History of the Siege of Boston,* p. 302.

36. Ibid., p. 307.

37. Stephen Kemble, *Journals of Lieut-Col. Stephen Kemble . . . and British Army
Orders . . .* (Boston: Gredd Press, 1972; originally published by the New-York
Historical Society), p. 318.

38. Van Tyne, *The Loyalists in the American Revolution,* pp. 57–58, quoting from letter in *Memorial History of Boston,* vol. 3, p. 164.

39. French, *The Siege of Boston,* pp. 423–424.

40. Thomas Jones, *History of New York During the Revolutionary War* (New-York Historical Society, 1879), vol. 1, p. 54.

41. All references to Allen's misadventures are from the Jolley Allen Minute Book, Massachusetts Historical Society. Parts of the Minute Book also appeared in the *Proceedings* of the society in February 1878 and as a book, *An Account of Part of the Sufferings and Losses of Jolley Allen, a Native of London* (Boston: Franklin Press/Rand, Aver & Co., 1883).

42. John Barker, "The Diary of Lieutenant John Barker," *Journal of the Society for Army Historical Research* (London) 7 (1928), p. 169.

43. James Thomas Flexner, *Washington: The Indispensable Man* (Boston: Little, Brown, 1974), pp. 8, 132.

44. Elizabeth A. Fenn, *Pox Americana* (New York: Hill and Wang, 2001), p. 90, quoting from Washington's General Orders, March 13 and March 14, 1776.

45. Ibid., p. 51.

46. French, *The Siege of Boston,* p. 428

47. Charles Francis Adams, *Familiar Letters of John Adams and His Wife Abigail during the Revolution* (New York: Hurd and Houghton, 1876), p. 142.

48. French, *The Siege of Boston,* pp. 430–431.

49. Frothingham, *History of the Siege of Boston,* p. 312.

50. Jack Coggins, *Ships and Seamen of the American Revolution* (Mineola, NY: Courier Dover Publications, 2002), p. 206.

51. Winthrop Sargent, ed., "Letters of John Andrews, Esq., of Boston, 1772–1776," *Proceedings of the Massachusetts Historical Society, 1864–1865* (Boston: Massachusetts Historical Society, 1866), p. 411. Andrews lived in Boston during the blockade.

52. Sabine, *Biographical Sketches,* vol. 1, p. 160.

53. Siebert. "Loyalist Troops of New England," p. 122. Also, John J. Duffy and Eugene A. Coyle, "Crean Brush vs. Ethan Allen: A Winner's Tale," *Vermont History,* p. 103.

54. Bakeless, *Turncoats, Traitors and Heroes,* pp. 10–22. A purported "journal" of a previously unknown spy surfaced in the nineteenth century, but it was a hoax. See http://massmoments.org/moment.cfm?mid=110; accessed 3/26/2010.

55. The encrypted letter (from "George Washington Papers at the Library of Congress, 1741–1799," series 4, General Correspondence, 1697–1799; Benjamin Church, Jr., to Maurice Cane, July 1775) can be seen at http://memory.loc.gov/cgi-bin/ampage?collId=mgw4&fileName=gwpage033.db&recNum=753. The full text of the decrypted version is available at http://www.history.org/

history/teaching/enewsletter/volume3/january05/primsource.cfm; accessed 3/26/2010.

56. Frank J. Rafalko, *A Counterintelligence Reader*. http://fas.org/irp/ops/ci/docs/ci1/ch1a.htm; accessed 4/25/2009.

57. William L. Clements Library, University of Michigan, Ann Arbor. See http://www.si.umich.edu/spies/gallery.html. See also, Bakeless, pp 9–19.

58. On November 9, 1775, members of the Continental Congress had sworn, under pain of expulsion, "not to divulge, directly or indirectly, any matter or thing agitated or debated in Congress, before the same shaft have been determined. . . ."

59. "Instructions for activity in France," document 19, March 3, 1776. Connecticut Historical Society Museum. http://www.silasdeaneonline.org/documents/doc19.htm; accessed 3/26/2010.

60. Thomas B. Allen, *George Washington, Spymaster* (Washington, DC: National Geographic Society, 2004), p. 82.

61. Streeter Bass, "Beaumarchais and the American Revolution," *Studies in Intelligence* 14 (Spring 1970), pp. 1–18.

62. Tom Paine, *Common Sense* (Oakton, VA: American Renaissance Books, 2009), p. 11.

63. Clarence F. Winsor, *The Memorial History of Boston*, p. 183.

64. Ibid. http://boston1775.blogspot.com/search/label/Thomas%20Crafts; accessed 11/22/2009.

65. Pauline Maier, *American Scripture* (New York: Vintage Books, 1998), p. 39.

66. Edward J. Lowell, *The Hessians and the Other German Auxiliaries of Great Britain in the Revolutionary War* (New York: Harper & Brothers, 1884), p. 26.

67. Griffith, *In Defense of the Public Liberty*, p. 468.

CHAPTER 8: BEATING THE SOUTHERN DRUMS

1. Letter, October 4, 1775, from Lieutenant Governor John Moultrie to General James Grant, former governor of East Florida, who was on the staff of General Gage. *American Archives,* "East Florida Correspondence, Miscellaneous Papers, Proceedings of Committees, &c." http://www.stanklos.net/?act=para&pid=5770&psname=CORRESPONDENCE%2C%20PROCEEDINGS%2C%20ETC; accessed 3/26/2010.

2. *Letter-book of Captain Alexander McDonald, of the Royal Highland Emigrants*, April 14, 1776. http://www.americanrevolution.org/mac77sep.html; accessed 11/22/2009.

3. Stark, *The Loyalists of Massachusetts*, p. 228.

4. Frank Moore, *Diary of the American Revolution* (New York: Charles Scribner, 1858), pp. 20–21.

5. Richard M. Ketchum, *Divided Loyalties* (New York: Henry Holt, 2002), p. 56.

6. History of Eighty-fourth Regiment of Foot, Second Battalion, Royal Highland Emigrants, http://www.kingsorangerangers.org/history2.html; accessed 3/26/2010. Musket cartridge information: Thomas B. Allen, *Remember Valley Forge* (Washington, DC: National Geographic Society, 2007), pp. 38–39.

7. David K. Wilson, *The Southern Strategy* (Columbia: University of South Carolina Press, 2005), p. 23. Also, Dallison, *Hope Restored*, p. 63.

8. Dallison, *Hope Restored,* p. 62.

9. *Letter-book of Captain Alexander McDonald, of the Royal Highland Emigrants*. Letter to Major Small from Halifax, January 27, 1776. http://www.american revolution.org/mac77sep.html; accessed 5/19/2009.

10. Dallison, *Hope Restored,* p. 63.

11. "Clansfolk of Clan MacDonald," http://www.chebucto.ns.ca/Heritage/FSC NS/Scots_NS/Clans/MacDonald/Clansfolk_MacDonald/Clansfolk_Mac Donald.html; accessed 5/19/2009.

12. "A History of the Royal Highland Emigrants," On-Line Institute for Advanced Loyalist Studies. http://www.royalprovincial.com/military/rhist/rhe/rhehist.htm; accessed 5/19/2009. See also Bruce L. Mouser, "Continuing British Interest in Coastal Guinea-Conakry and Fuuta Jaloo Highlands (1750 to 1850)," http://etudesafricaines.revues.org/index1465.html#tocto1n1; accessed 5/19/2009.

13. "Women in History of Scots Descent," Flora MacDonald, http://www.electricscotland.com/history/women/wih9.htm; accessed 5/19/2009.

14. Samuel Johnson and James Boswell, *Journey to the Hebrides,* edited by Ian McGowan (1785; reprint, Edinburgh: Canongate, 1996), p. 370.

15. William R. Brock, *Scotus Americanus: A Survey of the Sources for Links between Scotland and America in the Eighteenth Century* (Edinburgh: Edinburgh University Press, 1982), p. 68.

16. David Dobson, *Scottish Emigration to Colonial America, 1607–1785* (Athens: University of Georgia Press, 2004), p. 6.

17. Peter Wilson Coldham, *Emigrants in Chains* (Baltimore: Genealogical Publishing Co., 1994), p. 1. The transportation sentence was for exile to "foreign parts," but the implied destination was America. When the Revolution began, prisoners sentenced to transportation languished in British prisons and prison ships. Then, in 1787, British courts began shipping convicts to the new possession of Australia.

18. John R. Maass, " 'A Complicated Scene of Difficulties': North Carolina and the Revolutionary Settlement, 1776–1789" (Ph.D. diss., Ohio State University, 2007), p. 150.

19. See Cameron Flint, "To Secure To Themselves And Their Countrymen An Agreeable And Happy Retreat . . ." (M.A. thesis by Cameron Flint, University of Akron, 2006), pp. 5–6:

Since the union of Scotland and England in 1707, Highland Scots in North America had prospered and the wealth created by the growing transatlantic trade guided Highland Scots to the realization that the United Kingdom, with its established mercantile empire, could allow them to continue to prosper. When the war began in 1775 the likelihood of an American victory against the world's most dominant empire was small at best.

20. History of the Black Watch. http://www.theblackwatch.co.uk/index/raising-of-the-regiment; accessed 3/26/2010.

21. Flint. "To Secure To Themselves and Their Countrymen . . . ," p. 56.

22. Ibid., pp. 54, 58, 68, 76.

23. Andrews and Andrews, *Journal of a Lady of Quality*, p. 38.

24. Ferenc Morton Szasz, "Historians and the Scottish America Connection," *Scots in the North American West, 1790–1917* (Norman: University of Oklahoma Press, 2000), pp. 5–6, as quoted in Flint, "To Secure To Themselves and Their Countrymen . . . ," (hereafter, Flint).

25. Kevin Phillips, *The Cousins' Wars: Religion, Politics, and the Triumph of Anglo-America* (New York: Basic Books, 1999), p. 265, as quoted by Flint. The full phrase appears in Thomas Jefferson, *The Writings of Thomas Jefferson,* vol. 2, *1776–1781,* collected and edited by Paul Leicester Ford (New York: G. P. Putnam's Sons, 1893), p. 56.

26. Phillips, *The Cousins' Wars,* p. 202.

27. Ibid., pp. 201–203.

28. Jon Kukla, "The Proclamation Against Patrick Henry," *Early American Review*, vol. v, no. 2 (Summer/Fall 2004). http://www.earlyamerica.com/review/2004_summer_fall/proclamation.htm; accessed 5/20/2009.

CHAPTER 9: "BROADSWORDS AND KING GEORGE!"

1. Moore, *Diary of the American Revolution*, vol. 1, p. 169. The verse appeared in the *Middlesex* (New Jersey) *Journal*, January 30, 1776.

2. "Provincial Corps, Southern Campaign, American Revolution," British National Archives memo, September 1992. http://yourarchives.nationalarchives.gov.uk/index.php?title=British_Regiments%2C_Southern_Campaign%2C_American_Revolution; accessed 5/20/2009.

3. Marjoleine Kars, *Breaking Loose Together: The Regulator Rebellion in Pre-Revolutionary North Carolina* (Chapel Hill: University of North Carolina Press, 2002), p. 207; William Edward Fitch, *Some Neglected History of North Carolina: Being an Account of the Revolution of the Regulators and of the Battle of Alamance, the First Battle of the American Revolution* (New York: Neale Publishing, 1905), p 110. Also, Paul David Nelson, *William Tryon and the Course*

of Empire: A Life in British Imperial Service (Chapel Hill: University of North Carolina Press, 1990), p. 134.

4. Wallace Brown, *The King's Friends* (Providence, RI: Brown University Press, 1956), p. 206. Description of the oath comes in a letter from Martin to Dartmouth, November 12, 1775, in *Colonial Records* (of North Carolina), vol. x, p. 327.

5. Andrews, *Journal of a Lady of Quality*, pp. 149, 154.

6. Ibid., p. 317; Marshall De Lancey Haywood, *Governor William Tryon, and His Administration in the Province of North Carolina, 1765–1771* (Raleigh, NC: E. M. Uzzell, 1903), p. 166.

7. Andrews, *Journal of a Lady of Quality*, p. 167.

8. Ibid., pp. 190-191.

9. Ibid., p. 187.

10. Ibid., p. 198.

11. Ibid., p. 320.

12. Ibid., p. 326; Norton. *The British-Americans,* p. 28.

13. "Janet Schaw," *Dictionary of North Carolina Biography,* edited by William S. Powell (Chapel Hill: University of North Carolina Press, 1979–1996). Also Andrews, *Journal of a Lady of Quality*, p. 212.

14. Andrews, *Journal of a Lady of Quality*, p. 211.

15. Wilson, *The Southern Strategy,* pp. 1–2, quoting from letter from Martin to Dartmouth, June 30, 1775, in K. G. Davies, ed., *Documents of the American Revolution 1770–1783* (Shannon: Irish University Press, 1976), vol. 9, p. 213.

16. Hugh F. Rankin, "The Battle of Moore's Creek Bridge Campaign, 1776," *North Carolina Historical Review*, vol. 30, no. 1, 1953.

17. "An Introduction to North Carolina Loyalist Units," On-Line Institute for Advanced Loyalist Studies. http://www.royalprovincial.com/Military/rhist/ncindcoy/ncintro.htm; accessed 5/19/2009.

18. James G. Leyburn, *The Scotch Irish* (Chapel Hill: University of North Carolina Press, 1962), p. xvii.

19. *Pennsylvania Gazette,* November 20, 1729.

20. Edmund Burke, *Account of the European Settlements in America* (London: John Joseph Stockdalk, 1808). Vol. II, p. 250. Reprint of the 1757 edition. Also, Charles Augustus Hanna, *The Scotch-Irish* (New York: G. P. Putnam's Sons, 1902), vol. 1, p. 1.

21. Leyburn, *The Scotch-Irish*, p. 180. Other estimates go as high as 300,000.

22. Rev. Edward L. Parker, *The History of Londonderry, Comprising the Towns of Derry and Londonderry, N. H.* (Boston: Perkins and Whipple, 1851), p. 218.

23. *Gottlieb Mittelberg's Journey to Pennsylvania in the Year 1750 and Return to Germany in the Year 1754*, translated from the German by Carl Theo. Eben (Phila-

delphia: John Jos. McVey, 1898), pp. 19–29, http://www.vlib.us/amdocs/texts/
gottlieb_note.html; accessed 6/11/2009.

24. Leyburn, *The Scotch-Irish*, p. 181.

25. Christopher E. Hendricks and J. Edwin Hendricks, "Expanding to the West:
    Settlement of the Piedmont Region 1730 to 1775," *North Carolina Museum
    of History*. (2005). http://www.ncmuseumofhistory.org/collateral/articles/s95
    .expanding.west.pdf; accessed 3/26/2010.

26. Leyburn, *The Scotch-Irish*, p. 221.

27. Richard Hooker, ed., *The Carolina Backcountry on the Eve of the Revolution:
    The Journal and Other Writings of Charles Woodmason, Anglican Itinerant* (Cha-
    pel Hill: University of North Carolina Press, 1969), pp. xxv, xxvi.

28. David Caldwell, a Presbyterian minister quoted in Robert M. Calhoon, *Reli-
    gion and the American Revolution in North Carolina* (Raleigh: North Carolina
    Department of Cultural Resources, 1976), p. 9.

29. Leyburn, *The Scotch-Irish*, p. 301.

30. Captain Johann Heinrichs of the Hessian Jäger Corps to Herr H., January
    18, 1778, quoted in J. C. D. Clark, *The Language of Liberty, 1660–1832* (Cam-
    bridge, UK: Cambridge University Press, 1994), p. 362.

31. Duane Meyer, *The Highland Scots of North Carolina* (Raleigh: Carolina Char-
    ter Tercentenary Commission, 1963), chap. 4.

32. John Patterson MacLean, *An Historical Account of the Settlements of Scotch
    Highlanders in America Prior to the Peace of 1783*, (Cleveland: Helman-Taylor,
    1900), p. 117.

33. Henry Clinton, *The American Rebellion: Sir Henry Clinton's Narrative of His
    Campaign, 1775–1782*, William B. Willcox, ed. (New Haven: Yale University
    Press, 1954), p. 23.

34. Rankin, *The Battle of Moore's Creek Bridge Campaign, 1776*, p. 21.

35. Ibid., pp. 30–32.

36. Buchanan, *The Road to Guilford Courthouse* (New York: John Wiley & Sons,
    1997), pp. 4–5.

37. Moores Creek National Battlefield, Teachers Manual. http://www.nps.gov/
    archive/mocr/guide/covington.htm; accessed 6/16/2009.

38. Donald E. Graves, *Guide to Canadian Sources Related to Southern Revolution-
    ary War National Parks* (Carleton Place, ON: Ensign Heritage Consulting [for
    National Park Service], 2001), p. 50.

39. Buchanan, *The Road to Guilford Courthouse*, pp. 4–6; Rankin, *The Moore's
    Creek Bridge Campaign*, pp. 32–42, 45.

40. Wilson. *The Southern Strategy*, p. 31.

41. Ibid., pp. 36–37.

42. Walter Edgar, *Partisans & Redcoats* (New York: HarperCollins, 2001),
    pp. 35–36.

43. R. W. Gibbes, *Documentary History of the American Revolution: 1764–1776* (New York: D. Appleton & Co, 1855), p. 164.

44. Graves, *Guide to Canadian Sources,* p. 47.

45. National Park Service research reports, Ninety-Six National Historic Site, Ninety-Six, SC. See http://www.nps.gov/nisi/index.htm; accessed 7/2/2009.

46. Edgar, p. 33.

47. David R. Chesnutt, ed. *Papers of Henry Laurens* (Columbia, SC: University of South Carolina Press, 1968), vol. 11, pp. 51–52.

48. Robert M. Weir, *Colonial South Carolina: A History* (Columbia, SC: University of South Carolina Press, 1997), pp. 200–202; Robert Olwell, *Masters, Slaves, & Subjects: The Culture of Power in the South Carolina Low Country, 1740–1790* (Ithaca, NY: Cornell University Press, 1998), p. 235, quoting from Campbell letter to Dartmouth, August 18, 1775.

49. Hugh F. Rankin, *The North Carolina Continentals*, 2nd ed. (Chapel Hill: University of North Carolina Press, 2005), p. 71.

50. Wilson, *The Southern Strategy,* p. 50.

51. Ibid., p. 52.

52. Adolphus Egerton Ryerson, *The Loyalists of America and Their Times: From 1620 to 1816* (Toronto: William Briggs, 1880), vol. 1, pp. 465–466. Also, *On This Day: Legislative Moments in Virginia History* (Richmond: Virginia Historical Society, 2004), http://www.vahistorical.org/onthisday/42175.htm; accessed 6/10/2009.

53. Commager and Morris, *The Spirit of 'Seventy-Six,* vol. 1, p. 106.

54. Ibid., p. 111.

55. Cassandra Pybus, "Jefferson's Faulty Math: The Question of Slave Defections in the American Revolution," *The William and Mary Quarterly*, April 2005, vol. 62, no. 2, pp. 243–264, as reprinted by www.historycooperative.org. In 1786, Jefferson estimated that Virginia had lost about 30,000 slaves, an impossible figure that was frequently cited until Professor Pybus's research showed that her estimate of about 20,000 runaway slaves "can stand up against the documentary record."

56. "Slavery and the Making of America," Educational Broadcasting Corporation, 2004. http://www.pbs.org/wnet/slavery/timeline/1773.html; accessed 6/11/2009.

57. Patrick Charles, *Washington's Decision* (Charleston, SC: BookSurge, 2005), pp. 22–23, quoting from *American Archives,* vol. 2, Massachusetts Committee of Safety, July 8, 1775.

58. Gary B. Nash, *The Unknown American Revolution* (New York: Viking, 2006), pp. 228–229; Charles, *Washington's Decision*, pp. 70–78, 136.

59. Sidney Kaplan and Emma Nogrady Kaplan, *The Black Presence in the Era of the American Revolution* (Amherst: University of Massachusetts Press, 1989), p. 76.

60. Richard Podruchny, "The Battle of Great Bridge; A New Beginning for the Old Dominion," Military History on Line, http://www.militaryhistoryonline .com/18thcentury/articles/battleofgreatbridge.aspx; accessed 6/11/2009.

61. Robert A. Selig, "The Revolution's Black Soldiers," AmericanRevolution.org. http://www.americanrevolution.org/blk.html; accessed 6/11/2009.

62. *Virginia Gazette,* January 18, 1776. In September, Dunmore had sent a boatload of British soldiers and marines to Norfolk to seize the *Gazette*'s press, type, ink, and paper. On board one of Dunmore's ships, the paper published the deposed governor's version of news, including a claim that Rebels had set Norfolk afire. (Harry M. Ward, *The War for Independence and the Transformation of American Society* [Florence, KY: Routledge, 2003], p. 62.)

63. Colonial Williamsburg. http://research.history.org/pf/declaring/bio_dunmore .cfm; accessed 6/10/2009.

CHAPTER 10: WAR IN THE LOYAL PROVINCE

1. Jones, *History of New York During the Revolutionary War,* pp. 101–102.

2. General Gage in a letter to Cadwallander Colden, royal lieutenant governor of New York, on February 26, 1775. Military correspondence and headquarters papers of Lieutenant General Thomas Gage, Clements Library, University of Michigan, Ann Arbor. "Torytown": Letter of March 21, 1776, from John Eustace to Charles Lee; *Lee Papers* (New York: New-York Historical Society, 1871), vol. 2, p. 362.

3. Thomas Jefferson Wertenbaker, *Father Knickerbocker Rebels* (New York: Charles Scribner's Sons, 1948), p. 64.

4. Gage papers, Clements Library, August 14, 1775. Proof of the Tryon-Gage intelligence was not confirmed until the library purchased the papers in 1937. The plotting is described in John Campbell, *Minutes of a Conspiracy against the Liberties of America* (Philadelphia: John Campbell, 1865).

5. General Washington Proclamation, April 29, 1776. Ford, *The Writings of George Washington, 1776,* vol. 4, p. 25.

6. William Howard Adams, *Gouverneur Morris* (New Haven: Yale University Press. 2003), p. 48.

7. Jones, *History of New York During the Revolutionary War,* vol. 1, pp. 55–57; Ketchum, *Divided Loyalties,* p. 351.

8. Ford, *The Writings of George Washington,* pp. 498–499.

9. Paul David Nelson, *William Tryon and the Course of Empire: A Life in British Imperial Service* (Chapel Hill: University of North Carolina Press, 1990), pp. 4, 8, 9.

10. Wertenbaker, *Father Knickerbocker Rebels,* pp. 53, 59.

11. Mark V. Kwasny, *Washington's Partisan War, 1775–1783* (Kent, OH: Kent State University Press, 1996), pp. 29–30.

12. Stefan Bielinski, *An American Loyalist: The Ordeal of Frederick Philipse III* (Albany: New York State Museum, 1976), p. 2.

13. Wertenbaker, *Father Knickerbocker Rebels,* p. 66.

14. *American Archives,* series 4, vol. 5 (Correspondence, Miscellaneous Papers, Proceedings of Committees, &C.), New-York Committee of Safety, March 27, 1776. http://www.stanklos.net/?act=para&psname=&pid=8280; accessed 11/14/2008.

15. Henry Phelps Johnston, *The Campaign of 1776 around New York and Brooklyn,* part 2 (New York: S. W. Green, 1878), p. 108, quoting the diary for Thursday, June 13, 1776.

16. H. Morse Stephens, man. ed., *American Historical Review* 4, no. 1 (October 1898), p. 281.

17. I. W. Stuart, *Life of Jonathan Trumbull, Governor of Connecticut* (Boston: Crocker and Brewster, 1859), p. 220.

18. The Reverend Samuel Seabury, *Letters of a Westchester Farmer* (White Plains, NY: Westchester County Historical Society, 1930), November 16, 1774, p. 270.

19. Wertenbaker, *Father Knickerbocker Rebels,* pp. 64; Catherine Snell Crary, "The Tory and the Spy: The Double Life of James Rivington," *William and Mary Quarterly,* 3rd ser., 16, no. 1 (January 1959), p. 67; Ketchum, *Divided Loyalties,* pp. 355–356.

20. Wertenbaker, *Father Knickerbocker Rebels,* p. 65.

21. Bakeless. *Turncoats, Traitors and Heroes,* p. 100.

22. Collection of the Gilder Lehrman Institute of American History, GLC02437 .00344: David Mason to Henry Knox, June 22, 1776. http://www.gilderlehrman .org/collection/docs_archive/docs_archive_mutiny.html; accessed 3/26/2010.

23. Thomas Jones and Edward Floyd De Lancey, eds. *History of New York During the Revolutionary War,* vol. 1, p. 121.

24. Bakeless, *Turncoats, Traitors and Heroes,* p. 100.

25. William Edward Fitch, *Some Neglected History of North Carolina* (New York: Neale Pub. Co., 1905), p. 55.

26. Bakeless lays out the story of the conspiracy (*Turncoats, Traitors and Heroes,* pp. 98–109); Matthews, in London after the war, gave some details of the plot. A spurious document, published in London, became the basis for John Campbell, *Minutes of a Conspiracy Against the Liberties of America* (Philadelphia: John Campbell, 1865). This mixture of truth and falsehood indicates that more than forty suspects were questioned and accused of participation in the plot.

27. Letter from Dr. William Eustis, an army surgeon in New York, to Dr. Townsend in Boston, June 28, 1776. Reprinted in Henry P. Johnston, *The Campaign of 1776 around New York and Brooklyn* (New York: S. W. Green, 1878), part 2, p. 129.

28. Benjamin F. Thompson, *History of Long Island from Its Discovery and Settlement to the Present Time* (New York: Robert H. Dodd, 1918), vol. 1, p. 329.

29. Johnston, *The Campaign of 1776,* part 2, p. 171, n. 250.

30. Ibid., p. 82, n. 52.

31. Jones and De Lancey, eds. *History of New York During the Revolutionary War*, p. 108.

32. On-Line Institute for Advanced Loyalist Studies, http://www.royalprovincial.com/military/mems/ny/clmcrug.htm; accessed 3/3/2009.

33. Pauline Maier, *American Scripture* (New York: Vintage Books, 1998), p. 156.

34. King George's Head," *SAR Magazine* (Connecticut Sons of the Revolution), (Winter 1998). The exact count of bullets was 42,088, http://www.connecticutsar.org/articles/king_georges_head.htm; accessed 11/8/2008. The article notes that pieces began to appear in the early nineteenth century; by 1997 about fourteen hundred pounds were still unaccounted for.

35. Leah Reddy, "1776: Trinity Church and the American Revolution," *Trinity News,* July 3, 2008. http://www.trinitywallstreet.org/welcome/?article&id=996; accessed 12/1/2008.

36. Elias Boudinot, *Journal of Events in the Revolution* (New York: Arno Press, 1968), p. 3.

37. Letter from a Gentleman at Sandyhook, near New York, to his Friend in London, dated July 6, 1776. *Lloyd's Evening Post and British Chronicle,* August 14–16, 1776, as published in *Journal of the Society for Army Historical Research* 8 (1929), p. 132.

38. Barnet Schecter, *The Battle for New York* (New York: Walker, 2002), pp. 73, 80.

39. Henry Phelps Johnston, *Memoirs of the Long Island Historical Society: The Campaign of 1776 Around New York and Brooklyn* (Brooklyn, NY: Published by the Society, 1878), p. 100.

40. Jones, p. 157.

41. Thompson, *History of Long Island,* vol. 1, p. 282.

42. Siebert, "Loyalist Troops of New England," p. 129.

43. Alden. *A History of the American Revolution,* p. 267.

44. Isaac Q. Leake, *Memoir of the Life and Times of General John Lamb, an Officer of the Revolution* (Albany, NY: Joel Munsall, 1857; reprint, Alcester, Warwickshire: Read Books, 2008), p. 361.

45. Symonds and Clipson, *A Battlefield Atlas,* p. 27.

46. Ibid.

47. "George Washington to the President of Congress," September 2, 1776. George Washington Papers at the Library of Congress, http://rs5.loc.gov/learn/features/timeline/amrev/contarmy/prestwo.html; accessed 11/13/2008.

48. Johnston, *Memoirs of the Long Island Historical Society,* p. 264.

49. Wertenbaker, *Father Knickerbocker Rebels,* pp. 100–102.

50. On Shewkirk, see Johnston, *Memoirs of the Long Island Historical Society,* p. 108. On the *GR* painting, see Schecter, *The Battle for New York,* p. 176.

51. William A. Polf, *Garrison Town* (Albany: New York State American Revolution Bicentennial Commission, 1976), p. 9.

52. Ibid., pp. 9–10.

53. Ibid., pp. 11, 13.

54. Alden, *A History of the American Revolution,* pp. 503–504. Reprint. (New York: Da Capo Press, 1989).

55. The Loyalist Collection, University of New Brunswick, Winslow Papers, MIC-Loyalist FC LFR. W5E3P3.

56. Ann Gorman Condon, "The Mind in Exile: Loyalty in the Winslow Papers." http://www.lib.unb.ca/winslow/mind.html; accessed 3/30/2010.

57. Duval, "Edward Winslow, Portrait of a Loyalist," p. 28.

58. Wertenbaker, *Father Knickerbocker Rebels,* p. 22.

59. "Lord Dunmore's Ethiopian Regiment," *Online Encyclopedia of Significant People and Places in African American History,* http://www.blackpast.org/?q=aah/lord-dunmore-s-ethiopian-regiment (University of Washington); accessed 12/2/2008.

60. "A History of the Black Pioneers," On-Line Institute for Advanced Loyalist Studies, http://www.royalprovincial.com/military/rhist/blkpion/blkhist.htm; ccessed 11/10/2009.

61. Gary B. Nash, "Thomas Peters: Millwright and Deliverer," Revolutionary Essays: Michigan State University, http://revolution.h-net.msu.edu/essays/nash.html; accessed 12/2/2008; "A History of the Black Pioneers," On-Line Loyalist Institute for Advanced Loyalist Studies, http://www.royalprovincial.com/Military/rhist/blkpion/blkhist.htm;accessed 11/29/2008; Julie Hilvers, "Freedom Bound: Black Loyalists," *Freedom Chronicle,* Northern Kentucky University Institute for Freedom Studies, http://www.nku.edu/~freedomchronicle/OldSiteArchive/archive/issue4/studentscorner.php; accessed 12/4/2008.

62. Advertisement published in *Connecticut Courant,* June 1 and June 8, 1779; republished in *Baltimore Sun,* September 29, 2002, http://www.baltimoresun.com/news/specials/hc-dunmore.artsep29,0,119814.story; accessed 12/3/2008.

63. Wertenbaker, *Father Knickerbocker Rebels,* p. 54.

64. British lieutenant general Sir Henry Clinton and Baron Wilhelm von Knyphausen, the Hessian commander, later used the mansion as a headquarters. It still stands at Jumel Terrace, near West 160th Street.

65. Jared Sparks, *The Life of George Washington* (Auburn, NY: Derby & Miller, 1853), p. 98. "Her fate, how different, had she married Washington," Sabine (*Biographical Sketches,* vol. 2, p. 107) said in a conversation with Mary Morris's grandnephew in England. "You mistake, sir," he replied, saying, "my aunt Morris had immense influence over everybody; and, had she become the wife

of the Leader in the Rebellion which cost our family millions, *he* would not have been a Traitor; *she* would have prevented that, be assured, sir."

66. Robert A. East and Jacob Judd, eds., *The Loyalist Americans* (Tarrytown, NY: Sleepy Hollow Restorations, 1975), pp. 6, 98. (The book is a collection of original essays based upon papers presented at a conference on Loyalists, sponsored by Sleepy Hollow Restorations and the New York State American Bicentennial Commission at Tarrytown on November 2 and 3, 1973.)

67. Members of today's Army Rangers, Special Forces, and Delta Force trace themselves back to Knowlton's Rangers. The date 1776 on the seal of the army's intelligence service today refers to the creation of the Rangers.

68. Streeter Bass, "Nathan Hale's Mission," *Studies in Intelligence* (Winter 1973), pp. 67–74. Washington, DC: Central Intelligence Agency, Center for the Study of Intelligence. When the article appeared, *Studies in Intelligence* was classified. The article was one of many later declassified, as was *Studies* itself. The article can be found at http://www.cia.gov/library/center-for-the-study-of-intelligence/kent-csi/vol17no4/html/v17i4a03p_0001.htm; accessed 3/26/2010.

69. Ibid., p. 72.

70. A reproduction of the orderly book page appears opposite p. 110 in Wertenbaker, *Father Knickerbocker Rebels.*

71. James Hutson, "Nathan Hale Revisited," *Library of Congress Information Bulletin,* July/August 2003. http://www.loc.gov/loc/lcib/0307-8/hale.html; accessed 11/15/2008.

72. "Journal of Captain John Montrésor, July 1, 1777, to July 1, 1778, Chief Engineer of the British Army," *Pennsylvania Magazine of History and Biography* 5 (1881), p. 393. His Copley portrait is in the Detroit Institute of Art; the portrait of Mrs. Montrésor is in the collection of the U.S. Department of State.

73. "Journals of Capt. John Montrésor, 1757–1778," p. 123. *Collections of the New-York Historical Society,* 1881.

74. Bass, "Nathan Hale's Mission," p. 73.

75. Alexander Rose, *Washington's Spies* (New York: Bantam Books, 2006), p. 20; Robert Rogers, *The Journals of Major Robert Rogers* (Albany, NY: Josel Munsell's Sons, 1883), pp. 34, 101.

76. Sparks, *The Writings of George Washington,* pp 209–210.

77. Lieutenant-Colonel John Graves Simcoe, "The Raising of a Loyalist Corps in British Service," as quoted in Hart, *American History Told by Contemporaries,* vol. 2, pp. 511–513.

78. Sabine, *Biographical Sketches,* vol. 2, p. 236.

79. Memorandum from Major Patrick Ferguson, August 1, 1778; quoted in East and Judd, *The Loyalist Americans,* p. 7, citing Clinton Papers, Clements Library.

80. Sabine, *Biographical Sketches*, vol. 1, p. 254.

81. John R. Cuneo, "The Early Days of the Queen's Rangers, August 1776–February 1777," *Military Affairs* 22, no. 2 (Summer 1958), p. 68. Also, James L. Wells, Louis F. Haffen, and Josiah A. Briggs, eds., *The Bronx and Its People: a History, 1609–1927* (New York: Lewis Historical Pub. Co., 1927), p. 175.

82. Alexander Clarence Flick, ed., "Loyalism in New York during the American Revolution," *Studies in History, Economics and Public Law* (New York: Columbia University Press, 1901), p. 107.

83. *American Archives,* New York Committee of Safety to General Washington, Fishkill, NY, October 10, 1776. http://lincoln.lib.niu.edu/cgi-bin/amarch/getdoc.pl?/var/lib/philologic/databases/amarch/.24449; accessed 3/26/2010.

84. Brandt, *An American Aristocracy,* p. 112.

85. Robert Livingston's remark is from a letter to John Jay, quoted in Staughton Lynd, "The Tenant Rising at Livingston Manor, May 1777," by Staughton Lynd in *New-York Historical Society Quarterly* 48, no. 2 (April 1964), p. 167. Margaret Livingston's observation is in a letter to Robert Livingston, July 6, 1776. Livingston Family Papers, Broadside Collection, New York Public Library. They were cited by Clare Brandt in "Robert R. Livingston Jr., the Reluctant Revolutionary," at a symposium sponsored by the Friends of Clermont, Bard College/Hudson Valley Studies Program, and the New York State Office of Parks, Recreation, & Historic Preservation, Taconic Region, June 6-7, 1986.

86. East and Judd, *The Loyalist Americans,* pp. 30–37; Bielinski, *An American Loyalist,* p. 13.

87. William H. Nelson, *The American Tory*, p. 100.

88. "A History of the King's American Regiment," The On-Line Institute for Advanced Loyalist Studies, http://www.royalprovincial.com/military/rhist/kar/kar1hist.htm#karintro; accessed 11/19/2008. The Web site cites "Case of John Thompson (a Negro), June 10, 1788," Audit Office, Class 13, vol. 67, folio 340, Public Records Office, now UK National Archives. Seabury as chaplain: *Appleton's Cyclopaedia of American Biography*, James Grant Wilson and John Fiske, eds. (New York: D. Appleton, 1895), vol. 5, p. 445.

89. David H. Villers, "Loyalism in Connecticut, 1763–1783" (Ph.D. diss., University of Connecticut. 1976), pp. 310–314.

90. Thompson, *History of Long Island,* vol. 1, pp. 286–289.

91. Jones and De Lancey, eds., *History of New York During the Revolutionary War,* p. 362.

92. *Orderly Book of The Three Battalions of Loyalists commanded by Brigader-General Oliver De Lancey, 1776–1778,* compiled by William Kelby, New-York Historical Society, New York, 1917, pp. 59, 86.

93. Bakeless, *Turncoats, Traitors and Heroes,* p. 123.

94. Schecter, *The Battle for New York,* p. 206.

95. Symonds and Clipson, *A Battlefield Atlas of the American Revolution,* p. 29.

96. John Peter De Lancey, a British Army captain during the Revolution, lived in Mamaroneck. He was a son of James De Lancey, the patriarch. John's daughter, Susan, married James Fenimore Cooper.

97. Henry Barton Dawson, *Westchester County, New York, During the American Revolution* (Morrison, NY: Published by Author, 1886 [250 copies]), pp. 125–126. http://books.google.com/books?id=nl4EAAAAYAAJ&pg=PA9&dq=Loyalism+in+New+York+During+the+American+Revolution&lr=#PPP1 3,M1; accessed 11/20/2008.

98. Ibid., p. 252.

99. John R. Cuneo, "The Early Days of the Queen's Rangers," p. 72. Estimates of casualties came from Todd W. Braisted, founder of the On-Line Institute for Advanced Loyalist Studies, in a description of the Rangers that he prepared for this book.

100. Corey Slumkoski and David Bent, eds., Introduction to *Select Loyalist Memorials* by W. S. MacNutt. http://atlanticportal.hil.unb.ca/acva/loyalistwomen/en/context/articles/macnutt_manuscript.pdf; accessed 6/15/2009.

101. "Robert Rogers," *Dictionary of Canadian Biography.* http://www.biographi.ca/009004-119.01-e.php?&id_nbr=2149; accessed 6/17/2009.

102. The words came from a letter written by William Demont in London in 1792 and revealed in "Mount Washington and its Capture on the 16th of November, 1776," an article written by E. F. De Lancey, a descendant of the Loyalist family, in the February 1877 issue of the *Magazine of American History*, vol. 11, part 1, no. 2, pp. 65–90.

103. Thompson, *History of Long Island*, p. 166, quoting a letter from Governor Tryon to Lord George Germaine, December 24, 1776.

CHAPTER 11: TERROR ON THE NEUTRAL GROUND

1. *The American Crisis (No. 1).* On December 4, 1776, Thomas Paine published *The Crisis (No. 1)*, the first of a series of pamphlets. Library of Congress, http://memory.loc.gov/cgi-bin/query/r?ammem/rbpe:@field(DOCID+@lit(rbpe03902300)); accessed 11/22/2008.

2. "A History of the Guides & Pioneers," On-Line Loyalist Institute for Advanced Loyalist Studies, http://www.royalprovincial.com/military/rhist/g&p/gphist.htm; accessed 12/1/2008.

3. Ibid.; Richard M. Ketchum, *The Winter Soldiers* (New York: Macmillan, 1999), p. 136.

4. Carol Karels, *The Revolutionary War in Bergen County* (Charleston, SC: History Press, 2007), p. 26.

5. Ibid.; "A History of the Guides & Pioneers."

6. Fortescue, *The War of Independence,* p. 46.

7. Ketchum, *The Winter Soldiers,* pp. 138–139; Terry Lark, ed., *Hackensack— Heritage to Horizons* (Hackensack, NJ: Hackensack Bicentennial Committee, 1976), pp. 20–21; Sabine, p. 465.

8. Letter from Charles Lee to the President of the Massachusetts Council [James Bowdoin], November 22, 1776. *Lee Papers,* vol. 2 (New York: New-York Historical Society, 1872), p. 303.

9. "British Legion," http://www.royalprovincial.com/Military/rhist/britlegn/blin f1. htm. Accessed 11/22/2009.

10. John Buchanan, *The Road to Valley Forge* (New York: John Wiley & Sons, 2004), pp. 142–143; George H. Moore, *The Treason of Charles Lee* (New York: Charles Scribner, 1860), p. 64. Moore, the librarian of the New-York Historical Society, revealed the end-the-war plan that Lee composed for Howe.

11. Jones, *History of New York During the Revolutionary War*, p. 173.

12. Moore, *The Treason of Charles Lee,* pp. 86–87.

13. Alden, *A History of the American Revolution,* p. 277.

14. Adrian C. Leiby, *The Revolutionary War in the Hackensack Valley: The Jersey Dutch and the Neutral Ground, 1775–1783* (Piscataway, NJ: Rutgers University Press, 1992), p. 72.

15. Lark, *Hackensack,* p. 21.

16. George Washington to Lund Washington, December 10, 1776, John Rhodehamel, ed., *The American Revolution: Writings from the War of Independence* (New York: Penguin Putnam, Library of America, 2001), p. 236.

17. "A History of the 1st Battalion, New Jersey Volunteers," On-Line Institute for Advanced Loyalist Studies. http://www.royalprovincial.com/military/rhist/njv/1njvhist.htm; accessed 11/22/2008.

18. Leiby, *The Revolutionary War in the Hackensack Valley,* pp. 77–78.

19. Ibid., p. 99.

20. Jones, *History of New York During the Revolutionary War,* p. 171.

21. "History of the King's American Regiment," On-Line Institute for Advanced Loyalist Studies, http://www.royalprovincial.com/military/rhist/kar/kar5hist.htm#karraid; accessed 11/22/2008.

22. Sydney George Fisher, *The Struggle for American Independence* (New York: J. B. Lippincott, 1908), vol. 1, pp. 255–256.

23. Ford, *The Writings of George Washington,* vol. 7, p. 345.

24. Jones, *History of New York During the Revolutionary War*, vol. 2, p. 115.

25. Alexander Clarence Flick, *Loyalism in New York during the American Revolution* (New York: Columbia University Press, 1901), p. 99.

26. Samuel Eliot Morison and Henry Steele Commager, *The Growth of the American Republic* (New York: Oxford University Press, 1960), vol. 1, p. 204.

27. Leiby, *The Revolutionary War in the Hackensack Valley,* pp. 107–108.

28. Ira K. Morris, *Memorial History of Staten Island* (New York: Memorial Publishing Co., 1898), vol. 1, p. 244.

29. Epaphroditus Peck, *The Loyalists of Connecticut* (New Haven: Yale University Press, 1934), p. 12.

30. "Return of the Provincial Forces at Kingsbridge and Morrisania, 1st July 1777," Sir Henry Clinton Papers, vol. 21, item 24, William L. Clements Library, University of Michigan, as cited by the On-Line Institute for Advanced Loyalist Studies, http://www.royalprovincial.com/military/rhist/kar/kar1hist .htm; accessed 12/19/2008.

31. Sandra Riley, *Homeward Bound, A History of the Bahama Islands to 1850* (Miami, FL: Island Research, 2000), pp. 99–103. Also, *American War of Independence—at Sea,* "The New Providence Expedition: '. . . We thought Ourselves secure. . . .' " http://www.awiatsea.com/Narrative/New%20Providence%20Expedition.html; accessed 6/13/2009.

32. "From Confinement to Commandant," On-Line Institute for Advanced Loyalist Studies, http://www.royalprovincial.com/Military/rhist/pwar/pwarhist .htm; accessed 12/14/2008.

33. "A History of the Prince Of Wales' American Regiment." On-Line Institute for Advanced Loyalist Studies, http://www.royalprovincial.com/Military/ rhist/pwar/pwarhist.htm; accessed 6/17/2009.

34. Ibid.; W. O. Raymond, ed., *The Winslow Papers* (Saint John: New Brunswick Historical Society, 1901). Letter from Gov. Montfort Browne to Edward Winslow, June 22, 1777.

35. Bakeless, *Turncoats, Traitors and Heroes,* pp. 154–155.

36. "Cortlandt Skinner," *New Jersey in the American Revolution,* 1763–1783, http:// www.njstatelib.org/NJ_Information/Digital_Collections/NJInTheAmerican Revolution1763-1783/8.18.pdf; accessed 6/15/2009. Based on *The Royal Commission on the Losses and Services of American Loyalists, 1783–1785. . . . Notes of Mr. Daniel Parker Coke. . . .* (Oxford: Oxford University Press, 1915), pp. 113–115.

37. "New Jersey Volunteers," On-Line Institute for Advanced Loyalist Studies, http:// www.royalprovincial.com/military/rhist/njv/4njvhist.htm; accessed 6/12/2009.

38. W. Woodford Clayton and William Nelson, *History of Bergen and Passaic Counties, New Jersey* (Philadelphia: Everts & Peck [press of J. B Lippincott & Co.], 1882), p. 74.

39. William S. Stryker, *The New Jersey Volunteers (Loyalists) in the Revolutionary War* (Trenton, NJ: Naar, Day & Naar, 1887), pp. 4–5, http://www.archive.org/stream/ newjerseyvol00stryrich/newjerseyvol00stryrich_djvu.txt; accessed 11/30/2008.

40. Bakeless, *Turncoats, Traitors and Heroes,* p. 155.

41. Leiby, *The Revolutionary War in the Hackensack Valley,* p. 38.

42. Wells, Haffen, and Briggs, *The Bronx and its People,* p. 172.

43. Thomas S. Wermuth and James M. Johnson, "The American Revolution in the Hudson Valley—An Overview," *Hudson River Valley Review* (Summer 2003), pp. 41, 42.

44. Stryker, *The New Jersey Volunteers,* p. 28; "The Beginning of the End," chapter in a work-in-progress by Stefan Bielinski, director of the Colonial Albany Social History Project at the New York State Museum. See http://www.nysm.nysed.gov/albany/bios/d/stdl.html; accessed 12/1/2008.

45. Stryker, *The New Jersey Volunteers,* p. 43.

46. Charles Inglis, "The True Interest of America Impartially Stated, 1776," quoted in Gordon Wood, *The Creation of the American Republic 1776–1787* (Chapel Hill: University of North Carolina Press, 1998), p. 95.

47. C. M. Woolsey, *History of the Town of Marlborough, Ulster County, New York, from its Earliest Discovery* (Albany, NY: J. B. Lyon Co., 1908), pp. 123–128.

48. Simon Schama, *Rough Crossings* (London: BBC Books, 2006), pp. 136–139.

49. Flick, *Loyalism in New York during the American Revolution,* p. 101.

50. Kenneth Shefsiek, "A Suspected Loyalist in the Rural Hudson Valley: The Revolutionary War Experience of Roeloff Josiah Eltinge," *Hudson River Valley Review*, vol 10, no. 1 (Summer 2003), p. 42.

51. Crary, *The Price of Loyalty,* p. 184.

52. The committee officially requested Crosby to "use his utmost art to discover the designs, places of resort, and route, of certain disaffected persons." Minutes of the Committee, December 23, 1776. Also, James H. Pickering, "Enoch Crosby, Secret Agent of the Neutral Ground: His Own Story," *New York History* 47, no. 1 (January 1966), pp. 61–73.

53. Larry R. Gerlach, ed., *New Jersey in the American Revolution, 1763–1783* (Trenton: New Jersey Historical Commission, 1975), p. 139.

54. Richard P. McCormick, *New Jersey from Colony to State* (Princeton: D. Van Nostrand, 1964), p. 153.

55. Leiby, *The Revolutionary War in the Hackensack Valley,* pp. 103, 121.

56. Morison and Commager, *The Growth of the American Republic,* vol. 1, p. 854; Sharon McDonnell, "Revolutionary Martyrs," *American Spirit* (March–April 2007); p. 45. The estimate of deaths came from an oration at the Prison Ship Martyrs' Memorial in Fort Greene Park, Brooklyn, in 1801, according to the *New York Public Library Bulletin* 4 (January–December 1900), p. 217. When the Brooklyn Navy Yard was built on the filled-in bay, workers found hundreds of bones and skulls, which eventually were placed in the Martyrs' Tomb in the park. (A. J. Liebling, "The Yard," *The New Yorker,* July 2, 1938, p. 24.)

57. Danske Dandridge, *American Prisoners of the Revolution* (Charlottesville, VA: Michie Co., 1911), p. 133.

58. Jones, *History of New York During the Revolutionary War*, vol. 1, p. 351.

59. Orderly Book, Captain Henry Knight, Aide-de-Camp to General Howe, New-York Historical Society Collection, manuscript, p. 155 (as cited in *Brigade Dispatch*, Autumn 1992, "a publication of the 'brigade of the American Revolution,'" http://www.doublegv.com/ggv/battles/geary.html#16; accessed 12/8/2008.

60. John Frelinghuysen Hageman, *A History Of Princeton And Its Institutions* (Philadelphia: J. B. Lippincott, 1879), vol. 1, p. 155; "History of the Princeton Campus," http://etcweb.princeton.edu/CampusWWW/Companion/stockton_richard .html; accessed 12/17/2008; Frederick Bernays Wiener, "The Signer Who Recanted," *American Heritage,* June 1975.

61. Nelson, *William Tryon and the Course of Empire*, p. 150.

62. *New York Gazette,* October 16, 1777.

63. Jones, *History of New York During the Revolutionary War,* vol. 1, p. 54. The account came from Charlotte in 1835, when she was the wife of Field Marshal Sir David Dundas, commander in chief of British forces. The account was given to Edward Floyd De Lancey, editor of Jones's *History* and the grandson of Elizabeth Floyd.

64. Mark V. Kwasny, *Washington's Partisan War, 1775–1783* (Kent, OH: Kent State University Press, 1996), p. 121; Virtual American Biographies, http://www .famousamericans.net/returnjonathanmeigs/; accessed 10/12/2008. Meigs's unusual name comes from his father's courting story: After his marriage proposal was rejected, he sadly rode off. Then he heard a shout, "Return, Jonathan! Return!" He decided he would use the most joyful words he had ever heard as the name for their firstborn son.

65. Christopher Vail, *Journal 1775–1782*. The Library of Congress has a handwritten copy of Vail's journal. *Newsday* published an excerpt. http:// vailhistfdtn.com/christopher_vail.htm; accessed 5/25/2010.

66. Nelson, *William Tryon and the Course of Empire,* p. 157.

67. "Index to Emmerick's Chasseurs History," On-Line Institute for Advanced Loyalist Studies, http://www.royalprovincial.com/military/rhist/emmerick/emmhist .htm; accessed 12/11/2008. After the war Emmerich returned to England, where he wrote *The Partisan in War, of the Use of a Corps of Light Troops to an Army* and, as "Deputy Surveyor General of Royal Forests, Chases and Parks," *The Culture of Forests*.

68. Gideon Hiram Hollister, *The History of Connecticut* (Hartford: Case, Tiffany & Co., 1857), pp. 311–313.

69. The quotation, attributed to Galloway from "Reply to the Observations of General Howe," is on page 21, in the "Historical Essay" of the 1847 edition of Sabine, *Loyalists of the American Revolution.* (It does not appear in the essay published in

the 1864 edition.) Galloway, who moved to England in 1778, was sharply critical of General Howe's conduct of the war.

70. William A. Polf, *Garrison Town* (Albany: New York State American Revolution Bicentennial Commission, 1976), pp. 20–24.

71. Nelson, *William Tryon and the Course of Empire,* p. 158.

72. Leiby, *The Revolutionary War in the Hackensack Valley,* p. 303.

73. John Fiske, "Washington's Great Campaign of 1776," *Atlantic Monthly* 63, no. 375 (January 1889), pp. 20–37.

74. Commager and Morris, *The Spirit of 'Seventy-Six* (Indianapolis: Bobbs-Merrill, 1958), p. 497. (Probably from Francis, Lord Rawdon, to Robert Auchmuty, November 25, 1776.)

75. Washington Crossing Historic Park, Washington Crossing, Pennsylvania. http://www.ushistory.org/washingtoncrossing/history/durham.htm; accessed 12/10/2008.

76. Fitzpatrick, *The Writings of George Washington from the Original Manuscript Sources,* vol. 6, p. 398. George Washington to John A. Washington, December 18, 1776.

77. Buchanan, *The Road to Valley Forge,* pp. 159–165.

78. Leiby, *The Revolutionary War in the Hackensack Valley,* p. 109, quoting from *The Journals of Henry Melchior Muhlenberg* (Philadelphia: Lutheran Historical Society, Whipporwill Publications, 1982), vol. 2, p. 773.

79. Buchanan, *The Road to Valley Forge,* p. 185.

80. George Washington to John A. Washington, February 24, 1777, Fitzpatrick, *The Writings of George Washington from the Original Manuscript Sources.*

81. Richard M. Ketchum, *The Winter Soldiers* (New York: Anchor Books, 1973), p. 323. The figure of eight hundred is based on a muster on January 19, 1777.

82. Regimental Orders of December 18, 1776. Orderly Book of the King's American Regiment, William L. Clements Library, University of Michigan, as cited in "A History of the King's American Regiment," On-Line Institute for Advanced Loyalist Studies, http://www.royalprovincial.com/military/rhist/kar/kar1hist.htm; accessed 6/18/2009.

83. Ibid.

84. Ibid.

85. Eddy N. Smith et al., *Bristol, Connecticut* (Hartford: City Printing. Co., 1907), p. 157.

86. Peck, *The Loyalists of Connecticut,* pp. 22–27; *Connecticut Courant,* March 24, 1777; E. LeRoy Pond, *The Tories of Chippeny Hill, Connecticut* (New York: Grafton Press, 1909), chap. 7, which contains Dunbar's jail-cell autobiographical writing. Jones (vol. 1, p. 175) says that after Mrs. Dunbar was forced to witness the hanging she sought refuge in Middletown with paroled Loyalists, who included former New Jersey governor William Franklin. She was

ordered out of town but taken in by a Loyalist family. After giving birth, she and the baby, named after his father, escaped to New York.

87. *New York Gazette & Mercury,* June 23, 1777., as cited by Villers, "Loyalism in Connecticut, 1763–1783," p. 254.

88. "British Intelligence, New York," Memorandum Book of the British Army, 1778, unpaginated, Microfilm Reel 689, David Library of the American Revolution. Washington Crossing, PA.

89. The black-stripe claim, which also appears in accounts of torching during a Tryon raid in Fairfield, Connecticut, seems to be more than a legend. "Tory chimney. A house chimney that is painted white with a band of black around the top," appears among definitions in Henry Lionel Williams and Ottalie K. Williams, *How to Furnish Old American Houses* (New York: Pellegrini & Cudahy, 1949), p. 232.

90. Randall, *Benedict Arnold,* pp. 332–334; Silvio A. Bedini, *Ridgefield in Review* (New Haven, CT: Higginson Book Co., 1994).

91. Casualty estimates vary; these come from "The British Attack Danbury," extracted from Albert E. Van Dusen, *Connecticut* (New York: Random House, 1961). http://www.ctheritage.org/encyclopedia/ct1763_1818/british_danbury .htm; accessed 3/27/2010. http://www.skyweb.net/~channy/danraid.html; accessed 12/18/2008. Arnold's promotion: Randall, *Benedict Arnold,* p. 334.

92. Randall, *Benedict Arnold,* p. 342, quoting Washington's letter to Congress, July 10, 1777.

93. "Deeds of the Cow-boys," *New York Times,* November 23, 1879. (The newspaper's story was inspired by the discovery of muskets that may have belonged to Smith and his Cow-boys.)

94. Wanted poster published by the North Jersey Highlands Historical Society, http:// www.northjerseyhistory.org/history/smith/claudius.htm; accessed 12/8/2008. A cave used as one of his hideouts, near Tuxedo Park, New York, is on a hiking trail in Harriman State Park.

95. Leiby, *The Revolutionary War in the Hackensack Valley,* p. 193, quoting from Samuel W. Eager, *An Outline History of Orange County* (Newburgh, NY: S.T. Callahan, 1846–1847), pp. 554, 556. Also, "Deeds of the Cow-boys."

96. Leiby, *The Revolutionary War in the Hackensack Valley,* p. 195.

CHAPTER 12: "INDIANS MUST BE EMPLOYED"

1. May 21, 1775 meeting. Samuel Ludlow Frey, ed., *The Minute Book of the Committee of Safety of Tryon County, the Old New York Frontier* (New York: Dodd, Mead & Company, 1905), pp. 7, 8. Because Tryon County was named in honor of the royal governor, the New York legislature changed its name to Montgomery County in 1784.

2. The segments were labeled with abbreviations for New England, New York, New Jersey, Pennsylvania, Maryland, Virginia, North Carolina, and South Carolina. The drawing, made from a woodcut, appeared in the *Pennsylvania Gazette* on May 9, 1754. Franklin probably was inspired by the superstition that if a snake was cut in two, it would come back to life, provided the pieces were joined before sunset. Early in the Revolution the image became a flag motif that evolved into an intact rattlesnake and the motto "Dont Tread on Me." Margaret Sedeen, *Star-Spangled Banner* (Washington, DC: National Geographic Society, 1993), pp. 34–35.

3. From George Washington's "Journal to the Ohio," published in the *Maryland Gazette,* March 21 and 28, 1754.

4. George A. Bray III, "Scalping During the French and Indian War," *Early American Review*, vol. II, no. 3 (Spring/Summer 1998), http://www.earlyamerica.com/review/1998/scalping.html; accessed 3/30/2010, citing *Pennsylvania Archives,* vol. 3 (Philadelphia: Joseph Severns & Co., 1853), pp. 199–200. Also, Henry J. Young, "A Note on Scalp Bounties in Pennsylvania," *Pennsylvania History* 24 (1957).

5. "Treaty of Fort Stanwix," Ohio History Central, http://www.ohiohistory central.org/entry.php?rec=1420; accessed 6/28/2009.

6. "Lord Dunmore's War," Ohio History Central, http://www.google.com/search ?hl=en&q=http://www.ohiohistorycentral.org/entry.php%3Frec%3D514&aq =f&oq=&aqi=; accessed 6/25/2009.

7. Wilbur H. Siebert, "The Loyalists of Pennsylvania," *Contributions in History and Political Science* 24 (April 1920), p. 9.

8. "Speech to the Six Nations; July 13, 1775," *Journals of the Continental Congress,* http://avalon.law.yale.edu/18th_century/contcong_07-13-75.asp;accessed 6/25/2009.

9. William Sawyer, "The Six Nations Confederacy During the American Revolution," Fort Stanwix National Monument. http://www.nps.gov/fost/history culture/the-six-nations-confederacy-during-the-american-revolution.htm; accessed 6/25/2009.

10. Fintan O'Toole, *White Savage* (London: Faber and Faber, 2005), p. 36.

11. Arthur Pound and Richard E. Day, *Johnson of the Mohawks* (New York: Macmillan, 1930), p. 21.

12. O'Toole, *White Savage,* pp. 36–37.

13. Ibid., pp. 68, 69, 83.

14. Ibid., pp. 122–123.

15. Ibid., pp. 152–154.

16. Ibid., p. 7.

17. National Park Service, "Mary (Molly) Brandt, 1736–1796." Fort Stanwix National Monument biographies, http://www.nps.gov/fost/historyculture/tory-leaders-british-military-allied-indian.htm; accessed 6/25/2009.

18. O'Toole, *White Savage,* p. 169; "Sir William Johnson," Virtual American Biographies, http://www.famousamericans.net/sirwilliamjohnson/; accessed 6/27/2009.

19. Bradt (Bratt) Family History, http://home.cogeco.ca/~gzoskey/bradthistory .html; accessed 6/24/2009.

20. Oscar Jewell Harvey, *A History of Wilkes-Barré,* vol. 2 (Wilkes-Barre: Raeder Press, 1909), p. 929.

21. O'Toole, *White Savage,* pp. 39–47.

22. Gavin K. Watt, *The Burning of the Valleys* (Toronto: Dundurn Press, 1997), p. 28.

23. Pound and Day, *Johnson of the Mohawks,* p. 458.

24. "John Butler," *Dictionary of Canadian Biography Online.* http://www.biographi .ca/009004-119.01-e.php?&id_nbr=1785&interval=20&&PHPSESSID=ukoj32 f4touu7fpgk8od4r86o0; accessed 6/23/2009.

25. "Guy Johnson," *Dictionary of Canadian Biography Online,* http://www.biographi .ca/009004-119.01-e.php?&id_nbr=1973; accessed 6/22/2009.

26. William W. Campbell, *Annals of Tryon County; or, the Border Warfare of New York, During the Revolution* (New York: J. & J. Harper 1831), p. 37.

27. Robert S. Allen, *His Majesty's Indian Allies: British Indian Policy in the Defence of Canada, 1774–1815* (Toronto: Dundurn Press, 1992), p. 46; Ernest A. Cruikshank, *The Story of Butler's Rangers and the Settlement of Niagara* (Welland, ON: Tribune Printing House, 1893), pp. 11, 25.

28. O'Toole, *White Savage,* p. 330.

29. "John Butler," *Dictionary of Canadian Biography Online.*

30. National Gallery of Art, "American Portraits of the Late 1700s and Early 1800s." http://www.nga.gov/fcgi-bin/tinfo_f?object=569; accessed 6/19/2009.

31. O'Toole, *White Savage,* p. 330. The Romney portrait is in the National Gallery of Canada in Ottawa. The gorget is in the Joseph Brant Museum in Burlington, Ontario.

32. "Samuel Kirkland," Clinton (NY) Historical Society, http://www.clinton history.org/samuelkirkland.html; accessed 6/23/2009.

33. O'Toole, *White Savage,* p. 329.

34. Crary, *The Price of Loyalty,* pp. 78–80.

35. "Index to King's Royal Regiment of New York," On-Line Institute for Advanced Loyalist Studies, http://www.royalprovincial.com/Military/rhist/krrny/krrlist. htm; accessed 6/23/6009. The regiment was also known as Johnson's Greens, the Royal Greens, and the King's Royal Yorkers.

36. Cruikshank, *The Story of Butler's Rangers and the Settlement of Niagara*, p. 34.

37. Mary Beacock Fryer, *King's Men: the Soldier Founders of Ontario* (Toronto: Dundurn Press, 1980), p. 52.

38. "Original Communications," *The Gentleman's Magazine*, October 1828, p. 291.

39. Mark Jodoin, "Shadow Soldiers," *Esprit de Corps*, August 2008. http://find articles.com/p/articles/mi_6972/is_3_16/ai_n31586164/pg_7/?tag= content;col1; accessed 3/27/2010. Jodoin quotes from *Annotated Transcript of the Journal of Lieutenant Henry Simmons 1777–1778,* edited by H. C. Burleigh. The transcript is at the Lennox and Addington County Museum in Napanee, Ontario.

40. Francis Whiting Halsey, *The Old New York Frontier* (New York: Charles Scribner's Sons, 1902), p. 218. Don Chrysler, in *The Blue-Eyed Indians* (Zephyrhills, FL: Chrysler Books, 1999), says that one of the Crysler brothers, Balthus, was probably hanged in Albany. His wife and four children, according to family records found by Chrysler, remained in their farm home throughout and after the war.

41. Jeptha R. Simms, *History of Schoharie County and Border Wars of New York* (Albany: Munsell & Tanner, Printers, 1845), chap. 7.

42. John Russell Bartlett, ed., *Records of the Colony of Rhode Island and Providence Plantations, in New England* (Providence: Alfred Anthony, Printer to the State, 1864), vol. 9, pp. 247–251, Circular letter, General Washington to the Governor of Rhode Island, October 18, 1780, Headquarters, near Passaic, New Jersey.

43. Commager and Morris, *The Spirit of 'Seventy-Six,* pp. 545–547. In Parliament, Edmund Burke later gave a parody of Burgoyne's speech: "My gentle lions, my sentimental wolves, my tender-hearted hyenas, go forth, but take care not to hurt men, women, or children." (Paul Langford, ed., *The Writings and Speeches of Edmund Burke* [New York: Oxford University Press, 1999], vol. 3, p. 361.)

44. Sheldon S. Cohen, "Connecticut's Loyalist Gadfly: The Reverend Samuel Andrew Peters," *American Revolution Bicentennial Commission of Connecticut*, 17 (1976), p. 22; Sabine, *The American Loyalists*, pp. 531–535.

45. Mary Beacock Fryer, *Buckskin Pimpernel* (Toronto: Dundurn Press, 1981), p. 52.

46. Ibid., p. 46.

47. James L. Nelson, *Benedict Arnold's Navy* (Camden, ME: International Marine/McGraw-Hill, 2006), pp. 23–24.

48. Morton Borden and Penn Borden, eds., *The American Tory* (Englewood Cliffs, NJ: Prentice-Hall, 1972), p. 126.

49. Henry Hall, "Governor Philip Skene," paper presented to the Vermont Historical Society, Barre, VT, July 2, 1863, and published in *The Historical Magazine* 1 (1867), pp. 280–283.

50. Ibid., p. 283; Sabine, *The American Loyalists,* pp. 304–305.

51. "General Court Martial of Philip Wickware and Robert Dunbar," August 5, 1777, On-Line Institute for Advanced Loyalist Studies, http://www.royal

provincial.com/Military/courts/cmwick.htm; accessed 6/25/2009. They were sentenced to be hanged, but the record, as often happens, does not show whether the sentence was carried out.

52. Symonds and Clipson, *A Battlefield Atlas*, p. 41; North Callahan, *Royal Raiders: The Tories of the American Revolution* (Indianapolis: Bobbs-Merrill, 1963), p. 152.

53. Guy Johnson to the Tryon County Committee of Safety, June 25, 1775, quoted in Crary, *The Price of Loyalty,* pp. 71–72.

54. Letter from Colonel Peter Gansevoort to Colonel Goose [*sic*] Van Schaick, July 28, 1777, as cited by Simms, *History of Schoharie County*, p. 232.

55. On a monument to Jane McCrea at Fort Edward, July 27, 1777, was given as the date of her murder. The date itself has been questioned in some accounts, particularly Richard M. Ketchum, *Saratoga* (New York: Henry Holt, 1997), pp. 505–506.

56. "The History of Fort Edward," http://www.fortedwardnewyork.net/history .htm; accessed 6/25/2009.

57. William L. Stone, "The Jane McCrea Tragedy," *The Galaxy* 3, no. 1 (January 1, 1867), pp. 46–52; http://www.4peaks.com/fkmcrea.htm; accessed 12/23/2009. As Stone notes, the contemporary, propagandistic account was later challenged by fairly conclusive evidence that Jane McCrea had been shot (possibly by a Continental or Rebel militiaman) and then scalped by one of the arguing Indians to claim a reward. But, as a matter of honor, an Indian would not take a scalp from a victim killed by someone else. Jane McCrea's remains were disinterred and reburied in cemeteries in 1822 and 1852. Modern forensic archaeologists exhumed the remains and found that her bones were commingled with those of Sara McNeil. In 2005 they were able to reconstruct the face of Sara, who had died in 1799 of natural causes. But Jane's skull—which might have settled the tomahawk-or-bullet controversy—was missing, probably taken as a souvenir in 1852.

58. Christopher Hibbert, *Redcoats and Rebels* (New York: W. W. Norton, 2002), p. 173.

59. Symonds and Clipson, *A Battlefield Atlas*, p. 47. James Fenimore Cooper draws upon the Jane McCrea murder for his novel *The Last of the Mohicans*.

60. Alan Taylor, *The Divided Ground: Indians, Settlers and the Northern Borderland of the American Revolution* (New York: Alfred A. Knopf, 2006), p. 109.

61. The fort, built in 1758 during the French and Indian War, was named for a British officer. The British continued to call it by that name, while the Patriots began calling it after General Schuyler. Both names were used in contemporary accounts. The original name was later restored, the National Park Service endorsing it by maintaining the Fort Stanwix National Monument.

62. Halsey, *The Old New York Frontier,* p. 188. The National Park Service merely says that a "hand-made" flag was hoisted on August 3. http://www.nps.gov/fost/historyculture/the-1777-siege-of-fort-schuyler.htm; accessed 3/31/2010.

63. "Patriot Leaders of New York," Fort Stanwix National Monument, http://www.nps.gov/fost/historyculture/patriot-leader-of-new-york.htm; accessed 6/25/2009.

64. "The Valley Dwellers," the American Revolution on the New York Frontier, http://www.google.com/search?hl=en&q=http://www.nyhistory.net/~drums/herkimer.htm&aq=f&oq=&aqi=; accessed 6/24/2009. Also, Eugene W. Lyttle, "Nicholas Herkimer," *Proceedings of the New York State Historical Association* (Albany: Argus Co., 1904), http://www.nyhistory.net/~drums/herkimer; accessed 3/312010.

65. William L. Stone, *Border Wars of the American Revolution* (New York: A. L. Fowle, 1900). vol. 1, p. 191.

66. Mike Caldwell, superintendent of Valley Forge National Historic Park, welcoming the display of the painting, *The Oneidas at the Battle of Oriskany,* undated National Park Service announcement, 2008, http://www.nps.gov/vafo/parknews/upload/oriskanyPR-2.pdf; accessed 6/25/2009.

67. This account of the Battle of Oriskany, as it became known, is based on Hoffman Nickerson, *The Turning Point of the Revolution* (Boston: Houghton Mifflin, 1928), pp. 203–206; Campbell, *Annals of Tryon County,* chap.4; Symonds and Clipson, *A Battlefield Atlas*, p. 43; and William Sawyer, "The Battle at Oriska," Fort Stanwix National Monument, http://www.nps.gov/fost/historyculture/the-battle-at-oriska.htm; accessed 6/25/2009.

68. Symonds and Clipson, *A Battlefield Atlas*, p. 43.

69. James Thomas Flexner, "How a Madman Helped Save the Colonies," *American Heritage* 7, no. 2 (February 1956).

70. Randall, *Benedict Arnold* (New York: Barnes & Noble, 2003), p. 347; Stone, *Border Wars of the American Revolution,* p. 219.

71. H. Y. Smith and W. S. Rann, eds., *History of Rutland County Vermont* (Syracuse, NY: D. Mason & Co., 1886), chap. 29. http://www.bucklinsociety.net/Johns_Solomon.htm; accessed 6/25/2009.

72. Callahan, *Royal Raiders,* p. 152.

73. James H. Bassett, *Colonial Life in New Hampshire* (Boston: Ginn & Company, 1899), chap. 8.

74. Pierre Comtois, "Revolutionary War Upset at Bennington," *Military History* (August 2005).

75. Fryer, *King's Men,* p. 116.

76. Fryer, *Buckskin Pimpernel,* p. 62.

77. Comtois, "Revolutionary War Upset at Bennington."

78. Letter from Burgoyne to Lord Germain, August 20, 1777, Herbert Aptheker, *The American Revolution, 1763–1783* (New York: International Publisher, 1960), p. 119.

79. Richard M. Ketchum, "Bennington," *Military History Quarterly*, vol. 10, no. 1 (Autumn 1997), p. 110.

80. Randall, *Benedict Arnold,* p. 355.

81. Don Higginbotham, *Daniel Morgan* (Chapel Hill: University of North Carolina Press, 1979), pp. 56–57.

82. Commager and Morris, *The Spirit of 'Seventy-Six*, vol. 1, p. 581.

83. Symonds and Clipson, *A Battlefield Atlas*, p. 49.

84. Brandt, *An American Aristocracy,* p. 122.

85. The account of the Saratoga battles comes from Symonds and Clipson, *A Battlefield Atlas*, pp. 47–51; Smith. *A New Age Now Begins,* vol. 1, pp. 922–943; Alden. *A History of the American Revolution*, p. 327, and Randall, *Benedict Arnold,* pp. 349–368.

86. *Report on American Manuscripts in the Royal Institution of Great Britain* (Hereford: His Majesty's Stationery Office, 1907), vol. 3, pp. 314–315.

87. Wilbur H. Siebert, "American Loyalists in the Eastern Seigniories and Townships of the Province Of Quebec," *Transactions of the Royal Society of Canada* (Ottawa: Royal Society of Canada, 1913), p. 12.

88. Mary Beacock Fryer, *Buckskin Pimpernel*, p. 72.

89. Jodoin, "Shadow Soldiers."

90. Mary Beacock Fryer, *Buckskin Pimpernel*, p. 84. Carleton made the decision on June 1, 1778. Disenchanted with British policy, he had resigned on June 27, 1777. But, because his successor, Frederick Haldimand, was not to arrive in Canada for another year, Carleton remained in office until June 27, 1778.

91. Ibid., p. 73.

CHAPTER 13: TREASON ALONG THE CHESAPEAKE

1. Major General Greene to General Washington, February. 15. 1778, *The Papers of George Washington, Revolutionary War Series*, vol. 13, p. 546, //www.consource.org/index.asp?bid=582&fid=600&documentid=50265; accessed 11/18/2009.

2. Alden, *A History of the American Revolution*, pp. 296–297. Estimates of the size of Howe's army vary, some sources putting it at seventeen thousand.

3. Sabine, p. 157.

4. Charles P. Keith, "Andrew Allen," *Pennsylvania Magazine of History and Biography* 10, no. 4 (January 1887), pp. 361–365; Sabine, *Biographical Sketches*, vol. 1, p. 158.

5. "Penn in the 18th Century," University of Pennsylvania archives, http://www.archives.upenn.edu/histy/features/1700s/faculty.html; accessed 6/28/2009.

6. Edmund Cody Burnett, ed., *Letters of Members of the Continental Congress* (Washington, DC: Carnegie Institution, 1921), vol. 1, pp. 51–53.

7. John E. Ferling, "Joseph Galloway: A Reassessment of the Motivations of a Pennsylvania Loyalist," Pennsylvania History 39, no. 2 (April 1972), p. 183.

8. M. Christopher New, *Maryland Loyalists in the American Revolution* (Centreville, MD: Tidewater Publishers, 1996), p. 41.

9. Calhoon, *The Loyalists in Revolutionary America,* p. 470; Biographic Directory of the U. S. Congress, http://bioguide.congress.gov/scripts/biodisplay.pl?index=A000101; accessed 6/26/2009.

10. New, *Maryland Loyalists in the American Revolution*, pp. 40, 43.

11. John Thomas Scharf, *History of Maryland from the Earliest Period to the Present Day* (Baltimore: John B. Piet, 1879), vol. 2, p. 319.

12. Ibid., p. 296.

13. New, *Maryland Loyalists in the American Revolution*, p. 36.

14. Ibid.

15. Scharf, *History of Maryland,* vol. 2, p. 299.

16. Journals of the Continental Congress, 1774–1789; April 17, April 19, 1777, http://memory.loc.gov/cgi-bin/query/r?ammem/hlaw:@field(DOCID+@lit(jc00784)); accessed 6/25/2009.

17. Letter from Washington to Colonel John D. Thompson of the Maryland militia, August 28, 1777, Fitzpatrick, *The Writings of George Washington from the Original Manuscript Sources*, vol. 9, p. 140.

18. M. Christopher New, "James Chalmers and 'Plain Truth,'" *Early America Review* 1, no. 2 (Fall 1996), http://www.earlyamerica.com/review/fall96/loyalists.html; accessed 3/27/2010. Christopher Johnston, "The Key Family," *Maryland Historical Magazine* 5 (1910), pp. 196–197.

19. Siebert, *The Loyalists of Pennsylvania*, pp. 38–39.

20. General Sir William Howe's Orders, 1777, "The Kemble Papers," *Collections of the New-York Historical Society for the Year 1883* (New York: 1884), vol. I, p. 473, as cited by the On-Line Institute for Advanced Loyalist Studies, http://www.royalprovincial.com/military/rhist/paloyal/pal1hist.htm; accessed 12/20/2008.

21. New, *Maryland Loyalists in the American Revolution,* p. 42.

22. J. St. George Joyce, ed., *The Story of Philadelphia* (Philadelphia: Harry B. Joseph, 1919), pp. 156–157.

23. Sabine, *Biographical Sketches*, vol. 1, p. 502.

24. Stephen Jarvis, "An American's Experience in the British Army," *Connecticut Magazine* 11 (Summer/Autumn 1907), pp. 450–451.

25. Siebert, *The Loyalists of Pennsylvania*, p. 40.

26. Harry Stanton Tillotson, *The Beloved Spy* (Caldwell, ID: Caxton Printers, 1948), p. 34.

27. J. Smith Futhey, "The Massacre of Paoli," *Pennsylvania Magazine of History and Biography* 1, p. 305.

28. Thomas J. McGuire, *The Battle of Paoli* (Mechanicsburg, PA: Stackpole Books, 2006), p. 215.

29. Joseph Plumb Martin, *A Narrative of a Revolutionary Soldier* (New York: Signet Classics, 2001), p. 86.

30. Albigence Waldo, "Diary of Surgeon Albigence Waldo, of the Continental Line," *Pennsylvania Magazine of History and Biography* 21 (1897), p. 307.

31. Siebert, *The Loyalists of Pennsylvania*, p. 40. For more about the taking of the city, see "A History of the Provincial Corps of Pennsylvania Loyalists," On-Line Institute for Advanced Loyalist Studies, http://www.royalprovincial.com/military/rhist/paloyal/pal1hist.htm#palintro; accessed 6/27/2009.

32. Sabine, *Biographical Sketches*, vol. 2, p. 360.

33. W. W. H. Davis, *The History of Bucks County, Pennsylvania* (Doylestown, PA: Democrat Book and Job Office Printing, 1876), p. 638. Siebert, *The Loyalists of Pennsylvania*, p. 40.

34. Randall, *Benedict Arnold,* pp. 390–391.

35. Wertenbaker, *Father Knickerbocker Rebels*, p. 149.

36. "Regulations" poster, December 8, 1777, signed by Galloway, *Pennsylvania Magazine of History and Biography* 1 (1877), pp. 35–36.

37. Washington spoke of "Hundreds" of deserters. (Fitzpatrick, *The Writings of George Washington from the Original Manuscript Sources,* vol. 11, p. 417.) Galloway put the figure at twenty-three hundred. (Van Tyne, *The Loyalists in the American Revolution*, p. 157.)

38. Siebert, *The Loyalists of Pennsylvania*, p. 43.

39. Ibid., pp. 45, 47.

40. "British Legion Biographical Sketches, Cavalry Officers," On-Line Institute for Advanced Loyalist Studies. Richard Hovenden of Bucks County, Pennsylvania, formed the Philadelphia Light Dragoons on January 8, 1778; it was later absorbed into the British Legion. http://www.royalprovincial.com/Military/rhist/britlegn/blcav1.htm; accessed 6/27/2009.

41. Siebert, *The Loyalists of Pennsylvania*, pp. 41–42.

42. Letter to the President of Congress, December 23, 1777, Fitzpatrick, *The Writings of George Washington from the Original Manuscript Sources,* vol. 10, p. 193.

43. "Brigadier General John Lacey Jr.," Militia & Associated Companies of Bucks County, Pennsylvania, http://www.geocities.com/oldebucks/General_Lacey.htm; accessed 6/23/2009.

44. Paul Gouza, "Bird-in-Hand Raid Reenactment," *The Half-Moon* (publication of the Newtown Historic Association) 7, no. 2 (March 2008), http://www.newtonhistoric.org/NHANewsletter (03-08).pdf; accessed 3/27/2010. 7, no. 2, March 2008.

45. Davis, *The History of Bucks County,* p. 634.

46. Washington to Lacey, January 23, 1778, Fitzpatrick, *The Writings of George Washington from the Original Manuscript Sources,* vol. 10, p. 340; Ibid.

47. Allen, *The Loyal Americans.*

48. W. W. H. Davis, *History of the Battle of the Crooked Billet.* (Doylestown, PA: Democrat Book and Job Office Printing, 1894). http://goodyear-mascaro.org/ Warminster-History/billet.html; accessed 3/27/2010.

49. Sabine, *Biographical Sketches,* vol. 2, p. 416.

50. André's description of the Mischianza appears as a lengthy endnote to vol. 2, chap. 4, in Benson J. Lossing, *Pictorial Field Book of the Revolution* (New York: Harper & Brothers, 1859).

51. Fleming, *Now We Are Enemies,* pp. 269–270.

52. Randall, *Benedict Arnold,* pp. 387, 395. The Shippen family, down the years, insisted that Peggy and her two sisters had not appeared, despite the fact that André puts them in his account. He even sketched a portrait of Peggy in her "turban spangled and edged with gold or silver."

53. Wayne Bodle, *Valley Forge Winter* (University Park: Pennsylvania State University Press, 2002), pp. 228–229.

54. "Benedict Arnold's Oath of Allegiance, 05/30/1778," MLR Number A1 5A, Record Group 93, War Department Collection of Revolutionary War Records, 1709–1915. http://narademo.umiacs.umd.edu/cgi-bin/isadg/viewitem .pl?item=100956; accessed 3/27/2010.

55. Randall, *Benedict Arnold,* p. 406.

56. Many sources estimate three thousand as the number of fleeing Tories. The five thousand number comes from William Eden, an undersecretary in Britain's colonial department, who had arrived in Philadelphia just as the fleet was leaving. Eden, on an abortive peace-seeking mission, sailed in one of the ships carrying the Loyalists and wrote, "Near 5,000 of the Philadelphia inhabitants are attending us to New York." His observation is published in Wertenbaker, *Father Knickerbocker Rebels,* p. 149, who quotes Benjamin Franklin Stevens's *Facsimiles of Manuscripts in European Archives Relating to America 1773–1783* (London: Malby & Sons/Chiswick Press, 1889), vol. 5, p. 501.

57. Estimate and quote from Carl G. Karsch, "The Battle for Philadelphia," Independence Hall Association, http://www.ushistory.org/carpentershall/history/ battle.htm; accessed 6/27/2009.

58. Randall, *Benedict Arnold,* p. 397. André gave the portrait to his superior, General Grey. In 1906, on the two-hundredth anniversary of Franklin's birth, the Grey family returned the portrait, which was hung in the White House.

59. Noah Andre Trudeau, "Charles Lee's Disgrace at the Battle of Monmouth," *Military History Quarterly* (Fall 2006). The positioning of the Loyalist units is described by New, *Maryland Loyalists in the American Revolution,* pp. 58–59.

60. Siebert, *The Loyalists of Pennsylvania*, p. 53.

61. Randall, *Benedict Arnold,* pp. 408–409.

62. Siebert, *The Loyalists of Pennsylvania*, p. 74.

63. Ibid., pp. 56, 57.

64. Fleming, *Now We Are Enemies,* pp. 311–316.

65. Martin, *A Narrative of a Revolutionary Soldier* (New York: Signet Classics, 2001) pp. 110–111.

66. John O. Raum, *The History of New Jersey* (Philadelphia: John E. Potter and Co., 1877), vol. 2, p. 67.

67. The plan that Lee wrote for Howe was revealed by George H. Moore, librarian of the New-York Historical Society, who read his essay, "The Treason of Charles Lee," before the society on June 22, 1858. Moore subsequently used the same title in a book published in 1860 (New York: Charles Scribner). The plan had previously remained unpublished among the Howe papers. Lee died in Philadelphia in October 1782. His will directed that he not be buried in a church or churchyard, "for since I have resided in this country, I have kept so much bad company when living, that I do not chuse to continue it when dead." Despite his directive, he was buried in the churchyard of Christ Church, Philadelphia.

68. Letter to Joseph Kirkbride, county lieutenant of Bucks County, Pennsylvania, April 20, 1778, Fitzpatrick, *The Writings of George Washington from the Original Manuscript Sources*, vol. 11, p. 284.

69. Harvey, *A History of Wilkes-Barré* (Wilkes-Barre: Raeder Press, 1909), pp. 976–979.

70. Ibid., pp. 979–980.

CHAPTER 14: VENGEANCE IN THE VALLEYS

1. Letter from John Butler to Guy Carleton, May 15, 1778, Harvey, *A History of Wilkes-Barré*, p. 971, citing the Haldimand Papers in the British Museum; the papers are also on microfilm in the National Archives of Canada.

2. The quotations come from "The Butler Papers," National Archives of Canada, MG 31, G 36, as cited at http://www.iaw.on.ca/~awoolley/brang/brbp.html; accessed 2/2/2009. Other information from *The Loyalist Gazette*, March 2005, citing William A. Smy, *An Annotated Nominal Roll of Butler's Rangers 1777–1784 with Documentary Sources* (St. Catharines, ON: Friends of the Loyalist Collection at Brock University, 2004).

3. Harvey, *A History of Wilkes-Barré*, p. 971.

4. Cruikshank, *The Story of Butler's Rangers and the Settlement of Niagara,* p. 28; Stephen Davidson, "Five Spies Who Settled in New Brunswick," *Loyalist Trails* (June 2008).

5. Sarah B. Hood, "Life in Butler's Rangers," *Suite101*, a Canadian online magazine. http://canada-at-war.suite101.com/article.cfm/life_in_butlers_rangers; accessed 2/2/2009.

6. Harvey, *A History of Wilkes-Barré*, p. 971.

7. Information about forts is based primarily on John F. Luzader, "Fort Stanwix History, Historic Furnishing, And Historic Structure Reports," *Fort Stanwix* (Washington, DC: Office of Park Historic Preservation, National Park Service, 1976). Also, Dow Beekman, "Schoharie County in the Revolution," address to the Captain Christian Brown Chapter of the Daughters of the American Revolution, September 10, 1920, at Cobleskill, NY, http://www.rootsweb.ancestry .com/~nyschoha/cobleski.html; accessed 6/30/2009.

8. William E. Roscoe, *History of Schoharie County* (Syracuse, NY: D. Mason & Co., 1882), chap. 3, http://www.rootsweb.ancestry.com/~nyschoha/chap3.html; accessed 6/30/2009.

9. Ibid., chaps. 3, 11 ("History of the Town of Summit").

10. Jeptha R. Simms, *The Frontiersmen of New York* (Albany: Geo. C. Riggs, 1883), vol. 2, pp. 152–158; Richard Christman, "225 Years Ago: Torch, Tomahawk Ravage Cobleskill," *Schoharie County Historical Review* (Spring 2003). http://www.schohariehistory.net/Review/Spring2003/Christman.htm#_ ftn11; accessed 6/30/2009; Beekman, "Schoharie County in the Revolution." http://www.schohariehistory.net/Review/Spring2003/Christman.htm#Top; accessed 3/27/2010. A similar Masonic rescue of a wounded Patriot by a Tory had been reported at the battle of Bennington. See Robert Collins McBride, "Loyalists and Masons," *Loyalist Gazette* (Fall 2007). Brant, Butler, and Sir John Johnson were Masons, as were George Washington, Benjamin Franklin, Paul Revere, and John Hancock.

11. Alden, *A History of the American Revolution*, pp. 114–115.

12. Harvey, *A History of Wilkes-Barré*, pp. 988–989.

13. Ibid., p. 992.

14. History of Forty Fort, http://dsf.pacounties.org/fortyfort/cwp/view.asp?A=3& Q=470200; accessed 6/25/2009.

15. George Peck, *Wyoming; Its History, Stirring Incidents and Romantic Adventures* (New York: Harper & Brothers, 1858), pp. 38–42.

16. Harvey, *A History of Wilkes-Barré*, p. 893.

17. Ibid., p. 1013.

18. Ibid., p. 1024.

19. Denise Dennis, "Black Patriots of the 18th Century," address presented at Delaware Humanities Forum and Washington-Rochambeau Route Conference, September 28, 2006. The powder horn is in the Luzerne County Historical Society Museum in Wilkes-Barre.

20. Letter from Major John Butler to Lieutenant Colonel Mason Bolton, commandant of Fort Niagara, July 8, 1778, http://revwar75.com/battles/primary-docs/wiom1778.htm; accessed 6/25/2009.

21. James J. Talman, ed., "The Journal of Adam Crysler" (Toronto: Champlain Society in Canada, 1946), p. 58.

22. Butler on deaths of Indian and Rangers, letter to Bolton, "The Wyoming Massacre and Columbia County," Columbia County Historical and Genealogical Society. http://www.colcohist-gensoc.org/Essays/wyomingmassacre.htm; accessed 6/25/2009. Stories of the raid lived on in Rebel propaganda and in poetry. *Wyoming* (from the Delaware Indians' word for "at the big river flat") many years later became the name of a western state thanks to Rep. James Mitchell Ashley of Ohio. As chairman of the Committee on Territories, Ashley suggested it in 1868, after the defeat of a proposal to name the territory Lincoln. Ashley said he had been inspired by "Gertrude of Wyoming," a sentimental and thoroughly inaccurate 1809 ballad by Thomas Campbell, a Scottish poet. It begins "On Susquehanna's side, fair Wyoming! / Although the wild-flower on thy ruin'd wall, /And roofless homes, a sad remembrance bring, / Of what thy gentle people did befall." The immensely popular ballad called Brant the "monster" responsible for the atrocities when, in fact, Brant had not been present.

23. Peck, *Wyoming; Its History,* p. 44.

24. Letter from Major John Butler to Bolton. http://revwar75.com/battles/primary docs/wiom1778.htm; accessed 3/28/2010.

25. Harvey, *A History of Wilkes-Barré*, pp. 984, 985, 1016. Sources differ on the size of invader and defender forces and on the number of casualties. The numbers used here come from Alden, *A History of the American Revolution,* p. 433.

26. Horace Edwin Hayden, "Echoes of the Massacre of Wyoming," *Proceedings and Collections of the Wyoming Historical and Geological Society, for the years 1913–1914* (Wilkes-Barre, PA: Printed for the Society, 1914), vol. 113, pp. 124–130. The site of the reputed tomahawking is marked in Wilkes-Barre by "Bloody Rock," also known as "Queen Esther's Rock."

27. Lossing, *Pictorial Field Book of the Revolution,* vol. 1, chap. 15.

28. Frederic A. Godcharles, *Daily Stories of Pennsylvania* (Milton, PA: Published by the author, 1924), p. 460, quoting from a letter written by William Maclay of Paxtang, PA, on July 12, 1778.

29. Lossing, *Pictorial Field Book of the Revolution,* vol. 1, chap. 15.

30. Louise Welles Murray, *A History of Old Tioga Point and Early Athens, Pennsylvania* (Wilkes-Barre, PA: Raeder Press, 1908), p. 142.

31. Ibid., pp. 137, 138.

32. M. Paul Keesler, chap. 8, "Revolution," from *Discovering the Valley of the Crystals,*

a work in progress. I am grateful to Paul for permission to use information from the chapter as a source. http://www.paulkeeslerbooks.com/; accessed 6/30/2009.

33. "British Intelligence, New York," Memorandum Book of the British Army, 1778; film 689, David Library of the American Revolution, Washington Crossing, PA.

34. A. J. Berry, *A Time of Terror* (Victoria, BC: Trafford Publishing. 2005), pp. 44–45, 130–132.

35. Harvey, *A History of Wilkes-Barré*, p. 930.

36. "The Battle of Cherry Valley (Massacre)," http://www.myrevolutionarywar.com/battles/781111.htm; accessed 6/29/2009.

37. Watt, *The Burning of the Valleys*, p. 72.

38. Francis Whiting Halsey, *The Old New York Frontier* (New York: Charles Scribner's Sons, 1901), p. 241.

39. Benjamin Warren, "Diary of Captain Benjamin Warren at Massacre of Cherry Valley," *Journal of American History* (1909), http://www.newrivernotes.com/ny/cherryvalley.htm; accessed 6/30/2009.

40. Report from the scene, the day after the battle. http://www.myrevolutionarywar.com/battles/781111.htm; accessed 3/28/2010.

41. Berry, *A Time of Terror*, pp. 46, 131.

42. Sabine, *Biographical Sketches of Loyalists*, vol. 2, p. 119.

43. William Ketchum, *An Authentic and Comprehensive History of Buffalo* (Buffalo, NY: Rockwell, Baker & Hill, Printers, 1864), vol. 1, p. 314. Letter from Clinton to Butler, January 1, 1779.

44. Letter of instructions to General Sullivan, May 31, 1779, Ford, *The Writings of George Washington*, vol. 7, pp. 460–462.

45. Alden, *A History of the American Revolution*, pp. 434–435.

46. Simms, *The Frontiersmen of New York*, pp. 241–242; the letter from Clinton to his wife was written at Canajoharie on July 6, 1779.

47. Stanley J. Adamiak, "The 1779 Sullivan Campaign," *Early America Review* 2, no. 3 (Spring-Summer 1998). http://www.earlyamerica.com/review/1998/sullivan.html#end; accessed 6/25/2009.

48. Frederick Cook, ed., *Journals of the Military Expedition of Major General John Sullivan against the Six Nations of Indians in 1779* (Auburn, NY: Knapp, Peck & Thomson, Printers, 1887), p. 8. Cook was New York's secretary of state. The state legislature had ordered publication of the journals as part of a celebration of the centennial of the expedition.

49. Alden, *A History of the American Revolution*, p. 436.

50. Cook, *Journals of the Military Expedition of Major General John Sullivan*, p. 272.

51. Report of Colonel Goose Van Schaick, "The Van Schaick Expedition—April 1779," Fort Stanwix National Monument, http://www.nps.gov/fost/historyculture/the-van-schaick-expedition-april-1779.htm; accessed 2/6/2009.

52. Adamiak, "The 1779 Sullivan Campaign."

53. Halsey, *The Old New York Frontier,* p. 283; "Haldimand, Frederick," *Dictionary of Canadian Biography Online,* http://www.biographi.ca/009004-119.01-e.php?&id_nbr=2445&interval=25&&PHPSESSID=dr24vf4ons8pqadholf6fobbt5; accessed 6/30/2009.

54. William L. Stone, *Life of Joseph Brant,* vol 1. (Albany, NY: J. Munsell, 1865), pp. 6–7.

55. Louise Phelps Kellogg, ed., "Frontier Retreat on the Upper Ohio, 1779–1781," *Publications of the State Historical Society of Wisconsin* (Madison, WI: Published by the Society, 1917), vol. 21, pp. 22–24.

56. Halsey, *The Old New York Frontier,* pp. 280–283.

57. Ibid., p. 292.

58. Simms, *The Frontiersmen of New York,* pp. 327, 336.

59. Watt, *The Burning of the Valleys,* pp. 73–80, 90.

60. Mary Beacock Fryer, *Buckskin Pimpernel* (Toronto: Dundurn Press, 1981), pp. 95–96.

61. "Sir Frederick Haldimand" *Dictionary of Canadian Biography Online,* http://www.biographi.ca/009004-119.01-e.php?&id_nbr=2445; accessed 6/25/2009.

62. Watt, *The Burning of the Valleys,* pp. 160–164. The composition of the force was determined by Watt from many sources.

63. Ibid., pp. 109–112, 114.

64. Talman, "Reminiscences of Captain James Dittrick," *Loyalist Narratives from Upper Canada,* pp. 62–63.

65. Watt, *The Burning of the Valleys,* p. 171.

66. Ibid., pp. 171–172.

67. Simms, *The Frontiersmen of New York,* p. 420.

68. Watt, *The Burning of the Valleys,* pp. 173–174.

69. Simms, *The Frontiersmen of New York,* pp. 424–429.

70. Watt, *The Burning of the Valleys,* p. 184.

71. Nelson Greene, *The Story of Old Fort Plain and the Middle Mohawk Valley* (Fort Plain, NY: O'Connor Brothers Publishers, 1915), chap. 18.

72. The estimated size of Brown's force comes from careful assessment by Watt, *The Burning of the Valleys,* p. 206; other sources give Brown as few as one hundred men. The origin of Stone Arabia's name is a mystery. One theory has it come from a mispronunciation of *Stein,* the German word for "stone," and *Riegel,* meaning "bolt," but locally applied to rows of piled-up stones. "So immigrants . . . may have talked about their 'Stoina Riegel,' which eventually became attached to the village and transliterated into English as 'Stone Arabia.'" (Britta Schuelke Kling, http://threerivershms.com/saorigin.htm; accessed 6/30/2009.)

73. Watt, *The Burning of the Valleys,* pp. 206–208; Greene, *The Story of Old Fort Plain,* chap. 18.

74. Watt, *The Burning of the Valleys,* p. 192, 202.

75. Simms, *The Frontiersmen of New York,* p. 442.

76. Greene, *The Story of Old Fort Plain,* chap. 18.

77. Watt, *The Burning of the Valleys,* pp. 94–95.

78. Ibid., pp. 104, 128.

79. Ibid., pp. 106–107.

80. William L. Stone, *Reminiscences of Saratoga and Ballston* (New York: R. Worthington, 1880), chap. 34, "The Tory Invasion of 1780, and the Gonzalez Tragedy."

81. Watt, *The Burning of the Valleys,* pp. 109, 149.

82. Descriptions of the raids were compiled, primarily from pension applications, by James A. Morrison, http://morrisonspensions.org/index.htm; accessed 6/30/2009. Pension files are published in James A. Morrison and A. J. Berry, *Don't Shoot Until You See the Whites of Their Eyes* (Victoria, BC: Trafford Publishing, 2007). Morrison is also credited for research assistance in Gavin K. Watt's *The Burning of the Valleys.*

83. William M. Willett, *A Narrative of the Military Actions of Colonel Marinus Willett* (New York: G. & C. & H. Carvill, 1831), p. 75.

84. Samuel Ludlow Frey, introduction to *The Minute Book of the Committee of Safety of Tryon County, the Old New York Frontier* (New York: Dodd, Mead and Company, 1905), p. xiii.

85. Stone, *Life of Joseph Brant,* vol. 1, p. 337.

86. Watt, *The Burning of the Valleys,* p. 263.

87. Sparks, *The Writings of George Washington,* vol. 7, p. 282. Letter from Washington at Headquarters, Preakness, to Clinton, November 5, 1780.

88. Reuben G. Thwaites, ed., "Papers from the Canadian Archives," *Collections of the State Historical Society of Wisconsin* (Madison, WI: Democrat Printing Co., 1888), p. 176. Letter from Lord Germain to Sir Guy Carleton, enclosed in letter from Carleton to Lieutenant Governor Henry Hamilton, Quebec, May 21, 1777.

89. Ibid.

90. Henry Haymond, *History of Harrison County, West Virginia* (Morgantown, WV: Acme Publishing, 1910), p. 142. This is one of many sources using the label "Hair Buyer" for Hamilton, a canard that is questioned by many modern historians.

91. George E. Greene, *History of Old Vincennes and Knox County, Indiana.* Chicago: S. J. Clarke Publishing Co., 1911), p. 183.

92. "Simon Girty," *Historical Narratives of Early Canada,* http://www.uppercanada history.ca/ttuc/ttuc7.html; accessed 6/30/2009. Also, "Simon Girty," in Rich-

ard L. Blanco, ed., *The American Revolution 1775–1783* (New York: Garland Publishing, 1993).

93. "Simon Girty's Report of Colonel Crawford's Torture," Haldimand Papers, cited in Crary, *The Price of Loyalty,* pp. 256–257.

94. Wilbur E. Garrett, ed., *Historical Atlas of the United States* (Washington, DC: National Geographic Society, 1988), p. 96 (map).

95. "Henry Hamilton," *Dictionary of Canadian Biography Online.* http://www.biographi.ca/009004-119.01-e.php?&id_nbr=1931&&PHPSESSID=ychzfqkvzape; accessed 6/30/2009.

96. Colin G. Calloway, *The American Revolution in Indian Country* (New York: Cambridge University Press, 1995), p. 45. Also, Alden. *A History of the American Revolution*, p. 440.

97. Haymond, *History of Harrison County,* p. 173.

98. "Henry Hamilton," *Dictionary of Canadian Biography Online.* http://www.biographi.ca/009004-119.01-e.php?&id_nbr=1931&&PHPSESSID=ychzfqkvzape; accessed 3/28/2010.

99. Eugene H. Roseboom and Francis P. Weisenburger, *A History of Ohio* (Columbus: Ohio Historical Society, 1953), chap. 3, "The Struggle with the Indians 1763–1783."

CHAPTER 15: SEEKING SOUTHERN FRIENDS

1. Major General Greene, commander of the Southern Army, to Colonel William Davie, May 3, 1781, quoted in Charles Sumner, *Recent Speeches and Addresses* (Boston: Ticknor and Fields, 1856), p. 361.

2. Wilson, *The Southern Strategy,* p. 63, citing Davies, *Documents of the American Revolution*, vol. 15, p. 61.

3. Letter, October 4, 1775, from Lieutenant Governor John Moultrie to General James Grant, former governor of East Florida, who was on the staff of General Gage. *American Archives.* "East Florida Correspondence, Miscellaneous Papers, Proceedings of Committees, &c," http://www.stanklos.net/?act=para&pid=5770&psname=CORRESPONDENCE%2C%20PROCEEDINGS%2C%20ETC; accessed May 20, 2009.

4. Charles Alfred Risher, Jr., "Propaganda, Dissension, and Defeat: Loyalist Sentiment in Georgia, 1763–1783" (Ph.D. diss., Mississippi State University, 1976), pp. 91–92.

5. "James Wright," *New Georgia Encyclopedia.* http://www.georgiaencyclopedia.org/nge/Article.jsp?id=h-669; accessed 6/10/2009.

6. Wilson, *The Southern Strategy,* p. 79. Wilson methodically reconstructed the order of battle of both sides in southern battles.

7. Ibid., p. 77. Howe was the mysterious "Gentleman" who intrigued Janet Schaw during her stay on her brother's North Carolina plantation in 1775. Evangeline Walker Andrews and Charles McLean Andrews, eds., *Journal of a Lady of Quality*. (New Haven: Yale University Press, 1922), p. 167.

8. "Thomas Brown," *New Georgia Encyclopedia*, http://www.georgiaencyclopedia .org/nge/Article.jsp?id=h-1090; accessed 7/1/2009. The East Florida Rangers were also known as the King's Carolina Rangers or Brown's Rangers.

9. John Buchanan, *The Road to Guilford Courthouse* (New York: John Wiley & Sons, 1997), p. 96; "Revolutionary War in Georgia," *New Georgia Encyclopedia*. http://www.georgiaencyclopedia.org/nge/Article.jsp?id=h-2709; accessed 6/10/2009.

10. "East Florida Rangers Return," On-Line Institute for Advanced Loyalist Studies, http://www.royalprovincial.com/Military/rhist/eastfr/eastretn.htm; accessed 6/10/2009. Also, "Revolutionary War in Georgia" and "Thomas Brown," *New Georgia Encyclopedia*, http://www.georgiaencyclopedia.org;. accessed 6/30/2009.

11. From a report by James Simpson, former South Carolina attorney general. He had been asked by General Sir Henry Clinton for an assessment of Tory temperament, particularly in Charleston. The report is dated May 15, 1780, three days after Charleston fell to Clinton's forces. G. N. D. Evans,"Simpson's Report . . ," *Journal of Southern History* 21 (1955), pp. 518–519, as cited by Crary, *The Price of Loyalty*, pp. 277–278.

12. Wilson, *The Southern Strategy*, p. 84, refers to the unit as Carolina Loyalists; other sources call the unit the Carolina Royalists.

13. Risher, "Propaganda, Dissension, and Defeat," p. 172.

14. "An Introduction to North Carolina Loyalist Units," On-Line Institute for Advanced Loyalist Studies, http://www.royalprovincial.com/Military/rhist/ ncindcoy/ncintro.htm. Also, Robert Scott Davis, Jr., "Battle of Kettle Creek," *New Georgia Encyclopedia* http://www.georgiaencyclopedia.org/nge/Article .jsp?id=h-1088&hl=y; both accessed 7/4/2009.

15. Robert Scott Davis, Jr. "Battle of Kettle Creek," *New Georgia Encyclopedia*, http:// www.georgiaencyclopedia.org/nge/Article.jsp?id=h-1088; accessed 7/2/2009.

16. Wilson, *The Southern Strategy*, p. 86.

17. Jim Piecuch, *Three Peoples, One King* (Columbia: University of South Carolina Press, 2008), p. 140.

18. Wilson, *The Southern Strategy*, p. 89, citing a Campbell report to Clinton, March 4, 1779. Wilson also says that "cracker" apparently derives from the habit of some backcountry Georgians to "crack" boasts—a usage similar to one in contemporary Ireland—and to distill, or crack, whiskey (p. 297).

19. Letter from Captain T. W. Moore, aide-de-camp to General Augustine Prevost, to his wife, November 4, 1779, quoted by Wilson, *The Southern Strategy*, p. 170.

20. Symonds and Clipson, *A Battlefield Atlas of the American Revolution,* p. 85.

21. Todd W. Braisted, "The Prince of Wales' American Regiment," lecture, April 1998, online at Canadian Military Heritage Project. http://www.rootsweb.ancestry.com/~canmil/uel/pwar.htm; accessed 7/3/2009.

22. Crary, *The Price of Loyalty,* pp. 277–278, "Simpson's Report," May 15, 1780.

23. Alan S. Brown. "James Simpson's Reports on the Carolina Loyalists, 1779–1780," *Journal of Southern History* 21, no. 4 (November 1955), p. 514.

24. Banastre Tarleton, *A History of the Southern Campaigns of 1780 and 1781 in the Southern Provinces of North America* (Dublin: Colles, Exchaw et al., 1787), pp. 23–24.

25. Murtie June Clark, *Loyalists in the Southern Campaign of the Revolutionary War* (Baltimore: Genealogical Publishing Co., 1981), vol. 1, pp. xiii–xv, citing the United Kingdom National Archives, Cornwallis Papers, "Regulations for government of conquered towns, etc." and PRO (now National Archives of Britain), 30/11/2/44, "Clinton's instructions to Major Ferguson."

26. In a demonstration of his rifle before King George III, Ferguson said he could fire seven random shots a minute—"yet I could not undertake to bring down above five of his Majesties Enemys in that time. He laughed very heartily. . . ." (M. M. Gilchrist, *Patrick Ferguson* [Edinburgh: NMS Publishing, National Museums of Scotland, 2003], p. 29.)

27. Information on Ferguson derived from Gilchrist, *Patrick Ferguson*. http://www.silverwhistle.co.uk/lobsters/ferguson.html; accessed 3/21/2009.

28. M. M. Gilchrist, *Patrick Ferguson* (Edinburgh: National Museums of Scotland, 2003), pp. 60, 62.

29. Walter Edgar, *Partisans & Redcoats* (New York: William Morrow, 2001), pp. 52–53.

30. Don Higginbotham, "Some Reflections on the South in the American Revolution," *Journal of Southern History* 73, no. 3 (2007).

31. Governor Nash to the General Assembly, August 25, 1780, cited by John R. Maass, "A Complicated Scene of Difficulties" (Ph.D. diss., Ohio State University, 2007), p. 143.

32. "Regiment List: British and Loyalist," *Southern Campaigns, Revolutionary War, Final Report, June 2005* (Evans-Hatch & Associates, for the National Park Service, 2005).

33. Ann Taylor Andrus and Ruth M. Miller, *Charleston's Old Exchange Building* (Charleston, SC: History Press, 2005), pp. 34–35. Balfour had met his first Rebels in Massachusetts long ago when he was a guest on the Marshfield estate of Tory Nathaniel Ray Thomas.

34. Edward J. Cashin, *The King's Ranger* (New York: Fordham University Press, 1999), p. 118.

35. Steven J. Rauch, "Southern (Dis)Comfort," *Army History* (Spring 2009), pp. 44–45.

36. Pension Application of William Gipson, National Archives Microseries M804, Roll 1078, Application No. S17437. The application was found by Brent Brackett, curator of Tannenbaum Historic Country Park in North Carolina.

37. Michael E. Stevens, "The Hanging of Matthew Love," *South Carolina Historical Magazine*. South Carolina Historical Society, vol. 88, no. 1 (January 1987), pp. 55–58.

38. James Thacher, ed. *A Military Journal of the American Revolution* (Boston: Cohon & Barnard, 1827), p. 299. Cornwallis letter, August 1780, to Lieutenant Colonel Nisbet Balfour. Hanged Rebels: Edward McCrady, *The History of South Carolina in the Revolution, 1775–1780* (New York: Macmillan Company, 1901), vol. 3, p. 711.

39. Maass, "A Complicated Scene of Difficulties," pp. 144–147.

40. Thomas H. Raddall, "Tarleton's Legion," Collections of the Nova Scotia Historical Society, electronic version created by the Mersey Heritage Society in 2001 with the permission of Dalhousie University. http://www.mersey.ca/tarletons legion.htmlhttp://www.mersey.ca/tarletonslegion.html; accessed 3/28/2010.

41. J. Tracy Power, " 'The Virtue of Humanity was Totally Forgot': Buford's Massacre, May 29, 1780," *South Carolina Historical Magazine* 93 (1992), pp. 5–14. The march distance and duration is mentioned in Tarleton's report to Lieutenant General Cornwallis, May 30, 1780. http://www.royalprovincial.com/military/rhist/britlegn/bllet3.htm; accessed 7/4/2009. The number of Buford men: John Buchanan, *The Road to Guilford Courthouse* (New York: John Wiley & Sons, 1997), p. 82. Wilson, *The Southern Strategy*, says Buford had 380 infantrymen and forty dragoons. Buchanan and Wilson otherwise agree on most of the other details of the battle and subsequent massacre. Raddall (n. 40 above) does not believe there was a massacre.

42. Wilson, *The Southern Strategy,* pp. 256–260; Buchanan, *The Road to Guilford Courthouse*, pp. 83–85.

43. Samuel Cole Williams, *Tennessee During the Revolutionary War* (Knoxville: University of Tennessee Press, 1944), p. 140.

44. Ibid., p. 141.

45. Clark, *Loyalists in the Southern Campaign,* p. xviii.

46. Buchanan, *The Road to Guilford Courthouse,* pp. 206, 212–213.

47. Gilchrist, *Patrick Ferguson,* pp. 65–66.

48. Ibid., p. 66.

49. Estimate of Ferguson force: Buchanan, p. 229. Terrain: Author's notes and National Park Service Guide to King's Mountain National Military Park.

50. Gilchrist, *Patrick Ferguson,* p. 70. In 1845 an excavated grave yielded a female skeleton, believed to have been the remains of a young, red-haired woman reportedly shot during the battles (pp. 69–70).

51. Buchanan, quoting a Rebel witness, *The Road to Guilford Courthouse*, p. 234.

52. Isaac Shelby, one of the Rebel leaders, said there were "some who had heard that at Buford's death, the British had refused quarters. . .were willing to follow that bad example. . . ." (Ibid., p. 233)

53. Ibid., p. 236.

54. Ibid., pp. 238, 240; letter from Lieutenant Anthony Allaire, quoted in *The Royal Gazette* (New York), February 24, 1781, http://www.royalprovincial .com/history/battles/kingslet.shtml; accessed 3/22/2009.

55. George C. Mackenzie, *Kings Mountain National Military Park South Carolina*. Washington, DC: National Park Service, Historical Handbook Series No. 22, 1955. http://www.nps.gov/history/history/online_books/hh/22/hh22c.htm; accessed 11/20/2009.

CHAPTER 16: DESPAIR BEFORE THE DAWN

1. Bartlett, ed., *Records of the Colony of Rhode Island,* vol. 9, pp. 247–251. Circular letter, October 18, 1780, written at Washington's headquarters near Passaic, New Jersey. The circular letter is also addressed to New York governor George Clinton and is dated November 5, 1780. Headquarters, Preakness, New Jersey (Sparks, *The Writings of George Washington,* vol. 7, p. 282).

2. Randall, *Benedict Arnold,* p. 456.

3. Allen, *George Washington, Spymaster,* p. 113.

4. Winthrop Sargent, ed., *The Loyal Verses of Joseph Stansbury and Doctor Jonathan Odell; Relating to the American Revolution* (Albany: J. Munsell, 1860), p. 22.

5. Ibid., p. 24.

6. Randall, *Benedict Arnold,* p. 456.

7. Sabine, *Biographical Sketches of Loyalists,* vol. 1, p. 177.

8. Randall, *Benedict Arnold,* pp. 428–429, 440–442.

9. Ibid., p. 462.

10. Many of the letters can be seen at http://www.si.umich.edu/SPIES/, from the collection of the Clements Library at the University of Michigan.

11. John André to Joseph Stansbury, May 10, 1779, http://www.si.umich.edu/ spies/letter-1779may10-4.html; accessed 3/1/2009.

12. Clare Brandt, *The Man in the Mirror* (New York: Random House, 1994), p. 184.

13. Sparks, *The Writings of George Washington,* vol. 7, p. 533, letter to George Washington from Arnold, on the *Vulture,* September 25, 1780.

14. Randall, *Benedict Arnold,* pp. 500–501, quoting an undated letter written by Robinson sometime in early 1779.

15. Because the plot to take West Point failed, Arnold received only £6,000, plus £350 in expenses. Randall, *Benedict Arnold,* estimates that the payment

equaled about $200,000 in 1990 U.S. purchasing power. As a brigadier general he would be paid £650 a year plus a lifelong pension of £220 annually (p. 574).

16. Sargent, *The Loyal Verses,* p. 58.

17. Randall, *Benedict Arnold,* pp. 457–458, 464.

18. J. E. Morpurgo, *Treason at West Point* (New York: Mason/Charter, 1975), p. 117.

19. Smith, tried by a military court for aiding André, was acquitted because of lack of evidence. He was later arrested by civilian authorities and imprisoned. Slipping into a woman's dress, he escaped and fled to New York City as a Tory refugee. Randall, *Benedict Arnold,* p. 570.

20. Randall, *Benedict Arnold,* p. 553; Benson John Lossing, *Our Country* (New York: Amies Publishing Co., 1888), p. 1055.

21. Randall, *Benedict Arnold,* p. 570, quoting an 1817 letter from Tallmadge to Timothy Pickering, veteran of the Revolutionary War and President Washington's Cabinet.

22. Ibid., p. 567, quoting a letter from Arnold to Washington, October 1, 1780.

23. Ibid., p. 570.

CHAPTER 17: BLOODY DAYS OF RECKONING

1. *Continental Journal* (Boston), July 13, 1780, as cited by Philip Davidson, *Propaganda and the American Revolution 1763–1783* (Chapel Hill: University of North Carolina Press, 1941), p. 369.

2. Thomas Fleming, *The Battle of Springfield* (Trenton: New Jersey Historical Commission, 1975), p. 6.

3. Sheila L. Skemp, *William Franklin* (New York: Oxford University Press, 1990), pp. 4–11.

4. Ibid., pp. 192, 219; "William Franklin Papers," American Philosophical Society, http://www.amphilsoc.org/library/mole/f/franklin/franklinw.xml; accessed 11/21/2008.

5. Skemp, *Wiliam Franklin,* p. 171, quoting from the proceedings of the New Jersey General Assembly.

6. Ibid., pp. 38–39.

7. Srodes, *Franklin*, p. 259.

8. "The Loyalist Opposition," New Jersey State Library Digital Collection, http://www.njstatelib.org/NJ_Information/Digital_Collections/NJInTheAmerican Revolution1763-1783/8.7.pdf; accessed 3/4/2009.

9. His name does not appear on prison lists in Richard H. Phelps, *Newgate of Connecticut* (1876; reprint, Camden, ME: Picton Press, 1996). Nor does his

name appear in any prison record found by present-day researchers at the prison site, a National Historic Landmark and State Archaeological Preserve.

10. Edward W. Cooch, *The Battle of Cooch's Bridge Delaware September 3, 1777* (Wilmington, DE: William N. Cann, 1940).

11. Skemp, *William Franklin,* p. 234; Rivington's *Royal Gazette*, May 25, 1779.

12. Wertenbaker, *Father Knickerbocker Rebels,* p. 153.

13. Francis Bazley Lee, ed., *New Jersey as a Colony and as a State: One of the Original Thirteen* (New York: Publishing Society of New Jersey, 1903).

14. Clinton Proclamation, June 3, 1780.

15. Thomas J. Farnham, *Fairfield* (West Kennebunk, ME: Phoenix Publishing [for Fairfield Historical Society], 1988), p. 88.

16. Loyalist Institute: "A History of the King's American Regiment," part 2 of 8, http://www.royalprovincial.com/military/rhist/kar/kar2hist.htm; accessed 3/4/2009.

17. Judith Ann Schiff, "Naphtali Daggett: Pastor, Yale President, Sniper," *Yale Alumni Magazine*, July/August 2006. http://www.yalealumnimagazine.com/issues/2006_07/old_yale.html.

18. James Shepard, "The Tories of Connecticut," *Connecticut Quarterly* 1, no. 3 (1895); Stephen Eric Davidson, *The Burdens of Loyalty: Refugee Tales from the First American Civil War* (Saint John, NB: Trinity Enterprise, 2007); electronic book. The Daggett incident is described in R. D. French, *The Memorial Quadrangle: A Book About Yale* (New Haven: Yale University Press, 1929), pp. 36–37, as cited in "Resources on Yale's History," http://www.library.yale.edu/mssa/YHO/Daggett_bio.html; accessed 11/21/2009.

19. Schiff, "Naphtali Daggett: Pastor, Yale President, Sniper."

20. Shepard, "The Tories of Connecticut," *Connecticut Quarterly*, vol. 1, no. 3 (1895).

21. Rita Papazian, *Remembering Fairfield Connecticut* (Charleston, SC: History Press, 2007), pp. 13–14; Farnham, *Fairfield,* pp. 88–89.

22. Farnham, *Fairfield,* p. 92.

23. "Testimony of Eunice Burr, wife of Thaddeus Burr," a deposition taken three weeks after the raid, Royal R. Hinman's Historical Collection, Fairfield Museum and History Center. Her account, like all the others, describes the raiders as British and Hessians, making no mention of the King's American Regiment.

24. Ibid.

25. Ibid.

26. Personal correspondence from Raymond Longden of Fairfield, 3/24/2010.

27. "The Burning of Fairfield during the American Revolution," Fairfield Museum and History Center exhibit (2009); Farnham, *Fairfield,* pp. 92–93.

28. Compilation of Fairfield Museum and History Center.

29. Samuel Richards Weed, *Norwalk After Two Hundred and Fifty Years* (South Norwalk, CT: C. A. Freeman, 1902), pp. 289, 297.

30. Nelson, *William Tryon and the Course of Empire*, pp. 170–172.

31. Ibid., p. 168.

32. Ibid., p. 171; "The Battle of Stoney Point," http://www.myrevolutionarywar .com/battles/790716.htm; accessed 11/21/2009.

33. Nelson, *William Tryon and the Course of Empire,* p. 169.

34. Ibid., p. 172, quoting a letter from Tryon to Clinton, July 20, 1779.

35. Ibid.; Skemp, *William Franklin,* p. 235, quoting from William Smith's *Historical Memoirs, 1778–83.*

36. Ruth M. Keesey, "Loyalism in Bergen County, New Jersey," *William and Mary Quarterly,* 3rd ser., 18, no. 4 (October 1961), p. 559.

37. Dennis P. Ryan, *New Jersey's Loyalists* (Trenton, NJ: New Jersey Historical Commission, 1975), p. 6.

38. Nelson, *William Tryon and the Course of Empire,* p. 168. Between the creation of the six-battalion Volunteers and its disbanding in 1783, a total of 2,450 men actively served.

39. Wertenbaker, *Father Knickerebocker Rebels,* p. 229.

40. Jones, *History of New York During the Revolutionary War*, vol. 2, p. 482.

41. Ibid.

42. Jeffery P. Lucas, "Cooling by Degrees: Reintegraton of Loyalists in North Carolina, 1776–1790" (M.A. thesis, North Carolina State University, 2007), p. 12; Robert DeMond, *Loyalists in North Carolina During the Revolution* (Durham, NC: Duke University Press, 1940), p. 49.

43. Sabine, *American Loyalists*, p. 432.

44. Skemp, *William Franlin,* p. 239; "Loyal Associated Refugees Letters," On-Line Institute for Advanced Loyalist Studies, http://www.royalprovincial.com/ military/rhist/loyaref/lareflet2.htm; accessed 3/13/2009.

45. Johann Conrad Dohla, *A Hessian Diary of the American Revolution,* translated by Bruce E. Burgoyne (Norman: University of Oklahoma Press, 1993), p. 121.

46. Leiby, *The Revolutionary War in the Hackensack Valley,* pp. 239–244.

47. Martin, *A Narrative of a Revolutionary Soldier* (New York: Signet Classics, 2001), pp. 151–154.

48. *New Jersey Gazette*, February 3, 1779.

49. Harry M. Ward, *The War for Independence and the Transformation of American Society* (Florence, KY: Routledge, 2003), p. 73.

50. Barbara J. Mitnick, ed., *New Jersey in the American Revolution* (Piscataway, NJ: Rutgers University Press, 2005), pp. 57–58.

51. Fleming, *The Battle of Springfield* (Trenton: New Jersey Historical Commission, 1975), p. 6.

52. Martin, *A Narrative of a Revolutionary Soldier*, p. 148.

53. Fleming, *The Battle of Springfield*, p. 9.

54. Fitzpatrick, *The Writings of George Washington from the Original Manuscript Sources*, vol. 18, pp. 106, 107. Communications with Major General Robert Howe and Johann von Robaii, Baron de Kalb, March to June 1780.

55. Fleming, *The Battle of Springfield*, p. 12.

56. *New-Jersey Journal,* June 14, 1780.

57. Marian Meisner, *A History of Millburn Township,* electronic book jointly published by the Millburn/Short Hills Historical Society and the Millburn Free Public Library, 2002; chap. VIII. http://www.millburn.lib.nj.us/ebook/eBook.pdf; accessed 3/15/2009.

58. Ibid.

59. One of the executed men was the father-in-law of Barbara Frietschie, heroine of John Greenleaf Whittier's Civil War poem, "Barbara Frietchie." "Early History of the Frederick County Jail," http://www.frederickcountymd.gov/documents/Sheriff%27s%20Office/Adult%20Detention%20Center/Early%20History.pdf; accessed 3/30/2010. Also, Bernard C. Steiner, "Western Maryland in the Revolution," *Historical and Political Science* (Baltimore: Johns Hopkins Press, 1902), pp. 54–56.

60. "The Conviction and Penalty of Seagoe Potter for Treason in Delaware 1780. *Delaware Archives* (Wilmington, 1919), vol. 3, pp. 1302–1304, as published in Borden and Borden, *The American Tory*, pp. 89–90. Little is known about Potter beyond his grisly sentence, recorded in Delaware's extensive treason archives. As frequently happens, there is no record that the entire sentence was carried out. Drawing and quartering was customary in Britain for high treason and became part of the common law accepted in American colonies. But judges and legislatures considered simple hanging an acceptable form of execution.

61. His age and height come from his master's newspapers advertisement offering £3 for his capture. See http://www.pbs.org/wgbh/aia/part2/2h1b.html; accessed 3/10/2009.

62. Schama, *Rough Crossings: Britain, the Slaves and the American Revolution* (London: BBC Books, 2006), pp. 135, 138–140.

63. Robert Dallison, "A Case of Justifiable Homicide," *The Officers' Quarters* 23, no. 1 (2005), pp. 5–8.

64. "Revolution Day by Day," National Park Service, http://www.nps.gov/revwar/revolution_day_by_day/1780_main.html; accessed 3/17/2009.

65. Gary D. Saretzky, "The Joshua Huddy Era," exhibition catalog, Monmouth County Library, Manalapan, NJ, 2004. (Facsimiles of documents from the David Library of the American Revolution, Library of Congress, Monmouth County Archives, Monmouth County Historical Association, New Jersey His-

torical Society, New Jersey State Archives, Alexander Library at Rutgers University, and Salem County Historical Society.)

66. Ibid.

67. Gary D. Saretzky, "Documents of the American Revolution," exhibition at Monmouth County Library, Manalapan, New Jersey, October 2002. (Facsmiles of Revolutionary War Era documents from the Monmouth County Archives, New Jersey Historical Society, New Jersey State Archives, and Special Collections of Rutgers University at Alexander Library.) http://www.co.monmouth .nj.us/page.aspx?Id=1681; accessed 3/17/2009.

68. McCormick, *New Jersey from Colony to State* (Princeton, NJ: VanNostrand, 1964), p. 152.

69. Information about Huddy, his hanging, and the events that followed come from William Scudder Stryker, *The Capture of the Block House at Toms River* (Trenton, NJ: Naar, Day & Naar, 1888), pp. 4–14, 20–22; and "Proceedings of a General Court Martial held at the City Hall in New York in the Province of New York, from Friday the 3d May to Saturday the 22d June 1782 for the Tryal of Captain Richard Lippincott." http://personal.nbnet.nb.ca/halew/Lippincott .html; accessed 3/29/2010. Also, the Saretzky sources above.

70. Stryker, *The Capture of the Block House at Toms River,* pp. 23–27.

71. Ibid., p. 25, from the memoir of one of the officers.

72. Proceedings of a General Court Martial held at the City Hall in New York in the Province of New York, from Friday the 3d May to Saturday the 22d June 1782 for the Tryal of Captain Richard Lippincott. http://personal.nbnet.nb.ca/ halew/Lippincott.html; accessed 3/29/2010.

73. Letter to Major General Benjamin Lincoln, June 5, 1784, Massachusetts Historical Society. http://www.masshist.org/objects/2007october.cfm; accessed 3/19/2009.

74. Ibid.

75. Randall, *Benedict Arnold,* pp. 577–578.

76. "Battle of Groton Heights," http://www.fortgriswold.org/id5.html, quoting from "Brigadier General Benedict Arnold To General Henry Clinton," September 8, 1781, in *Diary of Frederick Mackenzie* (Cambridge: Harvard University Press, 1930), vol. 2, pp. 623–627).

77. Randall, *Benedict Arnold,* p. 589; "Stephen Hempsted's Account," http://www .fortgriswold.org/id5.html; accessed 3/20/2009.

78. Van Tyne, *The Loyalists in the American Revolution,* p. 269. Thirteen on the list were "tried and acquitted"; seventy-one "surrendered and were discharged." The fate of the rest is not known, but the reasonable assumption is that they exiled themselves.

79. Ibid., p. 271; Some minutes of meetings of the New York (State) Commissioners for Detecting and Defeating Conspiracies were published in 1924 by the

New York Historical Society. See http://www.archive.org/details/minutesof committt571newy; accessed 3/20/2009.

80. "Treason Files," Delaware State Archives, pp. 1281–1312. The man hanged was Cheney Clow, leader of a Loyalist force declared to be in rebellion against the state of Delaware.

81. Willett, *A Narrative of the Military Actions of Colonel Marinus Willett* (New York: G. & C. & H Carvill, 1831), p. 89.

CHAPTER 18: AND THEY BEGAN THE WORLD ANEW

1. "Songs of Rebels and Redcoats," American Adventure Recording, National Geographic Society, 1976; courtesy of John M. Lavery, director. By tradition, "The World Turned Upside Down" was played by the British military band at the surrender of General Cornwallis at Yorktown. The words and mournful music fit the occasion. But there is no direct evidence that this was the tune played. The articles of capitulation specified that the surrendering soldiers, marching out of the redoubt as prisoners of war with colors cased and arms grounded, had to play a German or British march. This was Washington's retribution for a similar article in the surrender agreement imposed on Maj. Gen. Benjamin Lincoln when he surrendered at Charleston in 1780. At Yorktown, Lincoln was given the honor of accepting Cornwallis's sword.

2. Articles of Capitulation, "Done in the trenches before York Town in Virginia October 19, 1781." Jerome A. Greene, *The Guns of Independence* (New York: Savas Beatie, 2005), pp. 351–355.

3. Greene, *The Guns of Independence*, pp. 352–355, "The Articles of Capitulation."

4. Thomas H. Raddall, "Tarleton's Legion," Collections of the Nova Scotia Historical Society, electonic version created by the Mersey Heritage Society in 2001 with the permission of Dalhousie University. http://www.mersey.ca/tarletonslegion.html; accessed 3/29/2010.

5. Josiah Atkins, *The Diary of Josiah Atkins,* edited by Steven E. Kagle (New York: Arno, 1975), p. 32. As quoted in Elizabeth A. Fenn, *Pox Americana* (New York: Hill and Wang, 2001), p. 129.

6. Martin, *A Narrative of a Revolutionary Soldier* (New York: Signet Classic, 2001), p. 207.

7. Fenn, *Pox Americana* says that the footnote containing the Leslie letter "has been excised from most extant copies" of a 1777 book written by a British officer suggesting the use of smallpox as a weapon (p. 314).

8. Carleton to Lieutenant General Alexander Leslie, July 15, 1782, cited in Crary, *The Price of Loyalty,* p. 357.

9. Risher, "Propaganda, Dissension, and Defeat," pp. 243–244.

10. "The British Evacuate Savannah Georgia," *Sons of the American Revolution Magazine,* vol. 101, no. 4 (Spring 2007). http://www.revolutionarywararchives.org/savannah.html; accessed 3/20/2009.

11. Risher, "Propaganda, Dissension, and Defeat," p. 252.

12. Van Tyne, *The Loyalists in the American Revolution,* p. 289.

13. Jones, *History of New York During the Revolutionary War,* vol. 2, pp. 235–236, as cited by Crary, *The Price of Loyalty,* p. 358.

14. Carole Watterson Troxler, "The Migration of Carolina and Georgia Loyalists to Nova Scotia and New Brunswick" (Ph.D. diss., University of North Carolina, 1974), p. 53. Under the treaty ending the French and Indian War, Britain had traded British-occupied Havana to Spain in exchange for Florida. The British divided the colony into East Florida and West Florida, with St. Augustine and Pensacola the respective capitals.

15. Ibid., table, p. 46.

16. Crary, *The Price of Loyalty,* p. 360.

17. *London Chronicle,* July 5–7, 1783, quoted in Wertenbaker, *Father Knickerbocker Rebels,* p. 256.

18. Norton, *The British-Americans,* p. 125.

19. Crary, "The Tory and the Spy," *The William and Mary Quarterly,* pp. 61–72.

20. Wertenbaker, *Father Knickerbocker Rebels,* p. 254.

21. Sabine, p. 265.

22. Christopher Moore, *The Loyalists* (Toronto: McClelland & Stewart, 1984), p. 150.

23. Details of the *Union*'s voyage come from Stephen Eric Davidson, *The Burdens of Loyalty,* an electronic book produced by Trinity Enterprise. http://www.loonielink.com/component/option.com_virtuemart/page,shop.product_details/flypage_ebo/ok/category_id,59/product_id, 261/Itemid,185; accessed 3/29/2010. He is also the author of "A Manifest Destiny" in *Canada's History* (formerly *The Beaver*), August–September 2008. The article also is a source for information about the *Union.*

24. Davidson, "A Manifest's Destiny," p. 39.

25. This and other Digby quotations come from his 1783 Naval Order Books, which a descendant of the admiral, Lady Digby, presented to the Digby Museum in the admiral's namesake Nova Scotia town in 2006. The museum allowed the author to view the Order Books.

26. "Boston King," *Dictionary of Canadian Biography Online,* http://www.biographi.ca/009004-119.01-e.php?&id_nbr=2489; accessed 7/5/2009. *Memoirs of the Life of Boston King, a Black Preacher,* published in *The Methodist Magazine* in four installments in 1798. From an electronic edition prepared for the Antislavery

Literature Project, Arizona State University, http://antislavery.eserver.org/narratives/boston_kingproof.pdf. Accessed 3/23/2010.

27. Ellen Gibson Wilson, *The Loyal Blacks* (New York: G. P. Putnam's Sons/Capricorn Books, 1976), p. 51.

28. Nova Scotia Archives and Records Management, published in "The Black Loyalists of Nova Scotia," Nova Scotia Museum publication. http://museum.gov.ns.ca/blackloyalists/who.htm; accessed 3/28/2010.

29. Wilson, *The Loyal Blacks,* p. 53, based on diary of Judge William Smith, who attended as an aide to Carleton.

30. Personal communication, Christa Dierksheide, 3/4/2009, Thomas Jefferson's Monticello.

31. Wilson, *The Loyal Blacks,* p. 52.

32. Ibid., pp. 53–54.

33. Fitzpatrick, *The Writings of George Washington from the Original Manuscript Sources,* vol. 26, p. 403. Substance of a Conference between General Washington and Sir Guy Carleton, May 6, 1783; ibid., letter to Carleton, p. 408.

34. "Black Loyalists: Our History, Our People," http://www.blackloyalist.com/canadiandigitalcollection/people/secular/blucke.htm; accessed 3/24/2009. Also, guide at Birchtown Museum, Birchtown, NS, 8/18/2006.

35. Paul A. Gilje and William Pencak, *New York in the Age of the Constitution* (Rutherford, NJ: Associated University Presses, 1992), pp. 20, 40.

Bibliography

Abbot, W. W. *The Royal Governors of Georgia 1754–1775*. Chapel Hill: University of North Carolina Press, 1959.

Adair, Douglass, and John A. Schutz. *Peter Oliver's Origin & Progress of the American Revolution*. Stanford, CA: Stanford University Press, 1961.

Adams, Charles Francis. *Familiar Letters of John Adams and His Wife Abigail during the Revolution*. New York: Hurd and Houghton, 1876.

———. *The Works of John Adams, Second President of the United States*. Vols. 2 and 10. Boston: Little, Brown, 1856.

Adams, William Howard. *Gouverneur Morris: An Independent Life*. New Haven: Yale University Press, 2003.

Ahlin, John Howard. *James Lyon, Patriot, Preacher, Psalmodist*. Machias, ME: Centre Street Congregational Church, 2005.

Albury, Paul. *The Story of the Bahamas*. London: Macmillan/Caribbean Education Ltd, 1975.

Alden, John R. *A History of the American Revolution*. New York: Alfred A. Knopf, 1969. Reprint, New York: DaCapo Press, 1989.

Allen, Ira. *The Natural and Political History of the State of Vermont*. Montpelier, VT: Library of Congress, 1870.

Allen, Jolley. *An Account of Part of the Sufferings and Losses of Jolley Allen, a Native of London*. Boston: Franklin Press/Rand, Aver & Co., 1883.

Allen, Robert S. *His Majesty's Indian Allies: British Indian Policy in the Defence of Canada, 1774–1815*. Toronto: Dundurn Press, 1992.

———, gen. ed. *The Loyal Americans: The Military Role of the Loyalist Provincial Corps and Their Settlement in British North America, 1775–1784* (publication for a traveling exhibit). Ottawa: National Museums of Canada, 1983.

Allen, Thomas B. *George Washington, Spymaster.* Washington, DC: National Geographic Society, 2004.

———. *Remember Valley Forge.* Washington, DC: National Geographic Society, 2007.

Anburey, Thomas. *Travels Through the Interior Parts of America, 1776–1781.* Boston: Houghton Mifflin, 1923.

Andrews, Evangeline Walker, ed. *Journal of a Lady of Quality.* New Haven: Yale University Press, 1921.

———, and Charles McLean Andrews, eds. *Journal of a Lady of Quality.* New Haven: Yale University Press, 1922.

Aptheker, Herbert. *The American Revolution, 1763–1783.* New York: International Publisher, 1960.

Archibald, Mary. *Gideon White, Loyalist.* Halifax: Petheric Press, 1975.

Atkins, Josiah. *The Diary of Josiah Atkins.* Edited by Steven E. Kagle. New York: Arno, 1975.

Augur, Helen. *The Secret War of Independence.* Boston: Little, Brown, 1955.

Babits, Lawrence E. *A Devil of a Whipping: The Battle of Cowpens.* Chapel Hill: University of North Carolina Press, 1998.

Bailey, Thomas A., and David M. Kennedy, eds. *The American Spirit.* Vol. 1. Lexington, MA: D. C. Heath and Co., 1994.

Bailyn, Bernard. *The Ideological Origins of the American Revolution.* Cambridge, MA: Belknap & Harvard, 1992.

———. *The Ordeal of Thomas Hutchinson.* Cambridge, MA: University Press, 1974.

Bakeless, John. *Turncoats, Traitors and Heroes.* Philadelphia: Lippincott, 1959.

Baker, William Spohn. *Itinerary of General Washington June 15, 1775 to December 23, 1783.* Philadelphia: J. B. Lippincott, 1892.

Balderston, Marion, and David Syrett, eds. *The Lost War: Letters from British Officers During the American Revolution.* New York: Horizon Press, 1975.

Bancroft, George. *Bancroft's History of the United States.* Vol. 7. London: Charles Bowen, 1834.

Banwell, Selwyn. *The Loyalist.* Toronto: Rous & Mann, 1934.

Barck, Oscar Theodore. *New York City during the War for Independence with Special Reference to the Period of British Occupation.* New York: Columbia University Press, 1931.

Barnett, Cleadie, and Elizabeth Sewell. *Loyalist Families.* Fredericton, NB: Federation Branch of the United Empire Loyalists' Association of Canada, 1983.

Bartlett, John Russell, ed., *Records of the Colony of Rhode Island and Providence Plantations, in New England.* Vol. 9. Providence: Alfred Anthony, Printer to the State, 1864.

Bass, Streeter. "Nathan Hale's Mission" (declassified 1994). Washington, DC: Central Intelligence Agency, Center for the Study of Intelligence.

Bassett, James H. *Colonial Life in New Hampshire.* Boston: Ginn & Company, 1899.

Batchelder, Samuel Francis. *The Life and Surprising Adventures of John Nutting, Cambridge Loyalist.* Cambridge, MA: Cambridge Historical Society, 1912.

Bates, Walter. *Kingston and the Loyalists of the "Spring Fleet" of 1783.* 1889. Reprint, Fredericton, NB: Nonentity Press, 1980.

Bauman, Richard. *For a Reputation of Truth: Politics, Religion and Conflict among the Pennsylvania Quakers, 1750–1800.* Baltimore: Johns Hopkins Press, 1971.

Bedini, Silvio A. *Ridgefield in Review.* New Haven, CT: Higginson Book Co., 1994.

Benton, William Allen. *Whig-Loyalism: An Aspect of Political Ideology in the American Revolutionary Era.* Rutherford, NJ: Fairleigh Dickinson University Press, 1969.

Berger, Carl. *Broadsides and Bayonets: The Propaganda War of the American Revolution.* Philadelphia: University of Pennsylvania Press, 1961.

Berkin, Carol, and Jonathan Sewall. *Odyssey of an American Loyalist.* New York: Columbia University Press, 1974.

Berling, Ira, and Ronald Hoffman, eds. *Slavery and Freedom in the Age of the American Revolution.* Charlottesville: University of Virginia Press, 1983.

Bernhardt, Albert Faust. *The German Element in the United States, with Special Reference to its Political, Moral, Social, and Educational Influence.* Boston: Houghton Mifflin, 1909.

Berry, A. J. *A Time of Terror.* Victoria, BC: Trafford Publishing, 2005.

Bicheno, Hugh. *Rebels & Redcoats: The American Revolutionary War.* London: HarperCollins, 2003.

Bielinski, Stefan. *An American Loyalist: The Ordeal of Frederick Philipse III.* Albany: New York State Museum, 1976.

Blakeley, Phyllis R., and John N. Grant, eds. *Eleven Exiles: Accounts of Loyalists in the American Revolution.* Toronto: Dundurn Press Ltd., 1982.

Blumenthal, Walter H. *Women Camp Followers of the American Revolution.* Philadelphia: G. S. MacManus, 1974.

Bodle, Wayne. *Valley Forge Winter.* University Park: Pennsylvania State University Press, 2002.

Bolton, Charles Knowles, ed. *Letters of Hugh, Earl Percy, from Boston and New York, 1774–1776.* Boston: Charles E. Goodspeed, 1902.

Borden, Morton, and Penn Borden, eds. *The American Tory.* Englewood Cliffs, NJ: Prentice-Hall, 1972.

Boss, William. *The Stormont, Dundas and Glengarry Highlanders 1783–1951.* Ottawa: Runge Press, 1952.

Boucher, Jonathan. *Reminiscences of an American Loyalist.* New York: Houghton Mifflin, 1925.

Boudinot, Elias. *Journal of Events in the Revolution.* New York: Arno Press, 1968. Reprint of 1894 edition.

Boyd, Julian P. *Anglo-American Union: Joseph Galloway's Plans to Preserve the British Empire*. Philadelphia: University of Pennsylvania Press, 1941. Reprint, London: Octagon Press, 1970.

Bradley, Arthur G. *The United Empire Loyalists: Founders of British Canada*. London: T. Butterworth Limited, 1932.

Bradley, Patricia, *Slavery, Propaganda, and the American Revolution*. Jackson: University Press of Mississippi, 1998.

Brandt, Clare. *An American Aristocracy: The Livingstons*. Garden City, NY: Doubleday & Co., 1986.

———. *The Man in the Mirror*. New York: Random House, 1994.

Brebner, John. *The Neutral Yankees of Nova Scotia: A Marginal Colony During the Revolutionary Years*. New York: Russell & Russell, 1937.

Brock, William R. *Scotus Americanus: A Survey of the Sources for Links between Scotland and America in the Eighteenth Century*. Edinburgh: Edinburgh University Press, 1982.

Brooke, John. *King George III*. London: Constable, 1972.

Brown, Lloyd A. *Loyalist Operations at New Haven*. Ann Arbor: William L. Clements Library, 1938.

Brown, Richard M. *The South Carolina Regulators*. Cambridge: Harvard University Press, 1963.

Brown, Wallace. *The Good Americans: The Loyalists in the American Revolution*. New York: William Morrow, 1969.

———. *The King's Friends: The Composition and Motives of the American Loyalist Claimants*. Providence: Brown University Press, 1965.

Buchanan, John. *The Road to Guilford Courthouse*. New York: John Wiley & Sons, 1997.

———. *The Road to Valley Forge*. Hoboken, NJ: John Wiley & Sons, 2004.

Bumstead, J. M. *Understanding the Loyalists*. Sackville, NB: Centre for Canadian Studies, Mount Allison University, 1986.

Bunnell, Paul J. *Thunder Over New England: Benjamin Bonnell, The Loyalist*. Hanover, MA: Willow Bend Books, 1988.

Burke, Edmund. *Account of the European Settlements in America*. Vol. 2. London: John Joseph Stockdalk, 1808.

Burnett, Edmund Cody, ed. *Letters of Members of the Continental Congress*. Vol. 1. Washington, DC: Carnegie Institution, 1921.

Burrows, Edwin G., and Mike Wallace. *Gotham: A History of New York City to 1898*. New York: Oxford University Press, 1999.

Burt, A. L. *The Old Province of Quebec*. Toronto: McGill-Queen's Press, 1933.

Calhoon, Robert McCluer. *Religion and the American Revolution in North Carolina*. Raleigh: North Carolina Department of Cultural Resources, 1976.

————. *The Loyalist Perception and Other Essays*. Columbia: University of South Carolina Press, 1989.

————. *The Loyalists in Revolutionary America, 1760–1781*. New York: Harcourt Brace Jovanovich, 1973.

Callahan, North. *Flight from the Republic: The Tories of the American Revolution*. Indianapolis: Bobbs-Merrill, 1967.

————. *Royal Raiders: The Tories of the American Revolution*. Indianapolis: Bobbs-Merrill, 1963.

Calloway, Colin G. *The American Revolution in Indian Country*. New York: Cambridge University Press, 1995.

Cambridge Historical Commission. *Cambridge During the Revolution*. Cambridge, MA: Cambridge Discovery, 1975.

Cameron, Kenneth W. *The Church of England in Pre-Revolutionary Connecticut: New Documents and Letters Concerning the Loyalist Clergy and the Plight of Their Surviving Church*. Hartford: Transcendental Books, 1976.

————, ed. *The papers of loyalist Samuel Peters: a survey of the contents of his notebooks—correspondence during his flight to England, exile, and the last years of his life*. Hartford: Transcendental Books, 1978.

Campbell, William W. *Annals of Tryon County or, the Border Warfare of New York, during the Revolution*. New York: J. & J. Harper, 1831.

Carmer, Carl. *The Susquehanna*. New York: Rinehart, 1955.

Carr, Jacqueline Barbara. *After the Siege*. Boston: Northeastern University Press, 2005.

Carter, Clarence E. *The Correspondence of General Thomas Gage with the Secretary of State, 1763–1775*. 2 vols. New Haven: Yale University Press, 1931–33.

Cashin, Edward J. *The King's Ranger: Thomas Brown and the American Revolution on the Southern Frontier*. Athens: University of Georgia Press, 1989.

Cayton, Mary, Elliot Gorn, and Peter Williams. *Encyclopedia of American Social History*. New York: Scribner, 1993.

Chadwick, Edward M. *Ontarian Families: Genealogies of United Empire Loyalist and Other Pioneer Families of Upper Canada*. Lambertville, NJ: Hunterdon House, 1983.

Charles, Patrick. *Washington's Decision*. Charleston, SC: Booksurge, 2005.

Chartrand, René. *American Loyalist Troops 1775–84*. Westminster, MD: Osprey Publishing, 2008.

Chesnutt, David R., ed. *Papers of Henry Laurens*. Vol. 11. Columbia: University of South Carolina Press, 1968.

Chidsey, Donald Barr. *The Loyalists: The Story of Those Americans Who Fought Against Independence*. New York: Crown Publishers, 1973.

Clark, Dora Mac. *British Opinion and the American Revolution*. New York: Oxford University Press, 1966.

Clark, J. C. D. *The Language of Liberty, 1660–1832: Political Discourse and Social Dynamics in the Anglo-American World*. Cambridge, UK: Cambridge University Press, 1994.

Clark, William Bell, ed. *Naval Documents of the American Revolution*. Vol. 2 Washington, DC: Naval Historical Center, 1966.

Clarke, M. St. Clair, and Peter Force. *American Archives: Containing a Documentary History of the United States of America: under the authority of Acts of Congress; Washington, D.C.: 1848–1851*. (Available at http://dig.lib.niu.edu/amarch/)

Clinton, Henry. *The American Rebellion: Sir Henry Clinton's Narrative of His Campaign, 1775–1782*. Edited by William B. Willcox. New Haven: Yale University Press, 1954.

Coburn, Frank Warren. *The Battle of April 19, 1775*. Lexington, MA: Published by author, 1912.

Coffin, Victor. *The Province of Quebec and the Early American Revolution*. Madison: University of Wisconsin Press, 1896.

Coggins, Jack. *Ships and Seamen of the American Revolution*. Mineola, NY: Courier Dover Publications, 2002.

Cohen, Sheldon S. *Connecticut's Loyalist Gadfly: The Reverend Samuel Andrew Peters*. Hartford: American Revolution Bicentennial Commission of Connecticut, 1977.

Coldham, Peter Wilson. *American Loyalist Claims Abstracted from the Public Records Office*. Washington, DC: National Genealogical Society, 1980.

———. *American Migrations 1765–1799: The lives, times and families of colonial Americans who remained loyal to the British Crown before, during and after the Revolutionary War, as related in their own words and through their correspondence*. Baltimore: Genealogical Publishing Co., Inc., 2000.

———. *Emigrants in Chains*. Baltimore: Genealogical Publishing Co., 1994.

Coleman, Kenneth. *The American Revolution in Georgia*. Athens: University of Georgia Press, 1958.

Commager, Henry Steele. *Documents of American History*. New York: F. S. Crofts & Co., 1940.

———, and Richard B. Morris, eds. *The Spirit of 'Seventy-Six*. Vol 1. Indianapolis: Bobbs-Merrill, 1958.

Cooch, Edward W. *The Battle of Cooch's Bridge Delaware September 3, 1777*. Wilmington, DE: William N. Cann, 1940.

Cook, Frederick, ed. *Journals of the Military Expedition of Major General John Sullivan against the Six Nations of Indians in 1779*. Auburn, NY: Knapp, Peck & Thomson, Printers, 1887.

Copley, John Singleton, and Henry Pelham. *Letters & Papers of John Singleton Copley and Henry Pelham, 1739–1776*. Boston: Massachusetts Historical Society, 1914.

Coupland, R. *The Quebec Act: A Study in Statesmanship.* New York: Oxford University Press, 1925.

Crary, Catherine S., ed. *The Price of Loyalty: Tory Writings from the Revolutionary Era.* New York: McGraw-Hill, 1973.

Craton, Michael. *A History of the Bahamas.* London: Collins, 1962.

Cruikshank, E. A. *The King's Royal Regiment of New York: with the addition of an Index, Appendices and a Master Muster Roll.* 1931. Reprint, Toronto: Ontario Historical Society, 1984.

————, and A. F. Hunter, eds. *Correspondence of the Honourable Peter Russell.* Toronto: Ontario Historical Society, 1923–26.

————. *The Story of Butler's Rangers and the Settlement of Niagara.* Welland, ON: Tribune Printing House, 1893.

Cuneo, John R. *Robert Rogers of the Rangers.* New York: Oxford University Press, 1959.

Cunningham, Anne Rowe, ed. *Letters and Diary of John Rowe.* Boston: W. B. Clarke Co., 1903.

Curwen, Samuel. *Journals and Letters, 1775–1784.* New York: C. S. Francis and Co., 1864.

Cuthbertson, Brian C. *The Loyalist Governor: Biography of Sir John Wentworth.* Halifax, NS: Petheric Press, 1983.

Dallison, Robert L. *Hope Restored: The American Revolution and the Founding of New Brunswick.* Fredericton, NB: Goose Lane Editions, 2003.

Dana, Elizabeth Ellery, ed. *The British in Boston: Being the Diary of Lieutenant John Barker of the King's Own Regiment from November 15, 1774 to May 31, 1776.* Cambridge, MA: Harvard University Press, 1924.

Dandridge, Danske. *American Prisoners of the Revolution.* Charlottesville, VA: Michie Co., 1911.

Dann, John C., ed. *The Revolution Remembered: Eyewitness Accounts of the War for Independence.* Chicago: University of Chicago Press, 1980.

Dartmouth, William Walter Legge, William Oxenham Hewlett, Benjamin Franklin Stevens, and William Page. *The Manuscripts of the Earl Of Dartmouth.* Vol. 2. London: H.M. Stationery Office, 1895.

Davidson, Philip. *Propaganda and the American Revolution, 1763–1783.* Chapel Hill: University of North Carolina Press, 1941.

Davis, William T. *Ancient Landmarks of Plymouth.* Boston: A. Williams and Company, 1883.

————. *History of the Town of Plymouth.* Philadelphia: J. W. Lewis & Co., 1885.

Davis, William Watts Hart. *The History of Bucks County, Pennsylvania.* Doylestown, PA: Democrat Book and Job Office Printing, 1876.

————. *History of the Battle of the Crooked Billet.* Doylestown, PA: Democrat Book and Job Office Printing, 1894.

Dawson, Henry Barton. *Westchester County, New York, During the American Revolution*. Morrison, NY: Published by author, 1886.

Destler, Chester M. *Connecticut: The Provisions State*. Chester, CT: Pequot Press, 1973.

Dickerson, Oliver M. *The Navigation Acts and the American Revolution*. New York: A. S. Barnes, 1963.

Dobson, David. *Scottish Emigration to Colonial America, 1607–1785*. Athens, GA: University of Georgia Press, 2004.

Dohla, Johann Conrad. *A Hessian Diary of the American Revolution*. Translated by Bruce E. Burgoyne. Norman: University of Oklahoma Press, 1993.

Drake, Francis S. *Tea Leaves: Being a Collection of Letters and Documents relating to the Shipment of Tea to the American Colonies in the year 1773. . . .* Boston: A. O. Crane, 1884.

Drayton, John. *Memoirs of the American Revolution: A concise description of the rise and progress of the American Revolution in South Carolina and neighboring states*. Charleston: Ayer Company Publisher, 1821. Reprint, Bedford, MA: Applewood Books, 2009.

Drisko, George Washington. *Narrative of the Town of Machias*. Machias, ME: Press of the Republican, 1904.

Dubeau, Sharon. *New Brunswick Loyalists: A Bicentennial Tribute*. Agincourt, ON: Generation Press, 1983.

East, Robert A. *Business Enterprise in the American Revolution*. 1938. Reprint, New York: Peter Smith, 1964.

———. *Connecticut's Loyalists*. Chester, CT: Pequot Press, 1974.

———, and Jacob Judd, eds. *The Loyalist Americans: A Focus on Greater New York*. Tarrytown, NY: Sleepy Hollow Restorations, 1975.

Eddis, William, and Aubrey C. Land. *Letters from America*. Cambridge, MA: Harvard University Press, 1969.

Edelberg, Cynthia Dubin. *Jonathan Odell, Loyalist Poet of the American Revolution*. Durham, NC: Duke University Press, 1987.

Edgar, Walter. *Partisans & Redcoats: The Southern Conflict That Turned the Tide of the American Revolution*. New York: HarperCollins, 2001.

Egerton, Hugh E. *The Causes and Character of the American Revolution*. Oxford, UK: Oxford University Press (American branch), 1923.

———. *The Royal Commission on the Losses and Services of American Loyalists, 1783–1785*. 1915. Reprint, New York: Arno Press, 1969.

Egnal, Marc. *A Mighty Empire: The Origins of the American Revolution*. Ithaca, NY: Cornell University Press, 1988.

Einstein, Lewis. *Divided Loyalties: Americans in England During the War of Independence*. Manchester, NH: Ayer Co., 1969. Reprint, Boston: Houghton Mifflin, 1933.

Ekirch, A. Roger. *Bound for America*. Oxford, UK: Clarendon Press, 1990.

Eller, Ernest McNeill. *Chesapeake Bay in the American Revolution*. Centreville, MD: Tidewater Publications, 1981.

———. *After the Revolution: Profiles of Early American Culture*. New York: W. W. Norton, 1979.

———. *American Creation: Triumphs and Tragedies at the Founding of the Republic*. New York: Alfred A. Knopf, 2007.

———. *His Excellency: George Washington*. New York: Alfred A. Knopf, 2004.

Endicott, Charles M. *Account of Leslie's Retreat at the North Bridge, on Sunday, Feb'ry 26, 1775*. Salem, MA: William Ives and George W. Pease, Printers, 1856.

Evans, G. N. D. *Allegiance in America: The Case of the Loyalists*. Reading, MA: Addison-Wesley, 1969.

Faltsits, Victor Hugo. *Minutes of the Commissioners for Detecting and Defeating Conspiracies in the State of New York: Albany Sessions, 1778–1781*. Albany, NY: J. B. Lyon Company, State Printers, 1909.

Fanning, David. *The Narrative of Colonel David Fanning*. Richmond, VA: Privately published, 1861.

Farnham, Thomas J. *Fairfield: The Biography of a Community, 1639–1989*. West Kennebunk, ME: Phoenix Publishing (for Fairfield [CT] Historical Society), 1988.

Fenn, Elizabeth A. *Pox Americana*. New York: Hill and Wang, 2001.

Ferling, John. *A Leap in the Dark: The Struggle to Create the American Republic*. New York: Oxford University Press, 2003.

———. *The Loyalist Mind: Joseph Galloway and the American Revolution*. University Park: Pennsylvania State University Press, 1977.

Ferris, Robert G., ed. *Signers of the Declaration: Historic Places Commemorating the Signing of the Declaration of Independence*. Washington, DC: U.S. Department of the Interior, National Park Service, 1975.

Fingerhut, Eugene R., and Joseph S. Tiedemann. *The Other New York*. New York: SUNY Press, 2005.

Fischer, David Hackett. *Washington's Crossing*. New York: Oxford University Press, 2004.

———. *Paul Revere's Ride*. New York: Oxford University Press, 2004.

Fisher, Sydney George. *The Struggle for American Independence*. New York: J. B. Lippincott, 1908.

Fiske, John. *The Beginnings of New England*. Boston: Houghton Mifflin, 1889.

———. *The Critical Period of American History, 1783–1789*. Boston: Houghton Mifflin, 1897.

Fitch, William Edwards. *Some Neglected History of North Carolina*. New York: Neale Publishing Co., 1905.

Fitzpatrick, John C., ed. *The Writings of George Washington from the Original Manuscript Sources 1745–1799*. Washington, DC: U.S. George Washington Bicentennial Commission, U.S. Government Printing Office, 1931.

Fleming, Thomas. *The Battle of Springfield*. Trenton: New Jersey Historical Commission, 1975.

———. *Now We Are Enemies: The Story of Bunker Hill*. New York: St. Martin's Press, 1960.

———. *Washington's Secret War: The Hidden History of Valley Forge*. Washington, DC: Smithsonian Books/Collins, 2005.

Flexner, James Thomas. *Washington: The Indispensable Man*. Boston: Little, Brown, 1969.

Flick, Alexander Clarence. *Loyalism in New York During the American Revolution*. New York: Columbia University Press, 1901.

Fogleman, Aaron Spencer. *Hopeful Journeys: German Immigration, Settlement, and Political Culture in Colonial America, 1717–1775*. Philadelphia: University of Pennsylvania Press, 1996.

Foner, Philip Sheldon. *Blacks in the American Revolution*. Westport, CT: Greenwood Press, 1976.

Force, Peter, and Matthew St. Clair Clarke. *American Archives 4th and 5th Series*. 9 vols. 1837–53. Reprint, Washington, DC: Johnson Reprint Corp., 1972.

Ford, Paul Leicester, ed. *The Writings of Thomas Jefferson*. Vol. 2, *1776–1781*. New York: G. P. Putnam's Sons, 1893.

Ford, Worthington Chauncey. *Writings of George Washington*. Vol. 4, *1776*. New York: G. P. Putnam's Sons, 1889.

———, et al. *Journals of the Continental Congress, 1774–1789*. Washington, DC: U.S. Government Printing Office, 1906.

Fortescue, John. *The War of Independence*. Mechanicsburg, PA: Stackpole Books, 2001.

Fraser, Alexander. *United Empire Loyalists: Enquiry Into the Losses and Services in Consequence of Their Loyalty; Evidence in the Canadian Claims*. 1905. Reprint, Baltimore: Genealogical Publishing Co., 1994.

Fredriksen, John C. *America's Military Adversaries*. Santa Barbara, CA: ABC-CLIO, 2001.

French, Allen. *General Gage's Informers: New Material Upon Lexington and Concord, Benjamin Thompson as Loyalist and the Treachery of Benjamin Church, Jr.* Ann Arbor: University of Michigan Press, 1932.

———. *The Siege of Boston*. New York: Macmillan Co., 1911.

Frey, Samuel Ludlow, ed. *The Minute Book of the Committee of Safety of Tryon County, the Old New York Frontier*. New York: Dodd, Mead and Company, 1905.

Frey, Sylvia R. *Water from the Rock: Black Resistance in a Revolutionary Age*. Princeton, NJ: Princeton University Press, 1991.

Frothingham, Richard, Jr. *History of the Siege of Boston, and of the Battles of Lexington, Concord, and Bunker Hill.* Boston: Charles C. Little and James Brown, 1851.

Fryer, Mary Beacock. *Buckskin Pimpernel: The Exploits of Justus Sherwood, Loyalist Spy.* Toronto: Dundurn Press, 1981.

———. *Allan Maclean, Jacobite General.* Toronto: Dundurn Press, 1987.

———. *King's Men: The Soldier Founders of Ontario.* Toronto: Dundurn Press, 1980.

Gabriel, Michael P. *Major General Richard Montgomery: The Making of an American Hero.* Madison, NJ: Fairleigh Dickinson University Press, 2002.

Galvin, John R. *The Minute Men.* Dulles, VA: Brassey's, 1996.

Garrett, E. Wilbur, ed. *Historical Atlas of the United States.* Washington, DC: National Geographic Society, 1988.

Gerlach, Larry R., ed. *New Jersey in the American Revolution, 1763–1783.* Trenton: New Jersey Historical Commission, 1975.

Gibbes, R. W. *Documentary History of the American Revolution: 1764–1776.* New York: D. Appleton & Co., 1855.

Gilje, Paul A., and William Pencak. *New York in the Age of the Constitution.* Rutherford, NJ: Associated University Presses, 1992.

Gipson, Lawrence Henry. *Jared Ingersoll: A Study of American Loyalism in Relation to British Colonial Government.* New Haven: Yale University Press, 1920.

Godcharles, Frederic A. *Daily Stories of Pennsylvania.* Milton, PA: Published by author, 1924.

Golway, Terry. *Washington's General: Nathanael Green and the Triumph of the American Revolution.* New York: Henry Holt and Company, 2005.

Goodhart, Sir Philip. *The Royal Americans.* Hernes Keep: Wilton 65, 2005.

Gordon, William. *History of the Rise, Progress, and Establishment, of the Independence of the United States of America.* Vol. 2. Printed for the author, 1788.

Grafton, John. *The American Revolution: A Picture Source Book, 411 Copyright-free Illustrations, Battles, Events, Documents, People of the Revolution.* New York: Dover, 1975.

Granger, Bruce I. *Political Satire in the American Revolution, 1763–1783.* Ithaca, NY: Cornell University Press, 1960.

Graves, Donald E. *Guide to Canadian Sources Related to Southern Revolutionary War National Parks.* Carleton Place, ON: Ensign Heritage Consulting (for National Park Service), 2001.

Graydon, Alexander, and John Stockton Littell, eds. *Memoirs of His Own Time: With Reminiscences of the Men and Events of the Revolution.* Philadelphia: Lindsay & Blakiston, 1846.

Greene, George E. *History of Old Vincennes and Knox County, Indiana.* Chicago: S. J. Clarke Publishing Co., 1911.

Greene, Jack P. *The Ambiguity of the American Revolution.* New York: Harper & Row, 1968.

Greene, Jack P., ed. *The Reinterpretation of the American Revolution, 1763–1789*. New York: Harper & Row, 1969.

Greene, Jerome A. *The Guns of Independence: The Siege of Yorktown, 1781*. New York: Savas Beatie LLC, 2005.

Greene, Nelson. *The Story of Old Fort Plain and the Middle Mohawk Valley*. Fort Plain, NY: O'Connor Brothers Publishers, 1915.

Greene, Robert Ewell. *Black Courage, 1775–1783: Documentation of Black Participation in the American Revolution*. Washington, DC: National Society of the Daughters of the American Revolution, 1984.

Griffith, Samuel B. *In Defense of the Public Liberty: Britain, America and the Struggle for Independence, From 1760 to the Surrender at Yorktown in 1781*. Garden City, NY: Doubleday & Co., 1976.

Gruber, Ira D. *The Howe Brothers and the American Revolution*. New York: Atheneum, 1972.

Guthorn, Peter J. *British Maps of the American Revolution*. Monmouth Beach, NJ: Philip Freneau Press, 1972.

Gwyn, Julian. *Ashore and Afloat: The British Navy and the Halifax Naval Yard Before 1820*. Ottawa: University of Ottawa Press, 2004.

———. *Frigates and Foremasts: The North American Squadron in Nova Scotia Waters, 1745–1815*. Vancouver, BC: University of British Columbia Press, 2004.

Hageman, John Frelinghuysen. *A History of Princeton and Its Institutions*. Vol. 1. Philadelphia: J. B. Lippincott & Co., 1879.

Hager, Robert E. *Mohawk River Boats and Navigation Before 1820*. Syracuse, NY: Canal Society of New York State, 1987.

Haliburton, Thomas C. *History of Nova Scotia*. Halifax: Joseph Howe, 1829. Facsimile edition, Belleville: Mika Publishing, 1973.

Hall, Leslie. *Land and Allegiance in Revolutionary Georgia*. Athens: University of Georgia Press, 2001.

Halsey, Francis Whiting. *A Tour of Four Great Rivers: The Hudson, Mohawk, Susquehanna and Delaware in 1769: Being the Journal of Richard Smith of Burlington, New Jersey*. New York: Charles Scribner's Sons, 1906.

———. *The Old New York Frontier*. New York: Charles Scribner's Sons, 1902.

Hammond, Otis Grant. *Tories of New Hampshire in the War of the Revolution*. Concord: New Hampshire Historical Society, 1917.

Hancock, Harold B. *The Delaware Loyalists*. Wilmington: Historical Society of Delaware, 1940.

———. *The Loyalists of Revolutionary Delaware*. Newark: University of Delaware Press, 1977.

Hanna, Charles Augustus. *The Scotch-Irish*. Vol. 1. New York: G. P. Putnam's Sons, 1902.

Hanson, Willis Tracy. *History of Schenectady during the American Revolution, to which is Appended a Contribution to the Individual Records of the Inhabitants of the*

Schenectady District During that Period. Salem: Privately printed [E. L. Hildreth & Co.], 1916.

Harper, J. R., *The Fraser Highlanders*. Bloomfield, ON: Museum Restoration Service, 1979.

Harrell, Isaac S. *Loyalism in Virginia*. Durham, NC: Duke University Press, 1926.

Harrington, Virginia D. *The New York Merchants on the Eve of the Revolution*. New York: Columbia University Press, 1935.

Hart, Albert Bushnell, ed. *American History Told by Contemporaries*. New York: Macmillan Co., 1901.

Harvey, Oscar Jewell. *A History of Wilkes-Barré*. Vol. 2. Wilkes-Barre, PA: Raeder Press, 1909.

Harvey, Robert. *A Few Bloody Noses: The Realities and Mythologies of the American Revolution*. Woodstock and New York: Overlook Press, 2001.

Hast, Adele. *Loyalism in Revolutionary Virginia: The Norfolk Area and the Eastern Shore*. Ann Arbor: UMI Research Press, 1982.

Haymond, Henry. *History of Harrison County, West Virginia*. Morgantown, WV: Acme Publishing, 1910.

Haywood, Marshall De Lancey. *Governor William Tryon, and His Administration in the Province of North Carolina, 1765–1771*. Raleigh, NC: E. M. Uzzell, 1903.

Hazard, Samuel. *Colonial Records of Pennsylvania*. Vols. 9–12. Harrisburg: 1838–53.

———. *Minutes of the Supreme Executive Council of Pennsylvania*. Vol. 11. Philadelphia: Theo Fenn & Co., 1852.

Helsley, Alexia Jones. *South Carolinians in the War for American Independence*. Columbia, SC: South Carolina Department of Archives and History, 2000.

Henderson, Ernest F. *Select Historical Documents of the Middle Ages: Laws of Richard I Concerning Crusaders Who Were to Go by Sea*. London: George Bell and Sons, 1896.

Her Majesty's Stationery Office. *Historical Manuscripts Commission*. London: Her Majesty's Stationery Office, 1895.

Hibbert, Christopher. *Redcoats and Rebels: The American Revolution Through British Eyes*. New York: Norton 2002.

Higginbotham, Don. *Daniel Morgan, Revolutionary Rifleman*. Chapel Hill: University of North Carolina Press, 1979.

———. *The War of American Independence: Military Attitudes, Policies, and Practice 1763–1789*. Boston: Northeastern University Press, 1971.

———. *The War of American Independence*. Boston: Northeastern University Press, 1983.

Hill, Isabel Louise. *Some Loyalists and Others*. Fredericton, NB: Louise Hill, 1977.

Hodges, Graham Russell. *The Black Loyalist Directory: African Americans in Exile after the American Revolution*. New York: Garland/New England Historic Genealogical Society, 1996.

————. *Root & Branch: African Americans in New York and East Jersey, 1613–1863*. Chapel Hill and London: University of North Carolina Press, 1999.

Hollister, Gideon Hiram. *The History of Connecticut*. Hartford: Case, Tiffany & Co., 1857.

Holmes, Theodore C. *Loyalists to Canada: The 1783 Settlement of Quakers and Others at Passamaquoddy*. Camden: Picton Press, 1992.

Holton, Woody. *Forced Founders: Indians, debtors, slaves, and the making of the American Revolution in Virginia*. Chapel Hill: University of North Carolina Press, 1999.

Hooker, Richard, ed. *The Carolina Backcountry on the Eve of the Revolution: The Journal and Other Writings of Charles Woodmason, Anglican Itinerant*. Chapel Hill: University of North Carolina Press, 1969.

Horry, P., and M. L. Weens. *The Life of General Francis Marion*. Winston-Salem, North Carolina: John F. Blair, 2000.

Hosmer, James K. *The Life of Thomas Hutchinson, Royal Governor of the Province of Massachusetts Bay*. Boston: Houghton Mifflin, 1896. Reprint, New York: Da Capo Press, 1972.

Hosmer, Jerome Carter. *The Narrative of General Gage's Spies, March, 1775*, with notes. Boston: Reprinted from Bostonian Society's Publications, 1912.

Howarth, Stephen. *To Shining Sea: A History of the United States Navy 1775–1991*. New York: Random House, 1991.

Hulton, Ann. *Letters of a Loyalist Lady: Being the Letters of Ann Hulton, Sister of Henry Hulton, Commissioner of Customs at Boston, 1767–1776*. 8 vols. Cambridge: Harvard University Press, 1927.

Hurd, D. Hamilton. *History of Plymouth County*. Philadelphia: J. W. Lewis, 1884.

Hutchison, Bruce. *The Struggle for the Border*. Toronto: Longmans, Green and Co., 1955.

Hutchinson, Peter Orlando, ed. *Diary and Letters of Thomas Hutchinson*. London: Sampson Low, Marston, Searle & Rivington, 1883.

Hyman, Harold. *To Try Men's Souls: Loyalty Tests in American History*. Berkeley, CA: University of California Press, 1959.

Jackson, John W. *With the British Army in Philadelphia 1777–1778*. San Rafael, CA, and London: Presidio Press, 1979.

Jameson, J. Franklin. *The American Revolution Considered as a Social Movement*. Princeton, NJ: Princeton University Press, 1968.

Jennings, Francis. *Empire of Fortunes: Crowns, Colonies and Tribes in the Seven Years War in America*. New York: W. W. Norton, 1988.

Jodoin, Mark. *Shadow Soldiers of the American Revolution: Loyalist Tales from New York to Canada*. Charleston, SC: History Press, 2009.

Johnson, Allen. *Readings in American Constitutional History, 1776–1876*. Boston: Houghton Mifflin Co., 1912.

Johnson, Arthur. *Myths and Facts of the American Revolution.* Toronto: William Briggs, 1908.

Johnson, Cecil. *British West Florida, 1763–1783.* New Haven: Yale University Press, 1943.

Johnson, Samuel, and James Boswell. *Journey to the Hebrides.* 1785. Reprint, edited by Ian McGowan. Edinburgh: Canongate, 1996.

Johnson, Uzal. *Uzal Johnson, Loyalist Surgeon: A Revolutionary War Diary.* Edited by Bobby Gilmer Moss. Blacksburg, SC: Scotia Hibernia Press, 2000.

Johnston, Henry Phelps. *Memoirs of the Long Island Historical Society: The Campaign of 1776 Around New York and Brooklyn.* Brooklyn, NY: Published by the Society, 1878.

Jones, Alfred E. *The Loyalists of Massachusetts: Their Memorials, Petitions and Claims.* Baltimore: Clearfield Co., 1930.

Jones, Thomas, and Edward Floyd De Lancey, eds. *History of New York During the Revolutionary War.* Vol. 1. New York: New-York Historical Society, 1879.

Kallich, Martin, and Andrew MacLeish, eds. *The American Revolution Through British Eyes.* New York: Harper & Row, 1962.

Kammen, Michael G. *Colonial New York: A History.* New York: Oxford University Press, 1996.

Kaplan, Sidney, and Emma Nogrady Kaplan. *The Black Presence in the Era of the American Revolution.* Amherst: University of Massachusetts Press, 1989.

Karels, Carol. *The Revolutionary War in Bergen County.* Charleston, SC: History Press, 2007.

Kars, Marjoleine. *Breaking Loose Together: The Regulator Rebellion in Pre-Revolutionary North Carolina.* Chapel Hill: University of North Carolina Press, 2002.

Keesler, M. Paul. *Mohawk: Discovering the Valley of the Crystals.* Utica, NY: North Country Books, 2006.

Kelby, William, ed. *Orderly Book of the Three Battalions of Loyalists Commanded by Brigadier-General Oliver De Lancey 1776–1778.* New York: New-York Historical Society, 1917.

Kemble, Stephen. *Journals of Lieut-Col. Stephen Kemble . . . and British Army Orders . . .* Boston: Gredd Press, 1972. Originally published by the New-York Historical Society, 1884.

Ketchum, Richard M. *Divided Loyalties: How the American Revolution Came to New York.* New York: Henry Holt and Company, 2002.

———. *Victory at Yorktown: The Campaign That Won the Revolution.* New York: Henry Holt and Company, 2004.

———. *The Winter Soldiers.* New York: Anchor Books, 1973.

Ketchum, William. *An Authentic and Comprehensive History of Buffalo.* Vol. 1. Buffalo, NY: Rockwell, Baker & Hill, Printers, 1864.

Kingsford, William. *The History of Canada*. Vol. 6. Toronto: Roswell & Hutchinson, 1893.

Knollenberg, Bernhard. *Origin of the American Revolution: 1759–1766*. New York: Macmillan Company, 1960.

Kwasny, Mark V. *Washington's Partisan War, 1775–1783*. Kent, OH: Kent State University Press, 1996.

Labaree, Benjamin W. *The Boston Tea Party*. New York: Oxford University Press, 1964.

Labaree, Leonard Woods, ed. *The Public Papers of the State of Connecticut for the Years 1783 and 1784, with the Journals of the Council of Safety from January 9, 1783 to November 15, 1783*. Hartford: State of Connecticut, 1943.

——. *Royal Government in America: A Study of the British Colonial System before 1783*. New Haven: Yale University Press, 1930.

Lambert, Robert Stansbury. *South Carolina Loyalists in the American Revolution*. Columbia: University of South Carolina Press, 1987.

Lanctot, Gustave. *Canada and the American Revolution, 1774–1783*. Cambridge, MA: Harvard University Press, 1967.

Land, Aubrey C. *The Dulanys of Maryland*. Baltimore: Johns Hopkins University Press, 1955.

Landrum, J. B. O. *Colonial and Revolutionary History of Upper South Carolina*. 1897. Reprint, Spartanburg, SC: Reprint Co., 1977.

Langguth, A. J. *Patriots: The Men Who Started the American Revolution*. New York: Simon & Schuster, 1988.

Lannings, Michael Lee. *African Americans in the Revolutionary War*. New York: Citadel Press Books, 2000.

Lark, Terry, ed. *Hackensack—Heritage to Horizons*. Hackensack, NJ: Hackensack Bicentennial Committee, 1976.

Launitz-Schürer, Leopold S., Jr. *Loyal Whigs and Revolutionaries: The Making of the Revolution in New York, 1765–1776*. New York: New York University Press, 1980.

Leake, Isaac Q. *Memoir of the Life and Times of General John Lamb, an Officer of the Revolution*. Albany, NY: Joel Munsall, 1857. http://books.google.com.

Leamon, James S. *Revolution Downeast: The War for American Independence in Maine*. Amherst: University of Massachusetts Press, 1995.

Leder, Lawrence H., ed. *The Colonial Legacy: The Loyalist Historians*. New York: Harper & Row, 1971.

Lee, Francis Bazley, ed. *New Jersey as a Colony and as a State: One of the Original Thirteen*. New York: Publishing Society of New Jersey, 1903.

Lefler, Hugh T., and Albert R. Newsome. *North Carolina: The History of a Southern State*. Chapel Hill: University of North Carolina Press, 1954.

Leiby, Adrian. *The Revolutionary War in the Hackensack Valley*. Piscataway, NJ: Rutgers University Press, 1992.

Leyburn, James G. *The Scotch-Irish*. Chapel Hill: University of North Carolina Press, 1962.

Lincoln, William. *The History of Worcester, Massachusetts, from Its Earliest Settlement to September, 1836*. Worcester: Charles Hersey, 1862.

Livesey, Robert. *The Loyal Refugees*. Toronto: Stoddart Kids, 1999.

Livingston, Edwin Brockholst. *The Livingstons of Livingston Manor*. New York: The Knickerbocker Press, 1910.

Lloyd, Alan. *The King Who Lost America: A Portrait of the Life and Times of George III*. Garden City, NY: Doubleday, 1971.

Lockhart, Paul. *The Drillmaster of Valley Forge: The Baron de Steuben and the Making of the American Army*. New York: Smithsonian Books, 2008.

Lossing, B. J. *Signers of the Declaration of Independence*. New York: George F. Cooledge & Brother, 1848.

———. *Pictorial Field Book of the Revolution*. New York: Harper & Brothers, 1860.

Lowell, Edward J. *Hessians and the Other German Auxiliaries of Great Britain in the Revolutionary War*. New York: Harper & Brothers, 1884.

Luzader, John F. *Fort Stanwix History, Historic Furnishing, and Historic Structure Reports*. Washington, DC: Office of Park Historic Preservation, 1976.

Lynn, Kenneth Schuyler. *A Divided People*. Westport, CT: Greenwood Press, 1977.

Lyttle, Eugene W. *Proceedings of the New York State Historical Association*. Albany, NY: Argus Co., 1904.

Maas, David Edward. *The Return of the Massachusetts Loyalists*. New York: Garland, 1989.

Mackenzie, Frederick. *A British Fusilier in Revolutionary Boston*. Edited by Allen French. Cambridge: Harvard University Press, 1926.

Mackesy, Piers. *The War for America: 1775–1783*. Lincoln: University of Nebraska Press, 1993.

MacKinnon, Neil. *This Unfriendly Soil: The Loyalist Experience in Nova Scotia, 1783–1791*. Ottawa: McGill-Queen's University Press, 1986.

MacLean, John Patterson. *An Historical Account of the Settlements of Scotch Highlanders in America Prior to the Peace of 1783*. Cleveland: Helman-Taylor, 1900.

Magee, Joan. *Loyalist Mosaic: A Multi-ethnic Heritage*. Toronto, ON, and Charlottetown, PEI: Dundurn Press, 1984.

Mahon, John K. *History of the Militia and the National Guard*. New York: Macmillan, 1983.

Maier, Pauline. *American Scripture: Making the Declaration of Independence*. New York: Vintage Books, 1998.

———. *From Resistance to Revolution: Colonial Radicals and the Development of American Opposition to Britain, 1765–1776*. New York: Alfred A. Knopf, 1972.

Main, Jackson Turner. *Rebel versus Tory: The Crisis of the Revolution, 1773–1776*. Chicago: Rand McNally, 1963.

Martin, James. *The Human Dimensions of Nation Making: Essays on Colonial and Revolutionary America*. Madison: State Historical Society of Wisconsin, 1976.

———. *Men in Rebellion: Higher Governmental Leaders and the Coming of the American Revolution*. New Brunswick, NJ: Rutgers University Press, 1973.

Martin, Joseph Plumb. *A Narrative of a Revolutionary Soldier*. New York: Signet Classics, 2001. Originally published anonymously in 1830.

Martin, Sir Theodore. *A Life of Lord Lyndhurst* [J. S. Copley's son]. London: John Murray, 1883.

Martyn, Charles. *The Life of Artemas Ward*. New York: Artemas Ward (the general's great-grandson), 1921.

Mather, David B. *Benjamin Lincoln and the American Revolution*. Columbia: University of South Carolina Press, 1995.

Mather, Frederic Gregory. *Refugees of 1776 from Long Island to Connecticut*. Albany, NY: J. B. Lyon Co., printers, 1913.

May, Robin. *The British Army in North America*. Westminster, MD/Oxford, UK: Osprey Publishing, 1997.

McCaughey, Elizabeth P. *From Loyalist to Founding Father: The Political Odyssey of William Samuel Johnson*. New York: Columbia University Press, 1980.

McCormick, Richard P. *New Jersey from Colony to State*. Princeton, NJ: D. Van Nostrand, 1964.

McCullough, David. *1776*. New York: Simon & Schuster, 2005.

———. *John Adams*. New York: Simon & Schuster, 2001.

McGrady, Edward. *History of South Carolina*. Vols. 2, 3, and 4. New York: Macmillan, 1899–1901.

McGuire, Thomas J. *The Battle of Paoli*. Mechanicsburg, PA: Stackpole Books, 2006.

Meisner, Marian. *A History of Millburn Township*. Millburn, NJ. Millburn/Short Hills Historical Society and the Millburn Free Public Library, 2002.

Meyer, Duane. *The Highland Scots of North Carolina*. Raleigh: Carolina Charter Tercentenary Commission, 1963.

Miller, A. E. *American Revolution*. Charleston, SC: Arno Press, 1821. Reprint, New York: Arno Press, 1969.

Miller, John C. *Origins of the American Revolution*. Boston: Little, Brown, 1943.

Miller, Kerby A, Bruce D. Boling, David N. Doyle, and Arnold Schrier. *Irish Immigrants in the Land of Canaan: Letters and Memoirs from Colonial and Revolutionary America, 1675–1815*. New York: Oxford University Press, 2003.

Miller, Lillian B. *"The Dye Is Now Cast."* Washington, DC: Smithsonian Institution Press, 1975.

Mitchell, Broadus. *The Price of Independence: A Realistic View of the American Revolution*. New York: Oxford University Press, 1974.

Mitnick, Barbara J., ed. *New Jersey in the American Revolution*. Piscataway, NJ: Rutgers University Press, 2005.

Mittelberg, Gottlieb. *Gottlieb Mittelberg's Journey to Pennsylvania in the Year 1750 and Return to Germany in the Year 1754*. Translated by Carl Theo. Eben. Philadelphia: John Jos. McVey, 1898.

Moody, James. *Lieut. James Moody's Narrative of His Exertions and Sufferings in the Cause of Government, Since the Year 1776; Authenticated by Proper Certificates*. 2nd ed. London: Richardson and Urquhart, 1783.

Moore, Christopher. *The Loyalists: Revolution, Exile and Settlement*. Toronto: Macmillan and Stewart, 1994.

Moore, George H. *The Treason of Charles Lee*. New York: Charles Scribner, 1860.

Moore, Frank, comp. *Diary of the American Revolution, 1775–1781*. London: Sampson Low, Son & Co, 1860.

Morgan, Edmond S., and Helen M. Morgan. *The Stamp Act Crisis*. Chapel Hill: University of North Carolina Press, 1995.

Morison, S. E. *Sources and Documents Illustrating the American Revolution, 1764–1788 and the Formation of the Federal Constitution*. Oxford, UK: Clarendon Press, 1923.

———, and Henry Steele Commager. *The Growth of the American Republic*. Vol. 1. New York: Oxford University Press, 1960.

Morpurgo, J. E. *Treason at West Point*. New York: Mason/Charter, 1975.

Morris, Ira K. *Memorial History of Staten Island*. Vol. 1. New York: Memorial Publishing Co., 1898.

Morris, Richard B. *The Peacemakers: The Great Powers and American Independence*. New York: Harper & Row, 1965.

———, and Jeffrey B. Morris, eds. *Encyclopedia of American History*. New York: HarperCollins, 1996.

Moss, Bobby Gilmer. *The Loyalists in the Siege of Fort Ninety Six*. Blacksburg, SC: Scotia-Hibernia Books, 1999.

Mowat, Charles L. *East Florida as a British Province, 1763–1784*. Berkeley: University of California Press, 1943.

Muhlenberg, Henry Melchior. *The Journals of Henry Melchior Muhlenberg*. Vol. 2. Philadelphia: Lutheran Historical Society, Whipporwill Publications, 1982. Reprint, translated from German. Originally published Philadelphia: Evangelical Lutheran Ministerium of Pennsylvania and Adjacent States, 1942–1958.

Murray, Louise Welles. *A History of Old Tioga Point and Early Athens, Pennsylvania*. Wilkes-Barre: Raeder Press, 1908.

Murtie, Jean Clark. *Loyalists in the Southern Campaigns of the Revolutionary War*. Baltimore: Genealogical Publishing Co., 1981.

Nash, Gary. *The Urban Crucible: Social Change, Political Consequences, and the Origins of the American Revolution*. Cambridge: Harvard University Press, 1979.

————. *The Unknown American Revolution*. New York: Viking, 2006.

Nay, James Han. *History of New Brunswick*. St. John, NB: John A. Bowes, 1909.

Neatby, Hilda. *The Quebec Act: Protest and Policy*. Scarborough, ON: Prentice-Hall of Canada, 1972.

Neimeyer, Charles. *America Goes to War: A Social History of the Continental Army*. New York: New York University Press, 1996.

Nell, William C. *The Colored Patriots of the American Revolution*. Boston: Robert F. Wallcut, 1855.

Nelson, James L. *Benedict Arnold's Navy*. Camden, ME: International Marine/McGraw-Hill, 2006.

Nelson, Paul David. *William Tryon and the Course of Empire: A Life in British Imperial Service*. Chapel Hill: University of North Carolina Press, 1990.

Nelson, William H. *The American Tory*. London: Oxford University Press, 1961.

New, M. Christopher. *Maryland Loyalists in the American Revolution*. Centreville, MD: Tidewater Publishers, 1996.

New York State. *Minutes of the Commissioners for Detecting and Defeating Conspiracies in the State of New York*. Albany: New York State, 1909.

Nicholson, Colin. *The "Infamas Govener."* Boston: Northeastern University Press, 2001.

Nickerson, Hoffman. *The Turning Point of the Revolution*. Boston: Houghton Mifflin, 1928.

Norton, Mary Beth. *The British-Americans: The Loyalist Exiles in England, 1774–1789*. Boston: Little, Brown, 1972.

Nova Scotia Museum. *The Black Loyalists of Nova Scotia*. Halifax: Nova Scotia Museum History Section, 2000.

Nye, Mary Greene, ed. *Sequestration, Confiscation and Sale of Estates*. Vol. 6. Montpelier: State Papers of Vermont, 1941.

O'Callaghan, E. B. *Documents Relative to the Colonial History of the State of New York. Albany, 1853–1887*. 15 vols. Albany, NY: Weed, Parsons and Co., 1883.

Oliver, Peter, Douglass Adair, and John Schutz, eds. *Peter Oliver's Origin and Progress of the American Revolution: A Tory View*. San Marino, CA: Huntington Library, 1961.

Olwell, Robert. *Masters, Slaves, and Subjects: The Culture of Power in the South Carolina Low Country, 1740–1790*. Ithaca, NY: Cornell University Press, 1998.

O'Toole, Fintan. *White Savage: William Johnson and the Invention of America*. London: Faber and Faber, 2005.

Ousterhout, Anne M. *A State Divided: Opposition in Pennsylvania to the American Revolution*. Westport, CT: Greenwood Press, 1987.

Paige, Lucius R. *History of Hardwick Massachusetts with a Genealogical Register*. Boston: Houghton Mifflin, 1883.

————. *History of Cambridge, Massachusetts. 1630–1877*. Boston: H. O. Houghton & Co., 1877.

Paine, Mary Standish. *Patriots and Tory Marshfield*. Marshfield, MA: Marshfield Historical Society, 1924.

Paine, Ralph D. *The Old Merchant Marine, A Chronicle of American Ships and Sailors*. New Haven: Yale University Press, 1921.

Palfrey, John Gorham, and Francis Winthrop Palfrey. *History of New England*. Boston: Little, Brown, 1890.

Palmer, Greg. *Biographical Sketches of American Loyalists*. New York: Meckler Publishing, 1982.

Paltsits, Victor Hugo, ed. *Minutes of the Commissioners for Detecting and Defeating Conspiracies in the State of New York*. Albany: State of New York (J. B. Lyon Company, State Printers), 1909.

Papazian, Rita. *Remembering Fairfield, Connecticut*. Charleston, SC: History Press, 2007.

Parker, Rev. Edward L. *The History of Londonderry, Comprising the Towns of Derry and Londonderry, N.H.* Boston: Perkins and Whipple, 1851.

Parsons, Charles Lathrop. *The Capture of Fort William and Mary, December 14 and 15, 1774*. Durham: New Hampshire Historical Society, 1903. Reprint, University of New Hampshire Library, 2009.

Peck, Epaphroditus. *A History of Bristol, Connecticut*. Hartford: Lewis Street Book Shop, 1932.

———. *The Loyalists of Connecticut*. New Haven: Yale University Press, 1934.

Peck, George. *Wyoming; Its History, Stirring Incidents and Romantic Adventures*. New York: Harper & Brothers, 1858.

Pennant, Thomas. *A Tour in Scotland and Voyage to the Hebrides, 1772*. London: Printed for Benj. White, 1776.

Penrose, Maryly B. *Compendium of Early Mohawk Valley Families*. Baltimore: Genealogical Publishing Company, 1990.

———. *Mohawk Valley in the Revolution, Committee of Safety Papers and Genealogical Compendium*. Franklin Park, NJ: Liberty Bell Associates, 1978.

Peterson, Harold L. *The Book of the Continental Soldier*. Harrison, PA: Stackpole Company, 1968.

Phelps, Richard H. *Newgate of Connecticut: Its Origin and Early History*. Camden, ME: Picton Press, 1996, reprint of 1876 edition.

Phillips, Kevin. *The Cousins' Wars: Religion, Politics, and the Triumph of Anglo-America*. New York: Basic Books, 1999.

Pipes, Ed, and Gail Bonsall Pipes. *Loyalists All: Stories Told About New Brunswick Loyalists by Their Descendants*. New Brunswick: New Brunswick Branch UEL Assn of Canada, 1985.

Polf, William A. *Garrison Town*. Albany: New York State American Revolution Bicentennial Commission, 1976.

Pond, E. LeRoy. *The Tories of Chippeny Hill*. New York: Grafton Press, 1909.

Pool, William, ed. *Landmarks of Niagara County, New York*. Buffalo: D. Mason & Co., 1897.

Poole, Edmund Duval. *Annals of Yarmouth and Barrington (Nova Scotia) in the Revolutionary War*. Yarmouth, NS: Reprinted from *Yarmouth Herald*, 1899.

Potter-MacKinnon, Janice. *The Liberty We Seek: Loyalist Ideology in Colonial New York and Massachusetts*. Cambridge: Harvard University Press, 1983.

———. *While the Women Only Wept: Loyalist Refugee Women*. Ottawa: McGill-Queen's University Press, 1993.

Pougher, Richard D. *"Averse . . . to Remaining Idle Spectators": The Emergence of Loyalist Privateering during the American Revolution, 1775–1778*. Orono: University of Maine Press, 2002.

Pound, Arthur, and Richard E. Day. *Johnson of the Mohawks*. New York: Macmillan, 1930.

Puls, Mark. *Samuel Adams: Father of the American Revolution*. New York: Palgrave Macmillan, 2006.

Purvis, Thomas L., ed. *Revolutionary America, 1763–1800*. New York: Facts on File, 1995.

Quarles, Benjamin. *The Negro in the American Revolution*. Chapel Hill: University of North Carolina Press, 1961.

———. "The Revolutionary War as a Black Declaration of Independence." In *Slavery and Freedom in the Age of the American Revolution,* edited by Ira Berling and Ronald Hoffman, Charlottesville: University of Virginia Press, 1983.

Randall, Willard Sterne. *Benedict Arnold, Patriot and Traitor*. New York: Barnes & Noble Books, 2003. Reprint, New York: William Morrow, 1990.

Rankin, Hugh F. *The North Carolina Continentals,* 2nd edition. Chapel Hill: University of North Carolina Press, 2005.

Ranlet, Philip. *The New York Loyalists*. Knoxville: University of Tennessee Press, 1986.

Raum, John O. *The History of New Jersey*. Vol. 2. Philadelphia: John E. Potter and Co., 1877.

Raymond, W. O. *The United Empire Loyalists*. Saint Stephen, New Brunswick: Saint Croix Printing and Publishing Co., 1893.

Raynor, George. *Patriots and Tories in Piedmont Carolina*. Salisbury, NC: *Salisbury Post,* 1990.

Reeks, Linsay. *Ontario Loyalist Ancestors*. Baltimore: Gateway Press, 1992.

Reese, Trevor R. *Colonial Georgia: A Study in British Imperial Policy in the Eighteenth Century*. Athens: University of Georgia Press, 1963.

Reid, William D. *The Loyalists in Ontario: The Sons and Daughters of the American Loyalists of Upper Canada*. Lambertville, NJ: Hunterdon House, 1973.

Rhodehamel, John H., ed. *The American Revolution: Writings from the War of Independence*. New York: Library of America, 2001.

Richards, Lysander Salmon. *History of Marshfield*. Vol. 1. Plymouth, MA: Memorial Press, 1901. Reprint, Salem, MA: Higginson Books, 1996.

Riddell, William Renwick. *Benjamin Franklin and Canada*. Toronto: Published by author, 1923.

Ridpath, John Clark. *James Otis, the Pre-Revolutionist*. Chicago: University Association, 1898. Reprint, Whitefish, MT: Kessinger Publishing, 2004.).

Riley, Sandra. *Homeward Bound, A History of the Bahama Islands to 1850*. Miami, FL: Island Research, 2000.

Ritcheson, Charles R. *British Politics and the American Revolution*. Norman; University of Oklahoma Press, 1954.

Roberts, Oliver Ayer. *History of the Military Company of the Massachusetts*. Vol. 3. Boston: Alfred Mudge & Son, Printers, 1898.

Robinson, Blackwell P. *The Five Royal Governors of North Carolina: 1729–1775*. Raleigh: Carolina Charter Tercentenary Commission, 1963.

———. *A History of Moore County, North Carolina 1747–1847*. Southern Pines, NC: Moore County Historical Association, 1956.

Robson, Eric. *The American Revolution in its Political and Military Aspects*. New York: W. W. Norton & Company, 1966.

Roeber, A. G. *Palatines, Liberty and Property: German Lutherans in Colonial British America*. Baltimore: Johns Hopkins University Press, 1998.

Rogers, Alan. *Empire and Liberty: American Resistance to British Authority, 1755–1763*. Berkeley: University of California Press, 1974.

Rogers, Robert. *The Journals of Major Robert Rogers*. Albany, NY: Josel Munsell's Sons, 1883. Reprint, New York: Cornith Books, 1961.

Rose, Alexander. *American Rifle: A Biography*. New York: Delacorte Press, 2008.

———. *Washington's Spies: The Story of America's First Spy Ring*. New York: Bantam Dell, 2006.

Roseboom, Eugene H., and Francis P. Weisenburger. *A History of Ohio*. Columbus: Ohio Historical Society, 1953.

Rosenberg, Bruce A. *The Neutral Ground: The André Affair and the Background of Cooper's The Spy*. Westport. CT: Greenwood Press, 1994.

Rowe, John. *The Diary of John Rowe, 1764–1779*. Boston: Massachusetts Historical Society, 1895.

Royster, Charles. *A Revolutionary People at War: The Continental Army & American Character, 1775–1783*. Chapel Hill: North Carolina University Press for the Omohundro Institute of Early American History and Culture, 1979.

Rubincam, Milton. *The Old United Empire Loyalist List*. Baltimore: Genealogical Publishing Co., 1969.

Ruch, John E., and Elizabeth Kipp. *Carleton's Loyalist Index*. Ottawa: Sir Guy Carleton Branch United Empire Loyalists' Association of Canada, 1996.

Ryan, D. Michael. *Concord and the Dawn of Revolution*. Salem, MA: History Press, 2007.

Ryan, Dennis P. *New Jersey's Loyalists*. Trenton: New Jersey Historical Commission, 1975.

Ryerson, Adolphus Egerton. *Loyalists of America and Their Time: From 1620 to 1816*. Toronto: William Briggs, 1880.

Sabine, Lorenzo. *The American Loyalists, or Biographical Sketches of Adherents to the British Crown in the American Revolution*. Boston: Charles C. Little and James Brown, 1847.

———. *Biographical Sketches of Loyalists of the American Revolution: With an Historical Essay*. Rev. ed. 2 vols. Boston: Little, Brown & Co., 1864.

Sargent, Winthrop, ed. *The Loyal Verses of Joseph Stansbury and Doctor Jonathan Odell; Relating to the American Revolution*. Albany: J. Munsell, 1860.

Sarles, Frank B., and Charles E. Shedd. *Colonials and Patriots*, Vol. 6. National Survey of Historic Sites and Buildings. Washington, DC: National Park Service, U.S. Department of the Interior, 1964.

Saunders, Gail. *Bahamian Loyalists and Their Slaves*. London: Macmillan Caribbean, 1983.

Schama, Simon. *Rough Crossing: Britain, the Slaves and the American Revolution*. London: BBC Books, 2006.

Scharf, John Thomas. *History of Maryland from the Earliest Period to the Present Day*. Vol. II. Baltimore: John B. Piet, 1879.

———. *History of Westchester County*. Philadelphia: L. E. Preston & Co., 1886.

Schecter, Barnet. *The Battle for New York: The City at the Heart of the American Revolution*. New York: Walker and Company, 2002.

Scheer, George F., and Hugh F. Rankin. *Rebels and Redcoats: The American Revolution Through the Eyes of Those Who Fought and Lived It*. New York: Da Capo Press, 1957.

Schlesinger, Arthur M. *The Colonial Merchants and the American Revolution*. New York: Columbia University Press, 1918.

———. *New Viewpoints in American History*. New York: Macmillan, 1923.

Scott, Kenneth, ed. *Rivington's New York Newspapers: Excerpts from a Loyalist Press, 1773–1785*. New York: New-York Historical Society, 1973.

Sedeen, Margaret. *Star-Spangled Banner*. Washington, DC: National Geographic Society, 1993.

Shelton, Hal T. *General Richard Montgomery and the American Revolution: From Redcoat to Rebel*. New York: New York University Press, 1996.

Shy, John. *A People Numerous and Armed: Reflections on the Military Struggle for American Independence*. Rev. ed. Ann Arbor: University of Michigan Press, 1990.

Sibley, John Langdon. *Biographical Sketches of Graduates of Harvard University in*

Cambridge, Massachusetts. Cambridge: Charles William Sever, University Bookstore, 1873.

Siebert, Wilbur Henry. *The American Loyalists in the Eastern Seigniories and Townships of the Province of Quebec*. Ottawa: Royal Society of Canada, 1913.

—. *The Exodus of the Loyalists from Penobscot to Passamaquoddy*. Columbus: Ohio State University, 1914.

—. *George Washington and the Loyalists*. Worcester: American Antiquarian Society, 1934.

—. *The Legacy of the American Revolution to the British West Indies and Bahamas: A Chapter Out of the History of the American Loyalists*. Columbus: Ohio State University, 1913.

—. *The Loyalist Refugees of New Hampshire*. Columbus: Ohio State University, 1916.

—. *Loyalists in East Florida, 1774 to 1785: The Most Important Documents Pertaining thereto, Edited with an Accompanying Narrative*. DeLand: Florida State Historical Society, 1929.

—. *The Loyalists of Pennsylvania*. Columbus: Ohio State University, 1920.

Simcoe, John Graves. *Simcoe's Military Journal: A History of the Operations of a Partisan Corps, Called the Queen's Rangers, Commanded by Lieut. Col. J. G. Simcoe, During the War of the American Revolution*. New York: Bartlett & Welford, 1844.

Simms, Jeptha R. *The Frontiersmen of New York*. Vol. 2. Albany: Geo. C. Riggs, 1883.

—. *History of Schoharie County and Border Wars of New York*. Albany: Munsell & Tanner, Printers, 1845.

Skemp, Sheila L. *William Franklin: Son of a Patriot, Servant of a King*. New York: Oxford University Press, 1990.

Skeoch, Alan. *United Empire Loyalists and the American Revolution*. Toronto: Grolier Ltd., 1982.

Smith, Albert Henry. *The Writings of Benjamin Franklin*. Vol. 10. New York: Macmillan Co., 1907.

Smith, Eddy N., et al. *Bristol, Connecticut* . . . Hartford: City Printing. Co., 1907.

Smith, H. Y., and W. S. Rann, eds. *The History of Rutland County Vermont*. Syracuse, NY: D. Mason & Co., 1886.

Smith, Page. *A New Age Now Begins*. 2 vols. New York: McGraw-Hill Co., 1976.

Smith, Paul H. *Loyalists and Redcoats: A Study in British Revolutionary Policy*. Chapel Hill: University of North Carolina Press, 1964.

Sparks, Jared. *The Life of George Washington*. Auburn, NY: Derby & Miller, 1853.

—, ed. *The Writings of George Washington*. Vol. 5. Boston: Ferdinand Andrews, 1838.

Spring, Matthew H. *With Zeal and with Bayonets Only: The British Army in North America, 1775–1783.* Norman: University of Oklahoma Press, 2008.

Srodes, James. *Franklin: The Essential Founding Father.* Washington, DC: Regnery Publishing, 2002.

Stanton, Harry. *The Beloved Spy.* Caldwell, ID: Caxton Printers, 1948.

Stark, James H. *The Loyalists of Massachusetts and the Other Side of the American Revolution.* Salem, MA: Salem Press, 1910.

Starr, Joseph Barton. *Tories, Dons, and Rebels: The American Revolution in British West Florida.* Gainesville: University Presses of Florida, 1976.

Stedman, C. *The history of the origin, progress, and termination of the American war.* Vol. 1. London: Printed for Author, 1794.

Steiner, Bruce. *Samuel Seabury, 1729–1796: A Study in the High Church Tradition.* Athens: Ohio University Press, 1972.

Steuben, Frederick William, Baron von. *Baron von Steuben's Revolutionary War Drill Manual.* 1794. Facsimile edition. New York: Cosimo, 2007.

Stewart, Walter. *True Blue: The Loyalist Legend.* Toronto: Collins, 1985.

Stoetzel, Donald I. *Encyclopedia of the French & Indian War in North America, 1754–1763.* Westminster, MD: Heritage Books, 2008.

Stone, William L. *Border Wars of the American Revolution.* Vol 1. New York: A. L. Fowle, 1900.

———. *The Campaign of Lieut. Gen. John Burgoyne.* Albany, NY: Joel Munsell, 1877.

———. *Life of Joseph Brant.* Albany, NY: J. Munsell, 1865.

———. *Reminiscences of Saratoga and Ballston.* New York: R. Worthington, 1880.

Stryker, William Scudder. *The New Jersey Volunteers (Loyalists) in the Revolutionary War.* Trenton: Naar, Day & Naar, 1887.

———. *The Capture of the Block House at Toms River.* Trenton, NJ: Naar, Day & Naar, 1888.

Stuart, I. W. *Life of Jonathan Trumbull, Governor of Connecticut.* Boston: Crocker and Brewster, 1859.

Sullivan, James. *Minutes of the Albany Committee of Correspondence, 1775–1778.* Albany, NY: University of the State of New York, 1923.

Swiggett, Howard. *War Out of Niagara: Walter Butler and the Tory Rangers.* New York: Columbia University Press, 1933.

Symonds, Craig L., and William J. Clipson. *A Battlefield Atlas of the American Revolution.* Mount Pleasant, SC: Nautical & Aviation Publishing Company of America, 1986.

Syrett, Harold C., ed. *The Papers of Alexander Hamilton.* New York: Columbia University Press, 1961.

Talman, James J., ed. *Loyalist Narratives from Upper Canada: The Journal of Adam Crysler.* Toronto: Champlain Society in Canada, 1946.

Taylor, Alan. *The Divided Ground: Indians, Settlers and the Northern Borderland of the American Revolution*. New York: Alfred A. Knopf, 2006.

Thacher, James. *A Military Journal of the American Revolution: From the Commencement to the Disbanding of the American Army*. Boston: Cottons & Barnard, 1827.

Thayer, William Roscoe. *George Washington*. Boston: Houghton Mifflin, 1922.

Thomas, William H. B. *Remarkable High Tories: Supporters of King and Parliament in Revolutionary Massachusetts*. Bowie, MD.: Heritage Books, 2001.

Thompson, Benjamin F. *History of Long Island from Its Discovery and Settlement to the Present Time*. Vol. 1. New York: Robert H. Dodd, 1918.

Thwaites, Reuben G. *Collections of the State Historical Society of Wisconsin*. Madison, WI: Democrat Printing Co., 1888.

Tiedemann, Joseph S. *Reluctant Revolutionaries: New York City and the Road to Independence*. Ithaca, NY: Cornell University Press, 1997.

———, and Eugene R. Fingerhut, eds. *The Other New York: The American Revolution Beyond New York City, 1763–1787*. Albany: State University of New York Press, 2005.

Tillotson, Harry Stanton, *The Beloved Spy*. Caldwell, ID: The Caxton Printers, 1948.

Todd, Charles B. *The History of Redding*. New York: Grafton Press, 1906.

Tourtellot, Arthur B. *Lexington and Concord: The Beginning of the War of the American Revolution*. New York: W. W. Norton & Company, 1959.

Troxler, Carol Watterson. *The Loyalist Experience in North Carolina*. Raleigh: North Carolina Bicentennial, 1976.

———. *Pyle's Defeat: Deception at the Racepath*. Graham, NC: Alamance County Historical Association, 2003.

Trueman, A. W. *The Story of the United Empire Loyalists*. Toronto: Copp Clark Co., 1946.

Trusell, John B. *Birthplace of an Army: A Study of the Valley Forge Encampment*. Harrisburg: Pennsylvania Historical and Museum Commission, 1976.

Tuchman, Barbara W. *The March of Folly: From Troy to Vietnam*. New York: Alfred A. Knopf, 1984.

Tyler, John W. *Connecticut Loyalists: An Analysis of Loyalist Land Confiscations in Greenwich, Stamford, and Norwalk*. New Orleans: Polyanthos, 1977.

Tyler, Moses Coit. *The Literary History of the American Revolution 1763–1783*. Vol. 2. New York: G. P. Putnam's Sons, 1897.

Upton, Leslie S. F., ed. *Revolutionary Versus Loyalist*. Waltham, MA: Blaisdell Publishing. Co., 1968.

Upton, Richard Francis. *Revolutionary New Hampshire: An Account of the Social and Political Forces Underlying the Transition from Royal Province to American Commonwealth*. Hanover, NH: Dartmouth College Publications, 1936.

Urban, Mark. *Fusiliers: The Saga of a British Regiment in the American Revolution.* New York: Walker and Company, 2007.

Van Alstyne, Richard W. *Empire and Independence: The International History of the American Revolution.* New York: John Wiley & Sons, 1965.

Van Buskirk, Judith L. *Generous Enemies: Patriots and Loyalists in Revolutionary New York.* Philadelphia: University of Pennsylvania Press, 2002.

Van Doren, Carl. *Secret History of the American Revolution.* New York: Viking Press, 1951.

Van Dusen, Albert E. *Connecticut.* New York: Random House, 1961.

Van Schaack, Henry C. *The Life of Peter Van Schaack.* New York: D. Appleton & Co., 1842.

Van Tyne, Claude Halstead. *The Loyalists in the American Revolution.* New York: Macmillan Co., 1902. Reprint, Gloucester, MA: Peter Smith, 1929.

Vance, Clarence H., ed. *Letters of a Westchester Farmer by Reverend Samuel Seabury.* White Plains, NY: Westchester County Historical Society, 1930.

Walker, James W. St. G. *The Black Loyalists: The Search for a Promised Land in Nova Scotia and Sierra Leone, 1783–1870.* New York: Africana Publishing Co., 1976.

Walker, Paul K. *Engineers of Independence: A Documentary History of the Army Engineers in the American Revolution, 1775–1783.* Washington, DC: U.S. Government Printing Office, 1981.

Wallace, Richard M. *Appeal to Arms: A Military History of the American Revolution.* New York: Harper and Brothers, 1951.

Wallace, W. Stewart. *The United Empire Loyalists.* 1914. Reprint, Whitefish, MT: Kessinger Publishing, 2004.

Ward, Christopher. *The War of the Revolution.* New York: Macmillan Co., 1952.

Ward, George Atkinson, ed. *Journal and Letters of Samuel Curwen, an American in England, from 1775 to 1783.* Boston: Little, Brown, 1864.

Ward, Harry M. *The War for Independence and the Transformation of American Society.* Florence, KY: Routledge, 2003.

Warfle, Richard T. *Connecticut's Western Colony: The Susquehannah Affair.* Hartford: American Revolution Bicentennial Commission of Connecticut, 1979.

Watt, Gavin K. *The Burning of the Valleys.* Toronto: Dundurn Press, 1997.

Weed, Samuel Richards. *Norwalk After Two Hundred and Fifty Years.* South Norwalk, CT: C. A. Freeman, 1902.

Wells, James L., Louis F. Haffen, and Josiah A. Briggs, eds. *The Bronx and Its People: a History, 1609–1927.* New York: Lewis Historical Pub. Co., 1927.

Wertenbaker, Thomas Jefferson. *Father Knickerbocker Rebels.* New York: Charles Scribner's Sons, 1948.

Weston, Thomas. *History of the Town of Middleboro, Massachusetts.* Boston: Houghton Mifflin, 1906.

Wheeler, Richard. *Voices of 1776.* New York: Fawcett Premier, 1972.

Whiteley, Peter. *Lord North: The Prime Minister Who Lost America*. London: Hambleton, 1996.

Willard, Margaret Wheeler, ed. *Letters on the American Revolution, 1774–1776*. Boston: Houghton Mifflin, 1925.

Willett, William M. *A Narrative of the Military Actions of Colonel Marinus Willett, Taken Chiefly from His Own Manuscript*. New York: G. & C. & H. Carvill, 1831.

Williams, Glenn F. *Year of the Hangman: George Washington's Campaign Against the Iroquois*. Yardley, PA: Westholme Publishing, 2005.

Williams, Henry Lionel, and Ottalie K. Williams. *How to Furnish Old American Houses*. New York: Pellegrini & Cudahy, 1949.

Williams, Samuel Cole. *Tennessee During the Revolutionary War*. Knoxville: University of Tennessee Press, 1944.

Wilson, David. *The Life of Jane McCrea*. New York: Baker, Godwin & Co., Printers, 1853.

Wilson, David K. *The Southern Strategy: Britain's Conquest of South Carolina and Georgia, 1775–1780*. Columbia: University of South Carolina Press, 2005.

Wilson, Ellen Gibson. *The Loyal Blacks*. New York: G. P. Putnam's Sons, 1976.

Wilson, James Grant, and John Fiske. *Appleton's Cyclopedia of American Biography*. New York: D. Appleton and Company, 1887–89.

Winsor, Clarence F. *The Memorial History of Boston*. Vol. 3. Boston: James R. Osgood, 1881.

Winsor, Justin. *History of the Town of Duxbury, Massachusetts, with Genealogical Registers*. Boston: Crosby & Nichols, 1849.

Wood, Gordon. *The Creation of the American Republic 1776–1787*. Chapel Hill: University of North Carolina Press, 1998.

———. *The Radicalism of the American Revolution*. New York: Alfred A. Knopf, 1992.

Wood, Louis Aubrey. *The War Chief of the Six Nations*. Toronto: Glasgow, Brook & Co. (Chronicles of Canada), 1920.

Wood, W. J. *Battles of the Revolutionary War, 1775–1781*. Chapel Hill: University of North Carolina Press, 1990.

Wood, William. *The Father of British Canada: A Chronicle of Carleton*. Toronto: Chronicles of Canada, 1916.

Woodford, Clayton W., and William Nelson. *History of Bergen and Passaic Counties, New Jersey*. Philadelphia: Everts & Peck (press of J. B Lippincott & Co.), 1882.

Woolsey, C. M. *History of the Town of Marlborough, Ulster County, New York, from its Earliest Discovery*. Albany, NY: J. B. Lyon Co., 1908.

Wright, Esmond. *Red, White and True Blue: the Loyalists in the Revolution*. New York: AMS Press, 1976.

———, ed. *The Fire of Liberty*. London: Folio Society, 1983.

Wright, Esther Clark. *The Loyalists of New Brunswick*. Fredericton, NB: n.p., 1981.

Wright, J. Leitch, Jr. *Florida in the American Revolution*. Gainesville: University Presses of Florida, 1975.

Wright, Robert K., Jr. *The Continental Army*. Washington, DC: Center of Military History, 1983.

Wrong, George M. *Canada and the American Revolution: The Disruption of the First British Empire*. New York: Macmillan, 1935.

Yosphe, Harry B. *The Disposition of Loyalist Estates in the Southern District of New York*. New York: AMS Press, 1967.

Young, Alfred F. *The Shoemaker and the Tea Party*. Boston: Beacon Press, 1999.

Zeichner, Oscar. *Connecticut's Years of Controversy*. Chapel Hill: University of North Carolina Press, 1949.

Index